T0180395

## Advanced Texts in Physics

This program of advanced texts covers a broad spectrum of topics which are of current and emerging interest in physics. Each book provides a comprehensive and yet accessible introduction to a field at the forefront of modern research. As such, these texts are intended for senior undergraduate and graduate students at the MS and PhD level; however, research scientists seeking an introduction to particular areas of physics will also benefit from the titles in this collection.

# Springer

*Berlin*
*Heidelberg*
*New York*
*Hong Kong*
*London*
*Milan*
*Paris*
*Tokyo*

**Physics and Astronomy**

ONLINE LIBRARY

http://www.springer.de/phys/

Helmut Wiedemann

# Synchrotron Radiation

With 89 Figures, 7 Tables,
100 Exercises and 55 Selected Solutions

 Springer

Professor Helmut Wiedemann
Applied Physics Department
and SSRL/SLAC
Stanford University
P.O. Box 20450
Stanford, CA 94309
USA

E-mail: wiedemann@SLAC.Stanford.edu

*Cover picture:*
Distortion of electrical field lines due to transverse acceleration of a charge (Figure 2.6)

ISBN 978-3-642-07777-7

Library of Congress Cataloging-in-Publication Data

Wiedemann, Helmut, 1938-
Synchrotron radiation / Helmut Wiedemann.
p.cm. – (Advanced texts in physics)
Includes bibliographical references and index.

1. Synchrotron radiation. I. Title. II. Series.
QC793.5.E627 W45 2002   539.7'35–dc21   2002021668

Springer-Verlag Berlin Heidelberg New York
a member of BertelsmannSpringer Science+Business Media GmbH

http://www.springer.de

© Springer-Verlag Berlin Heidelberg 2010
Printed in Germany

Cover design: *design & production* GmbH, Heidelberg
Printed on acid-free paper

To my family and students

# Preface

This book covers the physical aspects of synchrotron radiation generation and is designed as a textbook and reference for graduate students, teachers and scientists utilizing synchrotron radiation. It is my hope that this text may help especially students and young researchers entering this exciting field to gain insight into the characteristics of synchrotron radiation.

Discovered in 1945, synchrotron radiation has become the source of photons from the infrared to hard x-rays for a large community of researchers in basic and applied sciences. This process was particularly supported by the development of electron accelerators for basic research in high energy physics. Specifically, the development of the storage ring and associated technologies resulted in the availability of high brightness photon beams far exceeding other sources.

In this text, the physics of synchrotron radiation for a variety of magnets is derived from first principles resulting in useful formulas for the practitioner. Since the characteristics and quality of synchrotron radiation are intimately connected with the accelerator and electron beam producing this radiation, a short overview of relevant accelerator physics is included.

In the first four chapters radiation phenomena in general and synchrotron radiation in particular are introduced based on more visual and basic physical concepts. Where exact formulas are required, we borrow results from rigorous derivations in Chaps. 9 and 10. This way the physics of synchrotron radiation can be discussed without extensive deviations into mathematical manipulations, which can be quite elaborate although straightforward. The consequence for the reader, of this dual approach to synchrotron radiation is that, here and there, one will find some repetitive discussions, which the author hopes will provide easier reading and continuity in the train of thought.

Chapters 5 to 8 give an overview of beam dynamics in storage rings and guidance for the optimization of a storage ring for synchrotron radiation production. The theory of synchrotron radiation is derived rigorously in Chap. 9 and that of undulator or insertion device radiation in Chap. 10. Finally, in Chap. 11 the physics of a free electron laser is discussed.

Each chapter includes a set of exercises. For those exercises which are marked with the argument (S), solutions are provided in Appendix A. In support of the practitioner utilizing synchrotron radiation most relevant for-

mulas together with useful mathematical and physical formulae and constants
are compiled in Appendices B - D.

The author would like to thank the editorial staff at Springer Verlag and
especially Drs. H. Lotsch and C. Ascheron for suggesting the writing of this
book. The trained eyes of Dr. A. Lahee and Mrs. Dimler contributed much to
minimize typographical errors and to greatly improve the overall appearance
of the book. Special thanks goes to Professors J. Dorfan and K. Hodgson at
Stanford University for granting a sabbatical and to Professor T. Vilaithong
at the Chiang Mai University in Thailand for providing a quiet and peaceful
environment during the final stages of writing this book.


Chiang Mai,                                          *Helmut Wiedemann*
July 2002

# Contents

**1. Charges and Fields** .......................................... 1
   1.1 Radiation from Moving Charges ........................... 1
      1.1.1 Why do Charged Particles Radiate? ................ 2
      1.1.2 Spontaneous Synchrotron Radiation ................ 2
      1.1.3 Stimulated Radiation ............................ 4
      1.1.4 Electron Beam .................................. 5
   1.2 Maxwell's Equations ..................................... 6
      1.2.1 Conversion from cgs to MKS Units ................ 6
      1.2.2 Lorentz Force .................................. 8
   1.3 The Lorentz Transformations ............................ 10
      1.3.1 Lorentz Transformation of Coordinates ............. 11
      1.3.2 Energy and Momentum .......................... 13
   Exercises .................................................. 14

**2. Fundamental Processes** ..................................... 17
   2.1 Conservation Laws and Radiation ....................... 17
      2.1.1 Cherenkov Radiation ........................... 18
      2.1.2 Compton Radiation .............................. 20
   2.2 The Poynting Vector .................................... 20
   2.3 Electromagnetic Radiation ............................... 22
      2.3.1 Coulomb Regime ............................... 22
      2.3.2 Radiation Regime .............................. 23
   2.4 Spatial and Spectral Properties of Radiation .............. 26
   Exercises .................................................. 28

**3. Overview of Synchrotron Radiation** ...................... 31
   3.1 Radiation Power ....................................... 32
   3.2 Spectrum .............................................. 36
   3.3 Spatial Photon Distribution ............................. 41
   3.4 Fraunhofer Diffraction ................................. 42
   3.5 Spatial Coherence ..................................... 45
   3.6 Temporal Coherence ................................... 47
   3.7 Spectral Brightness .................................... 50
      3.7.1 Matching ...................................... 51
   Exercises .................................................. 52

**4.  Radiation Sources** ........................................... 55
    4.1  Bending Magnet Radiation .............................. 55
    4.2  Superbends ........................................... 56
    4.3  Wavelength Shifter ................................... 57
    4.4  Wiggler Magnet Radiation............................. 58
    4.5  Undulator Radiation .................................. 62
    4.6  Back Scattered Photons .............................. 68
        4.6.1  Photon Flux .................................. 68
    Exercises............................................... 70

**5.  Accelerator Physics** ...................................... 73
    Exercise .............................................. 76

**6.  Particle Beam Optics** ..................................... 77
    6.1  Deflection in Bending Magnets ........................ 77
    6.2  Beam Focusing........................................ 79
        6.2.1  Principle of Focusing ........................ 80
        6.2.2  Quadrupol Magnet ............................ 80
    6.3  Equation of Motion................................... 82
        6.3.1  Solutions of the Equations of Motion............. 84
        6.3.2  Matrix Formalism ............................ 84
        6.3.3  FODO Lattice ................................ 85
    6.4  Betatron Function ................................... 86
        6.4.1  Betatron Phase and Tune ..................... 87
        6.4.2  Beam Envelope .............................. 88
    6.5  Phase Ellipse ....................................... 88
    6.6  Beam Emittance ..................................... 89
        6.6.1  Variation of the Phase Ellipse ................. 90
        6.6.2  Transformation of Phase Ellipse ............... 91
    6.7  Dispersion Function ................................. 92
    6.8  Periodic Lattice Functions ........................... 93
        6.8.1  Periodic Betatron Function in a FODO Lattice....... 93
        6.8.2  Periodic Dispersion or $\eta$-Function ................. 95
        6.8.3  Beam Size ................................... 95
    Exercises............................................... 96

**7.  Radiation Effects** ....................................... 99
    7.1  Synchrotron Oscillations.............................. 99
        7.1.1  Longitudinal Phase Space Motion ................. 103
    7.2  Damping ............................................. 104
    7.3  Quantum Effects ..................................... 105
    7.4  Equilibrium Beam Parameters ......................... 106
        7.4.1  Equilibrium Energy Spread ..................... 106
        7.4.2  Bunch Length................................. 107
        7.4.3  Horizontal Beam Emittance .................... 108
        7.4.4  Vertical Beam Emittance ...................... 109

7.5    Transverse Beam Parameters............................ 110
       7.5.1   Beam Sizes ..................................... 111
       7.5.2   Beam Divergence ................................ 112
7.6    Beam Emittance and Wiggler Magnets ................... 112
       7.6.1   Damping Wigglers................................ 115
       7.6.2   Variation of the Damping Distribution ............. 117
       7.6.3   Can we Eliminate the Beam Energy Spread?......... 119
7.7    Photon Source Parameters.............................. 121
Exercises................................................... 122


8.  Storage Ring Design as a Synchrotron Light Source ...... 125
    8.1   Storage Ring Lattices ................................. 126
          8.1.1   FODO Lattice ................................. 126
    8.2   Optimization of a Storage Ring Lattice................... 127
          8.2.1   Minimum Beam Emittance ....................... 128
          8.2.2   The Double Bend Achromat (dba) Lattice .......... 131
          8.2.3   The Triple Bend Achromat (tba) Lattice ........... 134
          8.2.4   Limiting Effects ................................ 134


9.  Theory of Synchrotron Radiation ........................ 137
    9.1   Radiation Field ...................................... 137
    9.2   Total Radiation Power and Energy Loss .................. 144
          9.2.1   Transition Radiation............................ 144
          9.2.2   Synchrotron Radiation Power ..................... 147
    9.3   Radiation Lobes...................................... 150
    9.4   Synchrotron Radiation Spectrum ....................... 155
    9.5   Radiation Field in the Frequency Domain ................ 155
          9.5.1   Spectral Distribution in Space and Polarization ...... 160
          9.5.2   Spectral and Spatial Photon Flux ................. 163
          9.5.3   Harmonic Representation......................... 165
    9.6   Spatial Radiation Power Distribution .................... 165
          9.6.1   Asymptotic Solutions ........................... 167
    9.7   Angle-Integrated Spectrum ............................ 168
          9.7.1   Statistical Radiation Parameters .................. 174
Exercises................................................... 176


10. Insertion Device Radiation ............................. 177
    10.1  Periodic Magnetic Field ............................... 178
          10.1.1 Periodic Field Configuration ...................... 179
          10.1.2 Particle Dynamics in a Periodic Field Magnet ........ 182
          10.1.3 Focusing in a Wiggler Magnet ..................... 183
          10.1.4 Hard Edge Wiggler Model ........................ 186

10.2 Undulator Radiation . . . . . . . . . . . . . . . . . . . . . . . . . . . . . . . . . . . . 187
      10.2.1 Fundamental Wavelength . . . . . . . . . . . . . . . . . . . . . . . . . 188
      10.2.2 Radiation Power . . . . . . . . . . . . . . . . . . . . . . . . . . . . . . . . 189
      10.2.3 Spatial and Spectral Distribution . . . . . . . . . . . . . . . . . . 190
      10.2.4 Line Spectrum . . . . . . . . . . . . . . . . . . . . . . . . . . . . . . . . . 203
      10.2.5 Spectral Undulator Brightness . . . . . . . . . . . . . . . . . . . . 207
10.3 Elliptical Polarization . . . . . . . . . . . . . . . . . . . . . . . . . . . . . . . . . . 208
      10.3.1 Elliptical Polarization
             from Bending Magnet Radiation . . . . . . . . . . . . . . . . . . . 208
      10.3.2 Elliptical Polarization from Periodic Insertion Devices . 211
Exercises . . . . . . . . . . . . . . . . . . . . . . . . . . . . . . . . . . . . . . . . . . . . . . . . . 214

11. **Free Electron Lasers** . . . . . . . . . . . . . . . . . . . . . . . . . . . . . . . . . . . 217
11.1 Small Gain FEL . . . . . . . . . . . . . . . . . . . . . . . . . . . . . . . . . . . . . . . 220
      11.1.1 Energy Transfer . . . . . . . . . . . . . . . . . . . . . . . . . . . . . . . . 220
      11.1.2 Equation of Motion . . . . . . . . . . . . . . . . . . . . . . . . . . . . . 222
      11.1.3 FEL-Gain . . . . . . . . . . . . . . . . . . . . . . . . . . . . . . . . . . . . . 225
Exercises . . . . . . . . . . . . . . . . . . . . . . . . . . . . . . . . . . . . . . . . . . . . . . . . . 230

A. **Solutions to Exercises** . . . . . . . . . . . . . . . . . . . . . . . . . . . . . . . . . 231

B. **Mathematical Constants and Formulas** . . . . . . . . . . . . . . . . . . 243
    B.1  Constants . . . . . . . . . . . . . . . . . . . . . . . . . . . . . . . . . . . . . . . . . . 243
    B.2  Series Expansions . . . . . . . . . . . . . . . . . . . . . . . . . . . . . . . . . . . 243
    B.3  Multiple Vector Products . . . . . . . . . . . . . . . . . . . . . . . . . . . . . 244
    B.4  Differential Vector Expressions . . . . . . . . . . . . . . . . . . . . . . . . 244
    B.5  Theorems . . . . . . . . . . . . . . . . . . . . . . . . . . . . . . . . . . . . . . . . . 245
    B.6  Coordinate Systems . . . . . . . . . . . . . . . . . . . . . . . . . . . . . . . . . 245
    B.7  Gaussian Distribution . . . . . . . . . . . . . . . . . . . . . . . . . . . . . . . 247
    B.8  Miscellaneous Mathematical Formulas . . . . . . . . . . . . . . . . . . 248

C. **Physical Formulas and Parameters** . . . . . . . . . . . . . . . . . . . . . . 251
    C.1  Constants . . . . . . . . . . . . . . . . . . . . . . . . . . . . . . . . . . . . . . . . . 251
    C.2  Unit Conversion . . . . . . . . . . . . . . . . . . . . . . . . . . . . . . . . . . . . 252
    C.3  Relations of Fundamental Parameters . . . . . . . . . . . . . . . . . . . 253
    C.4  Energy Conversion . . . . . . . . . . . . . . . . . . . . . . . . . . . . . . . . . 253
    C.5  Maxwell's Equations . . . . . . . . . . . . . . . . . . . . . . . . . . . . . . . . 253
        C.5.1  Lorentz Force . . . . . . . . . . . . . . . . . . . . . . . . . . . . . . . 253
    C.6  Wave and Field Equations . . . . . . . . . . . . . . . . . . . . . . . . . . . . 254
    C.7  Relativistic Relations . . . . . . . . . . . . . . . . . . . . . . . . . . . . . . . . 254
    C.8  Four-Vectors . . . . . . . . . . . . . . . . . . . . . . . . . . . . . . . . . . . . . . 255

**D.  Electromagnetic Radiation** ............................. 257
  D.1  Radiation Constants ................................. 257
  D.2  Bending Magnet Radiation ........................... 258
  D.3  Periodic Insertion Devices ........................... 261
    D.3.1  Insertion Device Parameter ...................... 261
    D.3.2  Field Scaling for Hybrid Wiggler Magnets .......... 262
    D.3.3  Particle Beam Parameter ........................ 262
  D.4  Undulator Radiation ................................. 263
  D.5  Photon Beam Brightness ............................. 265
    D.5.1  Effective Source Parameter ...................... 265

**References** ................................................ 267

**Index** .................................................... 269

# 1. Charges and Fields

Ever since J.C. Maxwell formulated his unifying electromagnetic theory in 1873, the phenomenon of electromagnetic radiation has fascinated the minds of theorists as well as experimentalists. The idea of displacement currents was as radical as it was important to describe electromagnetic waves. It was only fourteen years later when G. Hertz in 1887 succeeded to generate, emit and receive again electromagnetic waves, thus, proving experimentally the existence of such waves as predicted by Maxwell's equations. The sources of the radiation are oscillating electric charges and currents in a system of metallic wires. In this text, we discuss the generation of electromagnetic radiation emitted by free electrons from first principles involving energy and momentum conservation as well as Maxwell's equations.

## 1.1 Radiation from Moving Charges

Analytical formulation of the emission of electromagnetic radiation posed a considerable challenge. Due to the finite speed of light one cannot make a snapshot to correlate the radiation field at the observer with the position of radiating charges. Rather, the radiation field depends on the position of the radiating charges some time earlier, at the retarded time, when the radiation was emitted. Already 1867 L. Lorenz included this situation into his formulation of the theory of electromagnetic fields and introduced the concept of retarded potentials. He did, however, not offer a solution to the retarded potentials of a point charge. Liénard [1] in 1898 and independently in 1900 Wiechert [2] derived for the first time expressions for retarded potentials of point charges like electrons. These potentials are now called the Liénard-Wiechert potentials relating the scalar and vector potential of electromagnetic fields at the observation point to the location of the emitting charges and currents at the time of emission. Using these potentials, Liénard was able to calculate the energy lost by electrons while circulating in a homogenous magnetic field.

In 1907 [3, 4] and 1912 [5] Schott formulated and published his classical theory of radiation from an orbiting electron. He was primarily interested in the spectral distribution of radiation and hoped to find an explanation for atomic radiation spectra. Verifying Liénard's conclusion on the energy loss, he derived the angular and spectral distribution and the polarization of radiation. Since this classical approach to explain atomic spectra was destined to fail, his paper was forgotten and only forty years later were many of his findings rediscovered.

### 1.1.1 Why do Charged Particles Radiate?

Before we dive into the theory of electromagnetic radiation in more detail we may first ask ourselves why do charged particles radiate at all? Emission of electromagnetic radiation from charged particle beams (microwaves or synchrotron radiation) is a direct consequence of the finite velocity of light. A charged particle in uniform motion through vacuum is the source of electric field lines emanating from the charge radially out to infinity. While the charged particle is at rest or moving uniformly these field lines also are at rest or in uniform motion together with the particle. Now, we consider a particle being suddenly accelerated for a short time. That means the field lines should also be accelerated. The fact that the particle has been accelerated is, however, still known only within the event horizon in a limited area close to the particle. The signal of acceleration travels away from the source (particle) only at the finite speed of light. Field lines close to the charged particle are directed radially toward the particle, but far away, the field lines still point to the location where the particle would be had it not been accelerated. Somewhere between those two regimes the field lines are distorted and it is this distortion travelling away from the particle at the speed of light what we call electromagnetic radiation. The magnitude of these field distortions is proportional to the acceleration.

In a linear accelerator, for example, electrons are accelerated along the linac axis and therefore radiate. The degree of actual acceleration, however, is very low because electrons in a linear accelerator travel close to the velocity of light. The closer the particle velocity is to the velocity of light the smaller is the actual acceleration gained from a given force, and the radiation intensity is very small. In a circular accelerator like a synchrotron , on the other hand, particles are deflected transversely to their direction of motion by magnetic fields. Orthogonal acceleration or the rate of change in transverse velocity is very large because the transverse particle velocity can increase from zero to very large values in a very short time while passing through the magnetic field. Consequently, the emitted radiation intensity is very large. Synchrotron radiation sources come therefore generally in form of circular synchrotrons. Linear accelerators can be the source of intense synchrotron radiation in conjunction with a transversely deflecting magnet.

### 1.1.2 Spontaneous Synchrotron Radiation

Charged particles do not radiate while in uniform motion, but during acceleration a rearrangement of its electric fields is required and this field perturbation, traveling away from the charge at the velocity of light, is what we observe as electromagnetic radiation. Free accelerated electrons radiate similarly to those in a radio antenna, although now the source (antenna) is moving. Radiation from a fast moving particle source appears to the observer in the laboratory as being all emitted in the general direction of motion of the particle. This forward collimation is particularly effective for highly relativistic electrons where most of the radiation is concentrated into a small cone around the forward direction with an opening angle of $1/\gamma$, typically 0.1 to 1 mrad, where $\gamma$ is the particle energy in units of its rest mass.

Radiation can be produced by magnetic deflection in a variety of ways. Whether it be a single kick-like deflection or a periodic right-left deflection, the radiation characteristics reflect the particular mode of deflection. Specific radiation characteristics can be gained through specific modes of deflections. Here, we will only shortly address the main processes of radiation generation and come back later for a much more detailed discussion of the physical dynamics.

In an *undulator* the electron beam is periodically deflected transversely to its direction of motion by weak sinusoidally varying magnetic fields, generating periodic perturbations of the electric field lines. A receiving electric field detector recognizes a periodic variation of the transverse electromagnetic field components and interprets this as quasi monochromatic radiation. In everyday life periodic acceleration of electrons occurs in radio and TV antennas and we may receive these periodic field perturbations with a radio or TV receiver tuned to the frequency of the periodic electron motion in the emitting antenna. The fact that we consider relativistic electrons is not fundamental, but we restrict ourselves in this text to high energy electrons only.

To the particle the wavelength of the emitted radiation is equal to the undulator period length ($\lambda_p$) divided by $\gamma$ due to relativistic Lorentz contraction. In a stationary laboratory system, this wavelength appears to the observer further reduced by another factor $2\gamma$ due to the Doppler effect. The undulator period length of the order of centimeters is thus reduced by a factor $\gamma^2$ ($10^6$ $10^8$) to yield short wavelength radiation in the VUV and x-ray regime. The spectral resolution of the radiation is proportional to the number of undulator periods $N_p$ and its wavelength can be shifted by varying the magnetic field. Most radiation is emitted within the small angle of $(\gamma\sqrt{N_p})^{-1}$.

Increasing the magnetic field strength causes the pure sinusoidal transverse motion of electrons in an undulator to become distorted due to relativistic effects generating higher harmonic perturbations of the electron trajectory. Consequently, the monochromatic undulator spectrum exhibits higher harmonics and changes into a line spectrum. For very strong fields, many harmonics are generated which eventually merge into a continuous spectrum from IR to hard x-rays. In this extreme, we call the source magnet a *wig-*

*gler magnet*. The spectral intensity varies little over a broad wavelength range and drops off exponentially at photon energies higher than the critical photon energy, $\varepsilon_{crit} \propto B\gamma^2$. Changing the magnetic field, one may vary the critical photon energy to suit experimental requirements. Compared to bending magnet radiation, wiggler radiation is enhanced by the number of magnet poles $N_p$ and is well collimated within an angle of $1/\gamma$ to say $10/\gamma$, or a few mrad.

A *bending magnet* is technically the most simple radiation source. Radiation is emitted tangentially to the orbit similar to a search light while well collimated in the non-deflecting, or vertical plane. The observer at the experimental station sees radiation from only a small fraction of the circular path which can be described as a piece of a distorted sinusoidal motion. The radiation spectrum is therefore similar to that of a wiggler magnet while the intensity is due to only one pole. Because bending magnets define the geometry of the electron beam  transport system or accelerator, it is not possible to freely choose the field strength and the critical photon energy is therefore fixed. Sometimes, specially in lower energy storage rings, it is desirable to extend the radiation spectrum to higher photon energies into the x-ray regime. This can be accomplished by replacing one or more conventional bending magnet with a superconducting magnet, or superbends, at much higher field strength. To preserve the ring geometry the length of these superbends must be chosen such that the deflection angle is the same as it was for the conventional magnet that has been replaced. Again, superbends are part of the ring geometry and therefore the field cannot be changed.

A more flexible version of a radiation hardening magnet is the wavelength shifter. This is a magnet which consists of a high field central pole and two weaker outside poles to compensate the deflection by the central pole. The total deflection angle is zero and therefore the field strength can be chosen freely to adjust the critical photon energy. It's design is mostly based on superconducting magnet technology, particularly in low energy accelerators, to extend (shift) the critical photon energy available from bending magnets to higher values.

A variety of more complicated magnetic field arrangements have been developed to primarily generate *circularly* or *elliptically polarized radiation*. In such magnets horizontal as well as vertical magnetic fields are sequentially employed to deflect electrons into some sort of helical motion giving raise to the desired polarization effect.

### 1.1.3 Stimulated Radiation

The well defined time structure and frequency of undulator radiation can be used to stimulate the emission of even more radiation. In an *optical klystron* [6] coherent radiation with a wavelength equal to the fundamental undulator wavelength enters an undulator together with the electron beam. Since the electron bunch length is much longer than the radiation wavelength, some electrons loose energy to the radiation field and some electrons gain energy from the radiation field while interacting with the radiation field. This energy modulation can be transformed into a density modulation by passing the modulated electron beam through a dispersive section. This section consists of deflecting magnetic fields arranged in such a way that the total path length through the dispersive section depends on the electron energy. The periodic energy modulation of the electron bunch then converts into a periodic density modulation. Now we have microbunches at a distance of the undulator radiation wavelength. This microbunched beam travels through a second undulator where again particles can loose or gain energy from the radiation field. Due to the microbunching, however, most particles are concentrated at phases where there is only energy transfer from the particle to the radiation field, thus providing a high gain of radiation intensity.

In a more efficient variation of this principle, radiation emitted by electrons passing through an undulator is recycled by optical mirrors in such a way that it passes through the same undulator again together with another electron bunch. The external field stimulates more emission of radiation from the electrons, and is again recycled to stimulate a subsequent electron bunch until there are no more bunches in the electron pulse. Generating from a linear accelerator a train of thousands of electron bunches one can generate a large number of interactions, leading to an exponential growth of electromagnetic radiation. Such a devise is called a *free electron laser, FEL*.

### 1.1.4 Electron Beam

In this text we consider radiation from relativistic electron beams. Such beams can be generated efficiently by acceleration in microwave fields. The oscillatory nature of microwaves makes it impossible to produce a uniform stream of particles, and the electron beam is modulated into bunches at the distance of the microwave wavelength. Typically the bunch length is a few percent of the wavelength. The circumference of the storage ring must be an integer multiple, the harmonic number, of the rf-wavelength. The rf-system actually provides potential wells, rf-buckets, which rotate around the ring. These buckets may or may not be filled with electrons and those electrons contained in a bucket are said to form an electron bunch. With special equipment in the injector it is possible to store any arbitrary pattern of electron bunches consistent with the equidistant distribution of the finite number of buckets equal to the harmonic number. Specifically, it is possible to operate the storage ring with all buckets filled or with just a single bunch or only a few bunches. The bunched nature of the electron beam and the fact that these bunches circulate in a storage ring determines the time structure and spectrum of the emitted radiation. Typically, the bunch length in storage rings is 30 100 ps at a distance of 2 3 ns depending on the rf-frequency.

During the storage time of the particle beam, the electrons radiate and it is this radiation that is extracted and used in experiments of basic and applied research. Considering, for example, only one bunch rotating in the storage ring, the experimenter would observe a light flash at a frequency equal to the revolution frequency $f_{\mathrm{rev}}$. Because of the extremely short duration of the light flash many harmonics of the revolution frequency appear in the light spectrum. At the low frequency end of this spectrum, however, no radiation can be emitted for wavelength longer than about the dimensions of the metallic vacuum chamber surrounding the electron beam. For long wavelengths the metallic boundary conditions for electromagnetic fields cannot be met prohibiting the emission of radiation. Practically, useful radiation is observed from storage rings only for wavelengths below the microwave regime, or for $\lambda \lesssim 1$ mm.

## 1.2 Maxwell's Equations

Radiation theory deals largely with the description of charged particle dynamics in the presence of external electromagnetic fields or of fields generated by other charged particles. We use Maxwell's equations in a vacuum or material environment with uniform permittivity $\epsilon$ and permeability $\mu$ to describe these fields. Furthermore, the Lorentz force provides the tool to formulate particle dynamics under the influence of these electromagnetic fields[1].

$$\boldsymbol{\nabla E} = \frac{4\pi}{[4\pi\epsilon_0]\,\epsilon_{\mathrm{r}}}\,\rho\,, \tag{1.1}$$

$$\boldsymbol{\nabla B} = 0\,, \tag{1.2}$$

Faraday's law :

$$\boldsymbol{\nabla}\times\boldsymbol{E} = -\frac{[c]}{c}\frac{\partial\boldsymbol{B}}{\partial t}\,, \tag{1.3}$$

Ampère's law :

$$\boldsymbol{\nabla}\times\boldsymbol{B} = \frac{4\pi}{c}\left[\frac{c}{4\pi}\right][\mu_0]\mu_{\mathrm{r}}\,\rho\boldsymbol{v} + [c\epsilon_0\mu_0]\frac{\epsilon_{\mathrm{r}}\,\mu_{\mathrm{r}}}{c}\frac{\partial\boldsymbol{E}}{\partial t}\,. \tag{1.4}$$

Here, $\rho$ is the charge density and $\boldsymbol{v}$ the velocity of the charged particle. In general, we are interested in particle dynamics in a material free environment and set therefore $\epsilon_{\mathrm{r}} = 1$ and $\mu_{\mathrm{r}} = 1$. For specific discussions we do, however, need to calculate fields in a material filled environment in which case we come back to this more general form of Maxwell's equations.

### 1.2.1 Conversion from cgs to MKS Units

In related literature we are faced with different choices of a system of units, mostly the cgs-or the MKS-system of units. While the cgs-system provides

---

[1] In this text we formulate equations both in the MKS-system (include factors in square brackets) and the cgs-system (ignore square brackets).

simple formulation of physical laws without the use of artificial factors, we cannot escape reality where we quantify and measure our results. In this text, we try to avoid differences in the formulation of physical laws between these systems whenever possible. Where this is not possible we include into the formulas factors in square brackets like $[4\pi\epsilon_0]$ which are to be used in case of MKS-units and to be ignored in case of cgs-units.

Generally, we use MKS-units to quote numerical results or expressing practical formulas. Sometimes, however, we find it necessary to perform numerical calculations with parameters given in different units or to compare with results given in another system of units. For such cases some helpful numerical conversions are compiled in Table 1.1.

**Table 1.1.** Numerical conversion factors

| quantity | label | replace cgs units | by SI units |
|---|---|---|---|
| voltage | $U$ | 1 esu | 300 V |
| electric field | $E$ | 1 esu | $3\ 10^4$ V/cm |
| current | $I$ | 1 esu | $10\ c = 2.9979\ 10^9$ A |
| charge | $q$ | 1 esu | $(10c)^{-1} = 3.3356\ 10^{-10}$ C |
| resistance | $R$ | 1 s/cm | $8.9876\ 10^{11}\ \Omega$ |
| capacitance | $C$ | 1 cm | $\frac{1}{8.9876}10^{-11}$ F |
| inductance | $L$ | 1 cm | $1\ 10^9$ Hy |
| magnetic induction | $B$ | 1 Gauss | $3\ 10^{-4}$ Tesla |
| magnetic field | $H$ | 1 Oersted | $\frac{1000}{4\pi} = 79.577$ A/m |
| force | $f$ | 1 dyn | $10^{-5}$ N |
| energy | $E$ | 1 erg | $10^{-7}$ J |

Analogous conversion factors can be derived for electromagnetic quantities in formulas. Table 1.2 includes some of the most frequently used conversions from cgs to MKS-units which were used in this text to define the square bracket factors.

The dielectric constant or permittivity of free space is

$$\epsilon_0 = \frac{10^7}{4\pi c^2}\frac{\text{C}}{\text{V m}} = 8.854187817 \times 10^{-12}\frac{\text{C}}{\text{V m}}, \tag{1.5}$$

and the magnetic permeability

$$\mu_0 = 4\pi \times 10^{-7}\frac{\text{V s}}{\text{A m}} = 1.2566370614 \times 10^{-6}\frac{\text{V s}}{\text{A m}}. \tag{1.6}$$

Both constants are related to the speed of light $v$ by

$$\epsilon_0\epsilon_{\text{r}}\mu_0\mu_{\text{r}}\,v^2 = 1\,, \tag{1.7}$$

**Table 1.2.** Equation conversion factors

| variable | replace cgs variable | by SI Variable |
|---|---|---|
| potential, voltage | $V_{cgs}$ | $\sqrt{4\pi\epsilon_0}\, V_{MKS}$ |
| electric field | $E_{cgs}$ | $\sqrt{4\pi\epsilon_0}\, E_{MKS}$ |
| current, current density | $I_{cgs}, j_{cgs}$ | $\frac{1}{\sqrt{4\pi\epsilon_0}}\, I_{MKS}, j_{MKS}$ |
| charge, charge density | $q, \rho$ | $\frac{1}{\sqrt{4\pi\epsilon_0}}\, q_{MKS}, \rho_{MKS}$ |
| resistance | $R_{cgs}$ | $\sqrt{4\pi\epsilon_0}\, R_{MKS}$ |
| capacitance | $C_{cgs}$ | $\frac{1}{\sqrt{4\pi\epsilon_0}}\, C_{MKS}$ |
| inductance | $L_{cgs}$ | $\sqrt{4\pi\epsilon_0}\, L_{MKS}$ |
| magnetic induction | $B_{cgs}$ | $\sqrt{\frac{4\pi}{\mu_0}}\, B_{MKS}$ |

or in vacuum by

$$\epsilon_0 \mu_0\, c^2 = 1\,. \tag{1.8}$$

### 1.2.2 Lorentz Force

Whatever the interaction of charged particles with electromagnetic fields and the reference system may be, in accelerator physics we depend on the invariance of the Lorentz force equations under coordinate transformations. All acceleration and beam guidance in accelerator physics will be derived from the Lorentz force which quantifies the force of an electric $\boldsymbol{E}$ and magnetic field $\boldsymbol{B}$ on a particle with charge $q$ by

$$\boldsymbol{F} = q\,\boldsymbol{E} + q\,\frac{[c]}{c}\,(\boldsymbol{v} \times \boldsymbol{B})\,. \tag{1.9}$$

Throughout this text, we use particles with one unit of electrical charge $e$ like electrons and protons unless otherwise noted. In case of multiply charged ions the single charge $e$ must be replaced by $eZ$ where $Z$ is the charge multiplicity of the ion. Both components of the Lorentz force are used in accelerator physics where the force due to the electrical field is mostly used to actually increase the particle energy while magnetic fields are mostly used to guide particle beams along desired beam transport lines. This separation of functions, however, is not exclusive as the example of the betatron accelerator shows where particles are accelerated by generating a time dependent magnetic field. Similarly, electrical fields are used in specific cases to guide or separate particle beams.

Integrating the Lorentz force over the interaction time of a particle with the field we get the change in its momentum,

$$\Delta\boldsymbol{p} = \int \boldsymbol{F}\,\mathrm{d}t\,. \tag{1.10}$$

On the other hand, if the Lorentz force is integrated with respect to the path length we get the change in kinetic energy $E_{\text{kin}}$ of the particle

$$\Delta E_{\text{kin}} = \int \boldsymbol{F} \, \mathrm{d}\boldsymbol{s} \, . \tag{1.11}$$

Comparing the last two equations we find with $\mathrm{d}\boldsymbol{s} = \boldsymbol{v} \mathrm{d}t$ the relation between the momentum and kinetic energy differentials

$$c\boldsymbol{\beta} \, \mathrm{d}\boldsymbol{p} = \mathrm{d}E_{\text{kin}} \, . \tag{1.12}$$

With the Lorentz force equation (1.9) and $\mathrm{d}\boldsymbol{s} = \boldsymbol{v} \mathrm{d}t$ in the second integral of (1.11) we get

$$\Delta E_{\text{kin}} = e \int \boldsymbol{E} \, \mathrm{d}\boldsymbol{s} + e \frac{[c]}{c} \int (\boldsymbol{v} \times \boldsymbol{B}) \, \boldsymbol{v} \, \mathrm{d}t \, . \tag{1.13}$$

The kinetic energy of a free particle increases whenever a finite electric field $\boldsymbol{E}$ component along the beam axis exists. This acceleration is independent of the particle velocity and acts even on a particle at rest $\boldsymbol{v} = 0$. Transverse field components do not affect the particle's kinetic energy, but do change its momentum vector (1.10). The second component of the Lorentz force, in contrast, depends on the particle velocity and is directed normal to the direction of propagation and normal to the magnetic field direction. The kinetic energy cannot be changed by the presence of magnetic fields since the scalar product $(\boldsymbol{v} \times \boldsymbol{B}) \boldsymbol{v} = 0$ vanishes. The magnetic field causes only a change in the transverse momentum (1.10) or a deflection of the particle trajectory.

The Lorentz force is used to derive the equation of motion of charged particles in the presence of electromagnetic fields

$$\frac{\mathrm{d}}{\mathrm{d}t} \boldsymbol{p} = \frac{\mathrm{d}}{\mathrm{d}t} (\gamma m \boldsymbol{v}) = e Z \boldsymbol{E} + e Z \frac{[c]}{c} (\boldsymbol{v} \times \boldsymbol{B}) \, , \tag{1.14}$$

noting that for ion accelerators the particle charge $e$ must be replaced by $eZ$. The fields can be derived from electrical and magnetic potentials

$$\boldsymbol{E} = -\frac{[c]}{c} \frac{\partial \boldsymbol{A}}{\partial t} - \nabla \phi \, , \tag{1.15}$$

$$\boldsymbol{B} = \nabla \times \boldsymbol{A} \, , \tag{1.16}$$

where $\phi$ is the scalar potential and $\boldsymbol{A}$ the vector potential. The particle momentum $\boldsymbol{p} = \gamma m \boldsymbol{v}$ and it's time derivative

$$\frac{\mathrm{d}\boldsymbol{p}}{\mathrm{d}t} = m\gamma \frac{\mathrm{d}\boldsymbol{v}}{\mathrm{d}t} + m\boldsymbol{v} \frac{\mathrm{d}\gamma}{\mathrm{d}t} \, . \tag{1.17}$$

With

$$\frac{d\gamma}{dt} = \frac{d}{d\beta} \frac{1}{\sqrt{1-\beta^2}} \frac{d\beta}{dt} = \gamma^3 \frac{\beta}{c} \frac{dv}{dt}$$

we get the equation of motion

$$\boldsymbol{F} = \frac{d\boldsymbol{p}}{dt} = m \left( \gamma \frac{d\boldsymbol{v}}{dt} + \gamma^3 \frac{\beta}{c} \frac{dv}{dt} \boldsymbol{v} \right). \tag{1.18}$$

For a force parallel to the particle propagation $\boldsymbol{v}$ we have $\dot{v}v = \dot{\boldsymbol{v}}v$ and

$$\frac{d\boldsymbol{p}_{\parallel}}{dt} = m\gamma \left( 1 + \gamma^2 \beta \frac{v}{c} \right) \frac{d\boldsymbol{v}_{\parallel}}{dt} = m\gamma^3 \frac{d\boldsymbol{v}_{\parallel}}{dt}. \tag{1.19}$$

On the other hand, if the force is directed normal to the particle propagation we have $\dot{v} = 0$ and (1.18) reduces to

$$\frac{d\boldsymbol{p}_{\perp}}{dt} = m\gamma \frac{d\boldsymbol{v}_{\perp}}{dt}. \tag{1.20}$$

It is evident from these results how differently the dynamics of particle motion is affected by the direction of the Lorentz force. Specifically, the dynamics of highly relativistic particles under the influence of electromagnetic fields depends greatly on the direction of the force with respect to the direction of particle propagation. The difference between parallel and perpendicular acceleration has a great impact on the design of electron accelerators. The ultimate achievable electron energy depends greatly on the type of accelerator due to the emission of synchrotron radiation. This limitation is most severe for electrons in circular accelerators where the magnetic forces act perpendicular to the propagation compared to the acceleration in linear accelerators where the accelerating fields are parallel to the particle propagation. This argument is also true for protons or for that matter, any charged particle, although the much larger mass renders the amount of synchrotron radiation negligible except for extremely high energies.

## 1.3 The Lorentz Transformations

Beam dynamics is expressed in a fixed laboratory system of coordinates but some specific problems are better discussed in the moving coordinate system of a single particle or of the center of charge for a collection of particles. We use therefore frequently transformations of coordinates as well as fields between the laboratory frame of reference and the particle rest frame or some other suitable moving reference frame.

### 1.3.1 Lorentz Transformation of Coordinates

To define the transformation of coordinates we consider two reference systems of which one is fixed to the laboratory $\mathcal{L}$, and the other $\mathcal{L}^*$ is attached to the particle moving with respect to $\mathcal{L}$. For simplicity, we assume that the the particle and with it the system $\mathcal{L}^*$ is moving with velocity $v_z$ along the positive $z$-axis of system $\mathcal{L}$. Transformation between the two reference systems is effected through a Lorentz transformation

$$
\begin{aligned}
x^* &= x \,, \\
y^* &= y \,, \\
z^* &= \frac{z - \beta_z\, ct}{\sqrt{1 - \beta_z^2}} = \gamma\,(z - \beta_z\, ct) \,, \\
ct^* &= \frac{ct - \beta_z\, z}{\sqrt{1 - \beta_z^2}} = \gamma\,(ct - \beta_z\, z) \,,
\end{aligned}
\tag{1.21}
$$

where $\beta_z = v_z/c$, $\gamma$ is the total particle energy $E$ in units of the particle rest energy $mc^2$

$$
\gamma = \frac{E}{m\,c^2} = \frac{1}{\sqrt{1 - \beta_z^2}} \,,
\tag{1.22}
$$

and quantities designated with $^*$ are measured in the moving system $\mathcal{L}^*$. These Lorentz transformations can be expressed in matrix formulation by

$$
\begin{pmatrix} x^* \\ y^* \\ z^* \\ ct^* \end{pmatrix} = \begin{pmatrix} 1 & 0 & 0 & 0 \\ 0 & 1 & 0 & 0 \\ 0 & 0 & \gamma & -\beta\gamma \\ 0 & 0 & -\beta\gamma & \gamma \end{pmatrix} \begin{pmatrix} x \\ y \\ z \\ ct \end{pmatrix} \,.
\tag{1.23}
$$

Characteristic for relativistic mechanics is the Lorentz contraction and time dilatation, both of which become significant in the description of particle dynamics. To describe the Lorentz contraction, we consider a rod at rest in $\mathcal{L}$ along the $z$-coordinate with the length $\Delta z = z_2 - z_1$. In the system $\mathcal{L}^*$ this rod appears to have the length $\Delta z^* = z_2^* - z_1^*$ related to the length in the $\mathcal{L}$−system by

$$
\Delta z = \gamma\,(z_2^* - v_z t^*) - \gamma\,(z_1^* - v_z t^*) = \gamma\,\Delta z^*
$$

or

$$
\Delta z = \gamma\,\Delta z^* \,.
\tag{1.24}
$$

The rod appears shorter in the moving particle system by the factor $\gamma$ and is longest in it's rest system.

Because of the Lorentz contraction, the volume of a body at rest in the system $\mathcal{L}$ appears also reduced in the moving system $\mathcal{L}^*$ and we have for the volume of a body in three dimensional space

$$V = \gamma V^* . \tag{1.25}$$

Only one dimension of this body is Lorentz contracted and therefore the volume scales only linearly with $\gamma$. As a consequence, the charge density $\rho$ of a particle bunch with the volume $V$ is lower in the laboratory system $\mathcal{L}$ compared to the density in the system $\mathcal{L}^*$ moving with this bunch and becomes

$$\rho = \frac{\rho^*}{\gamma} . \tag{1.26}$$

Similarly, we may derive the time dilatation or the elapsed time between two events occurring at the same point in both coordinate systems. From the Lorentz transformations we get with $z_2^* = z_1^*$

$$\Delta t = t_2 - t_1 = \gamma \left( t_2^* - \frac{\beta_z z_2^*}{c} \right) - \gamma \left( t_1^* - \frac{\beta_z z_1^*}{c} \right)$$

or

$$\Delta t = \gamma \, \Delta t^* . \tag{1.27}$$

For a particle at rest in the moving system $\mathcal{L}^*$ the time $t^*$ varies slower than the time $t$ in the laboratory system. This is the mathematical expression for the famous *twin paradox* where one of the twins moving in a space capsule at relativistic speed would age more slowly than the other twin who remains behind. This phenomenon becomes reality for unstable particles. For example, high-energy pion mesons, observed in the laboratory system, have a longer lifetime by the factor $\gamma$ compared to low-energy pions with $\gamma = 1$. As a consequence we are able to transport high-energy pion beams a longer distance than low energy pions before they decay. This effect is of great importance, for example, where a pion beam is used for cancer therapy and must be transported from the high radiation environment of the source to a safe location behind shielding walls where the patient can be placed.

**Lorentz transformation of fields.** Electromagnetic fields and the interaction of charged particles with these fields play an important role in accelerator physics. We find it often useful to express the fields in either the laboratory system or the particle system. Transformation of the fields from one to the other system is determined by the Lorentz transformation of electromagnetic fields. We assume again the coordinate system $\mathcal{L}^*$ to move with the particle at a velocity $v_z$ along the positive $z$-axis with respect to a right-handed $(x, y, z)$ reference frame $\mathcal{L}$. The electromagnetic fields in this moving reference frame

can be expressed in terms of the fields in the laboratory frame of reference
$\mathcal{L}$:

$$
\begin{pmatrix} E_x^* \\ E_y^* \\ E_z^* \\ [c]B_x^* \\ [c]B_y^* \\ [c]B_z^* \end{pmatrix} = \begin{pmatrix} \gamma & 0 & 0 & 0 & -\beta_z\gamma & 0 \\ 0 & \gamma & 0 & \beta_z\gamma & 0 & 0 \\ 0 & 0 & 1 & 0 & 0 & 0 \\ 0 & \beta_z\gamma & 0 & \gamma & 0 & 0 \\ -\beta_z\gamma & 0 & 0 & 0 & \gamma & 0 \\ 0 & 0 & 0 & 0 & 0 & 1 \end{pmatrix} \begin{pmatrix} E_x \\ E_y \\ E_z \\ [c]B_x \\ [c]B_y \\ [c]B_z \end{pmatrix}. \tag{1.28}
$$

These transformations exhibit interesting features for accelerator physics,
where we often use magnetic or electrical fields, which are pure magnetic
or pure electric fields when viewed in the laboratory system. For relativistic
particles, however, these pure fields become a combination of electric and
magnetic fields.

### 1.3.2 Energy and Momentum

The total energy of a particle is given by

$$
E = \gamma E_0 = \gamma mc^2, \tag{1.29}
$$

where $E_0 = mc^2$ is the rest energy of the particle. The kinetic energy is
defined as the total energy minus the rest energy

$$
E_{\text{kin}} = E - E_0 = (\gamma - 1)mc^2. \tag{1.30}
$$

In discussions of energy gain through acceleration we consider only energy
differences and need therefore not to distinguish between total and kinetic
energy. The particle momentum finally is defined by

$$
c^2 p^2 = E^2 - E_0^2 \tag{1.31}
$$

or

$$
cp = \sqrt{E^2 - E_0^2} = mc^2\sqrt{\gamma^2 - 1} = \beta\gamma mc^2 = \beta E, \tag{1.32}
$$

where $\beta = v/c$. The simultaneous use of the terms energy and momentum
might seem sometimes to be misleading. In this text, we use physically correct
quantities in mathematical formulations even though we sometimes use the
term "energy" for the quantity $cp$ rather than calling it the particle momentum. In high energy electron accelerators, the numerical distinction between
energy and momentum is insignificant and both "energy" and "momentum"
are used synonymous. For proton accelerators and even more so for heavy
ion accelerators the difference in both quantities is significant.

Often we need differential expressions or expressions for relative variations
of a quantity in terms of variations of another quantity. Such relations can
be derived from the definitions in this section. From the variation of $cp = mc^2\sqrt{\gamma^2 - 1}$ we get, for example,

$$cdp = \frac{mc^2}{\beta}\, d\gamma = \frac{dE}{\beta} = \frac{dE_{\text{kin}}}{\beta} \qquad (1.33)$$

and

$$\frac{cdp}{cp} = \frac{1}{\beta^2}\frac{d\gamma}{\gamma}. \qquad (1.34)$$

Varying $cp = \gamma\beta\, mc^2$ and eliminating $d\gamma$ we get

$$cdp = \gamma^3\, mc^2\, d\beta \qquad (1.35)$$

and

$$\frac{cdp}{cp} = \gamma^2 \frac{d\beta}{\beta}. \qquad (1.36)$$

In a similar way other relations can be derived.

## Exercises *

**Exercise 1.1 (S).** Use the definition for $\beta$, the momentum, the total and kinetic energy and derive expressions $p(\beta, E_{\text{kin}})$, $p(E_{\text{kin}})$, and $E_{\text{kin}}(\gamma)$.

**Exercise 1.2 (S).** Simplify the expressions obtained in exercise 1.1 for large energies, $\gamma \gg 1$. Derive from the relativistic expressions the classical nonrelativistic formulas.

**Exercise 1.3 (S).** Protons are accelerated to a kinetic energy of 200 MeV at the end of the Fermilab *Alvarez linear accelerator*. Calculate their total energy, their momentum and their velocity in units of the velocity of light.

**Exercise 1.4 (S).** Consider electrons to be accelerated in the 3 km long SLAC linear accelerator with a uniform gradient of 20 MeV/m. The electrons have a velocity $v = c/2$ at the beginning of the linac. What is the length of the linac in the rest frame of the electron? Assume the particles at the end of the 3 km long linac would enter another 3 km long tube and coast through it. How long would this tube appear to be to the electron?

**Exercise 1.5 (S).** An electron beam orbits in a circular accelerator with a circumference of 300 m at an average current of 250 mA and the beam consists of 500 equally spaced bunches each 1 cm long. How many particles are orbiting? How many electrons are there in each bunch? Assuming, the time structure of synchrotron radiation is the same as the electron beam time structure specify and plot the radiation time structure in a photon beam line.

---

* The argument (S) indicates an exercise for which a solution is given in Appendix A.

**Exercise 1.6 (S).** A $\pi^\pm$-meson is created at a kinetic energy of $E_{\mathrm{kin}} = 100$ MeV. Calculate it's velocity. What is the probability $P = \exp(-t/\tau_0)$ for the $\pi^\pm$-meson to travel 15 m before it decays. The pion half lifetime at rest is $\tau_0 = 26.0$ ns. This result is important for pion cancer therapy facilities. The pions are created in a highly radioactive environment where a high intensity proton beam strikes a target to produce pion-mesons. The beam line to carry the pions to the patient outside the thick radiation shielding wall is about 15 m long.

**Exercise 1.7 (S).** The half-life of muons $\mu$ at rest is $\tau_0 = 2.20$ $\mu$s. In 1941 the muon flux was measured on top of Mount Washington (2000 m above sea level) at 570 counts per hour. Another measurement at sea level detected 400 muons per hour. Estimate the kinetic energy of the cosmic ray muons $\left(m_\mu c^2 = 105.7 \text{ MeV}\right)$.

**Exercise 1.8 (S).** Consider a relativistic electron traveling along the $z$-axis. In its own system, the electrical field lines extend radially from the charge. Considering only the $xz$-plane, derive an expressions for the electrical field lines in the laboratory frame of reference. Sketch the field pattern in the electron rest frame and in the laboratory system of reference.

**Exercise 1.9 (S).** A circular accelerator with a circumference of 100 m contains a uniform distribution of singly charged particles orbiting with the speed of light. If the circulating current is 1 amp, how many particles are orbiting? We instantly turn on an ejection magnet so that all particles leave the accelerator during the time of one revolution. What is the peak current at the ejection point? How long is the current pulse duration? If the accelerator is a synchrotron accelerating particles at a rate of 100 acceleration cycles per second, what is the average ejected particle current?

**Exercise 1.10.** Verify the correctness of the unit conversion factors in Maxwell's equations (1.1) through (1.4).

**Exercise 1.11.** Show that $\frac{d\gamma}{dt} = \gamma^3 \beta \frac{d\beta}{dt}$.

**Exercise 1.12.** Determine the kinetic energy of an electron, a proton, and a $Ar^+$-ion which travel together at a speed of $0.95\,c$.

**Exercise 1.13.** How far will a 200 MeV pion travel before it has a 50% probability to decay?

**Exercise 1.14.** In a storage ring of 800 m circumference a total of $3 \times 10^{12}$ relativistic electrons are circulating. They are distributed into 300 bunches. Show that the circulating beam current is 180 mA and the bunch current is 0.6 mA. Sketch the temporal structure of the photon beam.

**Exercise 1.15.** A total of $2.34 \times 10^{12}$ electrons orbit in a circular accelerator with a revolution frequency of 1 MHz. Show that the circumference of the accelerator is 300 m and the total beam current 375 mA. The total beam be subdivided into 500 equally distant bunches. The instantaneous bunch current is 20.8 A. Show, that the bunch length is 1.2 cm and the distance between bunches 60 cm. Is the frequency of the rf-system 500 MHz? Why?

**Exercise 1.16.** Use the accelerator of exercise 1.15, but fill only every 5$^{\text{th}}$ bunch. Keep the total circulating beam current. Show that the temporal distance between photon pulses is 180 ps. Now fill the ring only with one bunch for timing experiments. What is the temporal distance between photon pulses?

**Exercise 1.17.** Assume a beam of $1.46 \times 10^{12}$ electrons circulating in a 243 m storage ring. Due to gas scattering, you loose a fraction $10^{-5}$ per second. Show that is takes 27.8 hours for the beam current to decay to 110 mA. What is the initial circulating beam current? How many electrons do you loose in the first turn? and how many per turn after 27.8 hours?

# 2. Fundamental Processes

The emission of electromagnetic radiation from free electrons is a classical phenomenon. We may therefore use a visual approach to gain some insight into conditions and mechanisms of radiation emission. First, we will discuss necessary conditions that must be met to allow an electron to emit or absorb a photon. Once such conditions are met, we derive from energy conservation a quantity, the Poynting vector, relating energy transport or radiation to electromagnetic fields. This will give us the basis for further theoretical definitions and discussions of radiation phenomena.

## 2.1 Conservation Laws and Radiation

The emission of electromagnetic radiation involves two components, the electron and the radiation field. For the combined system energy-momentum conservation must be fulfilled. These conservation laws impose very specific selection rules on the kind of emission processes possible. To demonstrate this, we plot the energy versus momentum for both electron and photon. In relativistic terms, we have the relation $\gamma = \sqrt{1 + (\beta\gamma)^2}$ between energy $\gamma$ and momentum $\beta\gamma$. For consistency in quantities used we normalize the photon energy to the electron rest energy, $\gamma_p = \varepsilon_p/mc^2$, where $\varepsilon_p$ is the photon energy and $mc^2$ the electron rest mass. Similarly, we express the speed of light by $\beta_p = c_p/c = 1/n$ where $n > 1$ is the refractive index of the medium surrounding the photon. With these definitions and assuming, for now, vacuum as the medium ($n = 1$) the location of a particle or photon in energy-momentum space is shown in Fig. 2.1(left).

Energy and momentum of a particle are related such that it must be located on the "particle"-line in Fig. 2.1(left). Similarly, a photon is always located on the "photon"-line. Transfer of energy between particle and photon must obey energy-momentum conservation. In Fig. 2.1(right) we apply this principle to a free electron in vacuum emitting (absorbing) a photon. To create a photon the electron would have to loose (gain) an amount of momentum which is numerically equal to the energy gained (lost) by the photon. Clearly, in this case the electron would end up at a location off the "particle"-line, thus violating momentum conservation. That cannot be, and

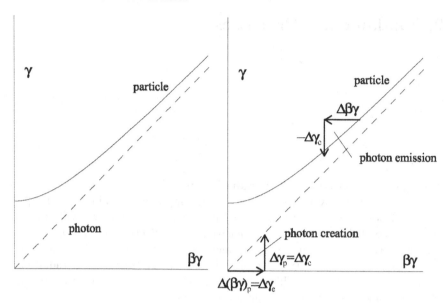

**Fig. 2.1.** Energy-momentum relationship for particles and photons (*left*). Violation of energy or momentum conservation during emission and absorption of electromagnetic radiation by a free electron travelling in perfect vacuum ($\beta_p = 1$) (*right*)

such a process is therefore not permitted. A free electron in vacuum cannot emit or absorb a photon without violating energy-momentum conservation.

### 2.1.1 Cherenkov Radiation

We have been careful to assume an electron in perfect vacuum. What happens in a material environment is shown in Fig. 2.2. Because the refractive index $n > 1$, the phase velocity of radiation is less than the velocity of light in vacuum and with $\beta = 1/n$, the "photon"-line is tilted towards the momentum axis.

Formally, we obtain this for a photon from the derivative $d\gamma/d(\beta\gamma)$, which we expand to $\frac{d\gamma}{d(\beta\gamma)} = \frac{d\gamma}{d\omega}\frac{d\omega}{dk}\frac{dk}{d(\beta\gamma)}$ and get with $\gamma = \hbar\omega/mc^2, k = n\frac{\omega}{c}$, and the momentum $\beta\gamma = \frac{\hbar}{mc}k$, the derivative

$$\frac{d\gamma_p}{d(\beta\gamma)_p} = \frac{1}{n} < 1, \tag{2.1}$$

where we have added the subscript $_p$ to differentiate between photon and electron parameters.

The dispersion function for a photon in a material environment has a slope less than unity as shown in Fig. 2.2. In this case, the numerical value of

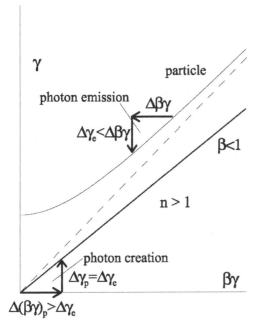

**Fig. 2.2.** Energy and momentum conservation in a refractive environment with $n > 1$

the photon momentum is less than the photon energy, analogous to the particle case. To create a photon of energy $\gamma_p$ we set $\gamma_p = -\Delta\gamma = -\beta\Delta\beta\gamma$ from (1.30), where from (2.1) the photon energy $\gamma_p = \frac{1}{n}(\beta\gamma)_p$ and get from both relations $(\beta\gamma)_p = -n\beta\Delta\beta\gamma$. Because of symmetry, no momentum transverse to the particle trajectory can be exchanged, which means radiation is emitted uniformly in azimuth. The change in longitudinal momentum along the trajectory is $-\Delta\beta\gamma = (\beta\gamma)_p \big|_{\parallel} = (\beta\gamma)_p \cos\theta$. In a dielectric environment, free electrons can indeed emit or absorb a photon although, only in a direction given by the angle $\theta$ with respect to the electron trajectory. This radiation is called Cherenkov radiation, and the Cherenkov angle $\theta$ is given by the Cherenkov condition

$$n\beta\cos\theta = 1. \tag{2.2}$$

Note that this condition is not the same as saying whenever an electron passes though a refractive medium with $n > 1$ there is Cherenkov radiation. The Cherenkov condition requires that $n\beta > 1$ which is, for example, not the case for an electron beam of less than 20 MeV traveling through air.

### 2.1.2 Compton Radiation

To generate electromagnetic radiation from free electrons in vacuum without violating energy-momentum conservation, we may employ the Compton effect which is the scattering of an incoming photon by the electron. In energy-momentum space this process is shown in Fig. 2.3. The electron, colliding head-on with an incoming photon absorbs this photon and emits again a photon of different energy. In this process it gains energy but looses momentum bringing the electron in the energy-momentum space to an intermediate point, from where it can reach its final state on the "particle"-line by emitting a photon as shown in Fig. 2.3. This is the process involved in the generation of synchrotron radiation. Static magnetic fields in the laboratory system appear as electromagnetic fields like an incoming (virtual) photon in the electron system with which the electron can collide. Energy-momentum conservation give us the fundamental and necessary conditions under which a free charged particle can emit or absorb a photon. We turn our attention now to the actual interaction of charged particles with an electromagnetic field.

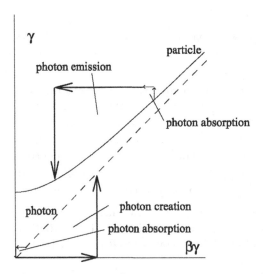

**Fig. 2.3.** Energy and momentum conservation for Compton scattering process

## 2.2 The Poynting Vector

The rate of work done in a charged particle-field environment is defined by the Lorentz force and the particle velocity

$$\boldsymbol{F}_{\mathrm{L}}\boldsymbol{v} = \left( e\boldsymbol{E} + e\frac{[c]}{c}\,[\boldsymbol{v}\times\boldsymbol{B}] \right)\boldsymbol{v}\,. \tag{2.3}$$

Noting that $[\boldsymbol{v}\times\boldsymbol{B}]\,\boldsymbol{v} = 0$ we set $e\boldsymbol{E}\boldsymbol{v} = \boldsymbol{j}\boldsymbol{E}$ and the total rate of work done by all particles and fields is the integral over all particles and fields

$$\int \boldsymbol{j}\boldsymbol{E}\mathrm{d}V = \frac{c}{4\pi}\,[4\pi\epsilon_0]\int \left( [c]\,\boldsymbol{\nabla}\times\boldsymbol{B} - \frac{1}{c}\dot{\boldsymbol{E}} \right)\boldsymbol{E}\,\mathrm{d}V\,. \tag{2.4}$$

With the vector relation $(B.17)$ we get

$$\int \boldsymbol{j}\boldsymbol{E}\,\mathrm{d}V = \frac{c}{4\pi}\,[4\pi\epsilon_0]\int \left[ [c]\,\boldsymbol{B}\underbrace{\boldsymbol{\nabla}\times\boldsymbol{E}}_{=-\frac{1}{c}\dot{\boldsymbol{B}}} - [c]\,\boldsymbol{\nabla}\,(\boldsymbol{E}\times\boldsymbol{B}) - \frac{1}{c}\dot{\boldsymbol{E}}\boldsymbol{E} \right]\mathrm{d}V$$

$$= \int \left[ \frac{\mathrm{d}\,u}{\mathrm{d}\,t} + \frac{c}{4\pi}\,[4\pi c\epsilon_0]\,\boldsymbol{\nabla}\,(\boldsymbol{E}\times\boldsymbol{B}) \right]\mathrm{d}V\,, \tag{2.5}$$

where $u = \frac{[4\pi\epsilon_0]}{8\pi}\left( E^2 + [c^2]\,B^2 \right)$ is the field energy density. Applying Gauss's theorem $(B.28)$ to the vector product we get an expression for the energy conservation of the complete particle-field system

$$\underbrace{\frac{\mathrm{d}}{\mathrm{d}t}\int u\,\mathrm{d}V}_{\substack{\text{change of}\\\text{field energy}}} + \underbrace{\int \boldsymbol{j}\boldsymbol{E}\,\mathrm{d}V}_{\substack{\text{particle energy}\\\text{loss or gain}}} + \underbrace{\oint \boldsymbol{S}\boldsymbol{n}\,\mathrm{d}s}_{\substack{\text{radiation loss through}\\\text{closed surface } s}} = 0\,, \tag{2.6}$$

where the Poynting vector is defined by

$$\boldsymbol{S} = \frac{c}{4\pi}\,[4\pi c\epsilon_0]\,(\boldsymbol{E}\times\boldsymbol{B})\,. \tag{2.7}$$

Equation (2.7) exhibits characteristic features of electromagnetic radiation. Both electric and magnetic fields are orthogonal to each other $(\boldsymbol{E}\perp\boldsymbol{B})$, orthogonal to the direction of propagation $(\boldsymbol{E}\perp\boldsymbol{n}, \boldsymbol{B}\perp\boldsymbol{n})$, and the vectors $\boldsymbol{E}, \boldsymbol{B}, \boldsymbol{S}$ form a right handed orthogonal system. For plane waves,

$$\boldsymbol{n}\times\boldsymbol{E} = [c]\,\boldsymbol{B}\,, \tag{2.8}$$

(Exercise 2.7) and (2.7) reduces to

$$\boldsymbol{S} = \frac{c}{4\pi}\,[4\pi\epsilon_0]\,E^2\,\boldsymbol{n}\,. \tag{2.9}$$

The Poynting vector is defined as the radiation energy flow through a unit surface area in the direction $\boldsymbol{n}$ and scales proportionally to the square of the electric radiation field.

## 2.3 Electromagnetic Radiation

Phenomenologically, synchrotron radiation is the consequence of a finite value for the velocity of light. Electric fields extend infinitely into space from charged particles in uniform motion. When charged particles become accelerated, however, parts of these fields cannot catch up with the particle anymore and give rise to synchrotron radiation. This happens more so as the particle velocity approaches the velocity of light.

The emission of light can be described by applying Maxwell's equations to moving charged particles. The mathematical derivation of the theory of radiation from Maxwell's equations is straightforward although mathematically somewhat elaborate and we will postpone this to Chap. 9 and 10 of this text. Here we follow a more intuitive discussion [7] which displays visually the physics of synchrotron radiation from basic physical principles.

Electromagnetic radiation occurs wherever electric and magnetic fields exist with components orthogonal to each other such that the Poynting vector

$$\boldsymbol{S} = \frac{c}{4\pi} \left[ 4\pi c \epsilon_0 \right] \left[ \boldsymbol{E} \times \boldsymbol{B} \right] \neq 0 \,. \tag{2.10}$$

It is interesting to ask what happens if we have a static electric and magnetic field such that $[\boldsymbol{E} \times \boldsymbol{B}] \neq 0$. We know there is no radiation but the Poynting vector is nonzero. Applying (2.6), we find the first two terms to be zero which renders the third term zero as well. For a static electric and magnetic field the integral defining the radiation loss or absorption is equal to zero and therefore no radiation or energy transport occurs.

Similarly, in case of a stationary electrostatic charge, we note that the electrostatic fields extend radially from the charge to infinity which violates the requirement that the field be orthogonal to the direction of observation or energy flow. Furthermore, the charge is stationary and therefore there is no magnetic field.

### 2.3.1 Coulomb Regime

Next we consider a charge in uniform motion. In the rest frame of the moving charge we have no radiation since the charge is at rest as just discussed. In the laboratory system, however, the field components are different. Since the charge is moving, it constitutes an electric current which generates a magnetic field. Formulating the Poynting vector in the laboratory system we express the fields by the pure electric field in the particle rest frame $\mathcal{L}^*$. That we accomplish by an inverse Lorentz transformations to (1.28), where the laboratory system $\mathcal{L}$ now moves with the velocity $-\beta_z$ with respect to $\mathcal{L}^*$ and $\beta_z$ in (1.28) must be replaced by $-\beta_z$ for

$$\begin{pmatrix} E_x \\ E_y \\ E_z \\ [c]B_x \\ [c]B_y \\ [c]B_z \end{pmatrix} = \begin{pmatrix} \gamma & 0 & 0 & 0 & \beta_z\gamma & 0 \\ 0 & \gamma & 0 & -\beta_z\gamma & 0 & 0 \\ 0 & 0 & 1 & 0 & 0 & 0 \\ 0 & -\beta_z\gamma & 0 & \gamma & 0 & 0 \\ \beta_z\gamma & 0 & 0 & 0 & \gamma & 0 \\ 0 & 0 & 0 & 0 & 0 & 1 \end{pmatrix} \begin{pmatrix} E_x^* \\ E_y^* \\ E_z^* \\ [c]B_x^* \\ [c]B_y^* \\ [c]B_z^* \end{pmatrix} . \tag{2.11}$$

In the laboratory system $\mathcal{L}$, the components of the Poynting vector (2.10) become then

$$\frac{4\pi}{c}S_x = [4\pi\epsilon_0]\gamma\beta_z E_x^* E_z^* ,$$

$$\frac{4\pi}{c}S_y = [4\pi\epsilon_0]\gamma\beta_z E_y^* E_z^* , \tag{2.12}$$

$$\frac{4\pi}{c}S_z = -[4\pi\epsilon_0]\gamma^2\beta_z \left(E_x^{*2} + E_y^{*2}\right) ,$$

where $^*$ indicates quantities in the moving system $\mathcal{L}^*$, and $\beta_z = v_z/c$. The Poynting vector is nonzero and describes the flow of field energy in the environment of a moving charged particle. The fields drop off rapidly with distance from the particle and the "radiation" is therefore confined close to the location of the particle. Specifically, the fields are attached to the charge and travel in the vicinity and with the charge. This part of electromagnetic radiation is called the Coulomb regime in contrast to the radiation regime and is, for example, responsible for the transport of electric energy along electrical wires and transmission lines.

We will ignore this regime in our further discussion of synchrotron radiation because we are interested only in free radiation which is not anymore connected to electric charges. It should be noted, however, that measurements of radiation parameters close to radiating charges may be affected by the presence of the Coulomb radiation regime. Such situations occur, for example, when radiation is observed close to the source point. Related theories deal with this mixing by specifying a formation length defining the minimum distance from the source required to sufficiently separate the Coulomb regime from the radiation regime.

### 2.3.2 Radiation Regime

In this text we are only interested in the radiation regime and therefore ignore from now on the Coulomb regime. To describe the physics of emission of radiation, we consider a coordinate system moving with a constant velocity equal to that of the charged particle and associated electric fields. The charge is at rest in the moving reference system, the electric field lines extend radially out to infinity, and there is no radiation as discussed before. Acceleration of the charge causes it to move with respect to this reference system generating a distortion of the purely radial electric fields of a uniformly moving charge

(Fig. 2.4). This distortion, resulting in a rearrangement of field lines to the new charge position, travels outward at the velocity of light giving rise to what we call radiation.

To be more specific, we consider a positive charge in uniform motion for $t \leqslant 0$, apply an accelerating force at time $t = 0$ for a time $\Delta T$ and observe the charged particle and its fields in the uniformly moving frame of reference. Due to acceleration the charge moves in this reference system during the time $\Delta T$ from point $A$ to point $B$ and as a consequence the field lines become distorted within a radius $c\Delta T$ from the original location $A$ of the particle. It is this distortion, travelling away from the source at the speed of light, that we call radiation.

The effects on the fields are shown schematically in Fig. 2.4 for an acceleration of a positive charge along its direction of motion. At time $t = 0$ all electric field lines extend radially from the charge located at point $A$ to infinity. During acceleration fieldlines emerge from the charge now at locations between $A$ and $B$. The new field lines must join the old field lines which, due to the finite velocity of light, are still unperturbed at distances larger than $c\Delta T$. As long as the acceleration lasts, a nonradial field component, parallel and opposite to the acceleration, is created. Furthermore, the moving charge creates an azimuthal magnetic field $B_\varphi^*(t)$ and the Poynting vector becomes nonzero causing the emission of radiation from an accelerated electrical charge.

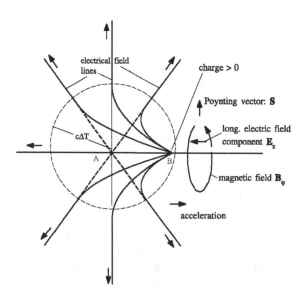

**Fig. 2.4.** Distortion of fields due to longitudinal acceleration

Obviously, acceleration would not result in any radiation if the velocity of propagation for electromagnetic fields were infinite, $c \to \infty$. In this case the radial fields at all distances from the charge would instantly move in synchrony with the movement of the charge. Only the Coulomb regime would exist.

The electrical field perturbation is proportional to the electrical charge $q$ and the acceleration $a^*$. Acceleration along the $z$-axis generates an electric field $E_z^* \neq 0$ and its component normal to the direction of observation scales like $\sin \Theta^*$, where $\Theta^*$ is the angle between the line of observation and the direction of particle acceleration. During the acceleration a fixed amount of field energy is created which propagates radially outward from the source. Since the total radiation energy must stay constant and the volume of the expanding spherical sheath of field perturbation increases like $R^2$, the field strength decays linear with distance $R$. With this, we make the ansatz

$$ \boldsymbol{E}_\parallel^* = - \left[ \frac{1}{4\pi\epsilon_0} \right] \frac{ea^*}{c^2 R} \sin \Theta^* \tag{2.13} $$

for the electric field, where we have added a factor $c^2$ in the denominator to be dimensionally correct. For an electron, $e < 0$, and the field perturbation points in the direction of the acceleration. As expected from the definition of the Poynting vector, the radiation is emitted predominantly orthogonal to the direction of acceleration and is highly polarized in the direction of acceleration. From (2.9)

$$ \boldsymbol{S} = \frac{c}{4\pi} [4\pi\epsilon_0] \, E_\parallel^{*2} \boldsymbol{n}^* \, , \tag{2.14} $$

where $\boldsymbol{n}^*$ is the unit vector in the direction of observation from the observer toward the radiation source. The result is consistent with our earlier finding that no free radiation is emitted from a charge at rest or uniform motion ($a^* \to 0$). The spatial radiation distribution is from (2.13) and (2.14) characterized by a $\sin^2 \Theta^*$-distribution resembling the shape of a doughnut as shown in Fig. 2.5, where the acceleration occurs along the $x$-axis.

Acceleration may not only occur in the longitudinal direction but also in the direction transverse to the velocity of the particle as shown in Fig. 2.6. The distortion of field lines in this case creates primarily transverse or radial field components. The radiation field component transverse to the direction of observation is

$$ \boldsymbol{E}_\perp^* = - \left[ \frac{1}{4\pi\epsilon_0} \right] \frac{e\,a^*}{c^2 R} \cos \Theta^* \, . \tag{2.15} $$

This case of transverse acceleration describes the appearance of synchrotron radiation created by charged particles being deflected in magnetic fields. Similar to (2.14) the Poynting vector for transverse acceleration is

$$ \boldsymbol{S} = \frac{c}{4\pi} [4\pi\epsilon_0] \, E_\perp^{*2} \boldsymbol{n}^* \, . \tag{2.16} $$

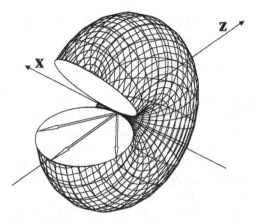

**Fig. 2.5.** Spatial radiation distribution in the rest frame of the radiating charge

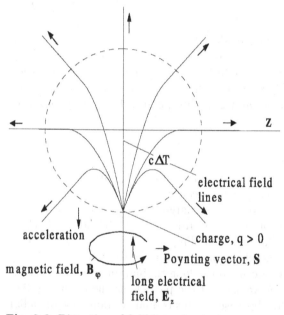

**Fig. 2.6.** Distortion of field lines due to transverse acceleration

## 2.4 Spatial and Spectral Properties of Radiation

Although the acceleration and the creation of radiation fields is not periodic, we may Fourier-decompose the radiation pulse and obtain a spectrum of plane waves

$$E^* = E_0^* e^{i\Phi^*} , \tag{2.17}$$

where the phase is defined by

$$\Phi^* = \omega^* \left[ t^* - \tfrac{1}{c} \left( n_x^* x^* + n_y^* y^* + n_z^* z^* \right) \right] . \tag{2.18}$$

The phase of an electromagnetic wave is proportional to the product of the momentum-energy and space-time 4-vectors. Like any other product of two 4-vectors this product is invariant under Lorentz transformations. We have therefore the equality $\Phi^* = \Phi$ or

$$\omega^* \left[ ct^* - n_x^* x^* - n_y^* y^* - n_z^* z^* \right] = \omega \left[ ct - n_x x - n_y y - n_z z \right] \tag{2.19}$$

between the phases as measured in both the laboratory $\mathcal{L}$ and the particle frame of reference $\mathcal{L}^*$. To derive the relationships between similar quantities in both systems, we use the Lorentz transformations noting that the particle reference frame is the frame, where the particle or radiation source is at rest. We use the Lorentz transformations (1.23) to replace the coordinates $(x^*, y^*, z^*, ct^*)$ in (2.19) by those in the laboratory system. Since the space-time coordinates are independent we may equate their coefficients on either side of the equation separately. In so doing, we get from the $ct$-coefficients for the oscillation frequency

$$\omega^* \gamma \left( 1 + \beta_z \, n_z^* \right) = \omega , \tag{2.20}$$

which expresses the relativistic Doppler effect. Looking parallel to the direction of particle motion $n_z^* = 1$ the observed oscillation frequency is increased by the factor $(1 + \beta_z) \, \gamma \approx 2\gamma$ for highly relativistic particles. The Doppler effect is reduced if the radiation is viewed at some finite angle or even normal to the direction of motion when $n_z^* = 0$. For viewing angles in between these two extremes we set $n_z^* = \cos \Theta^*$.

Similarly, we obtain from (2.19) also the transformation of spatial directions

$$n_x = \frac{n_x^*}{\gamma \left( 1 + \beta_z \, n_z^* \right)} , \tag{2.21}$$

$$n_y = \frac{n_y^*}{\gamma \left( 1 + \beta_z \, n_z^* \right)} , \tag{2.22}$$

$$n_z = \frac{\beta_z + n_z^*}{\left( 1 + \beta_z \, n_z^* \right)} . \tag{2.23}$$

These transformations define the spatial distribution of radiation in the laboratory system. In case of transverse acceleration the radiation in the particle rest frame is distributed like $\cos^2 \Theta^*$ about the direction of motion. This distribution becomes greatly collimated into the forward direction in the laboratory system. With $n_x^{*2} + n_y^{*2} = \sin^2 \Theta^*$ and $n_x^2 + n_y^2 = \sin^2 \Theta \approx \Theta^2$ and $n_z^* = \cos \Theta^*$ we find

$$\Theta \approx \frac{\sin \Theta^*}{\gamma (1 + \beta \cos \Theta^*)} . \tag{2.24}$$

In other words, radiation from relativistic particles, emitted in the particle system into an angle $-\pi/2 < \Theta^* < \pi/2$ appears in the laboratory system highly collimated in the forward direction within an angle of

$$\Theta \approx \frac{1}{\gamma}. \tag{2.25}$$

This angle is very small for highly relativistic electrons like those in a storage ring, where $\gamma$ is of the order of $10^3 - 10^4$.

# Exercises *

**Exercise 2.1 (S).** Use a 10 MeV electron beam passing through atmospheric air. Can you observe Cherenkov radiation and if so at what angle? Answer the same questions also for a 50 MeV electron beam. Describe and explain with Fig. 2.2 the fundamental difference of your results ($n_{air} = 1.0002769$ for $\lambda = 5600$ Å).

**Exercise 2.2 (S).** A 10 MeV electron beam passes with normal incidence through a plate of translucent plastic ($n = 1.7$). Is there any Cherenkov radiation and if so at what angle? Where does this radiation escape the plate?

**Exercise 2.3 (S).** Show that the product of two 4-vectors is Lorentz invariant.

**Exercise 2.4 (S).** Show that the product of the 4-momentum and 4-spacetime of a photon is proportional to the phase of the electromagnetic wave.

**Exercise 2.5 (S).** Derive from (2.20) the formula for the classical Doppler effect valid for sound waves emitted at a frequency $f_s$ from a source moving with velocity $v$ and received at an angle $\vartheta$.

**Exercise 2.6 (S).** From Heisenberg's uncertainty relation construct a "characteristic volume" of a photon with energy $\varepsilon_{ph} = \hbar\omega$. What is the average electric field in this volume for a 1 eV photon and an x-ray photon of 10 keV?

**Exercise 2.7 (S).** Prove that $n \times E = [c]\, B$ for plane waves.

**Exercise 2.8 (S).** Show that equations (2.7) and (2.9) are the same for electromagnetic waves.

**Exercise 2.9.** Resketch Fig. 2.4 to show the electric field lines from the charge to infinity at a time $t \gtrsim \Delta t$ after the acceleration has stopped.

---

* The argument (S) indicates an exercise for which a solution is given in Appendix A.

**Exercise 2.10.** Derive the equations of transformation for the frequency (2.20), and the direction of observations (2.21 2.23), from (2.19).

**Exercise 2.11.** Verify the equality of (C.24) and (C.25).

**Exercise 2.12.** Consider a beam of 123.8 meV and 10 keV photons, both at a power density of 100 Watt/mm$^2$. How many photons occupy their respective "characteristic volumes"? . Show that the photon flux density is $1.875 \times 10^{10}$ photons(100 meV)/mm$^3$ and $1.875 \times 10^5$ photons(10 keV)/mm$^3$. Verify that, 61.07 photons(123.8 meV) and $1.44 \times 10^{-18}$ photons (10 keV) occupy, on average, each characteristic volume in a 100 W/mm$^2$ beam. The x-ray photon distribution is indeed sparse among it's characteristic volume. What are the respective characteristic volumes?

**Exercise 2.13.** Verify that for a 10 MeV electron colliding head-on with a Ti-Saphire laser ($\lambda = 0.8$ $\mu$m) the wavelength in it's own system is $\lambda^* = 40.88$ nm. Also show that the wavelength of the backscattered photon in the laboratory system is $\lambda_\gamma = 10.4$ Å. What electron beam energy do you need to produce 1 Å radiation? What is the maximum acceptance angle allowable to still get a photon beam with a band width of 10% or less? Show that the acceptance angle is $\pm 18.15$ mrad.

# 3. Overview of Synchrotron Radiation

After Schott's [3] unsuccessful attempt to explain atomic radiation with his electromagnetic theory no further progress was made for some 40 years mainly because of lack of interest. Only in the mid forties did the theory of electromagnetic radiation from free electrons become interesting again with the successful development of circular high-energy electron accelerators. At this time powerful betatrons [8] have been put into operation and it was Ivanenko and Pomeranchouk [9], who first in 1944 pointed out a possible limit to the betatron principle and maximum energy due to energy loss from emission of electromagnetic radiation. This prediction was used by Blewett [10] to calculate the radiation energy loss per turn in a newly constructed 100 MeV betatron at General Electric. In 1946 he measured the shrinkage of the orbit due to radiation losses and the results agreed with predictions. On April 24, 1947 visible radiation was observed for the first time at the 70 MeV synchrotron built at General Electric [11, 12, 13]. Since then, this radiation is called synchrotron radiation.

The energy loss of particles to synchrotron radiation causes technical and economic limits for circular electron or positron accelerators. As the particle energy is driven higher and higher, more and more rf-power must be supplied to the beam not only to accelerate particles but also to overcome energy losses due to synchrotron radiation. The limit is reached when the radiation power grows to high enough levels exceeding technical cooling capabilities or exceeding the funds available to pay for the high cost of electrical power. To somewhat ameliorate this limit, high-energy electron accelerators have been constructed with ever increasing circumference to allow a more gentle bending of the particle beam. Since the synchrotron radiation power scales like the square of the particle energy (assuming constant magnetic fields) the circumference must scale similar for a constant amount of rf-power. Usually, a compromise is reached by increasing the circumference less and adding more rf-power in spaces along the ring lattice made available by the increased circumference. In general the maximum energy in large circular electron accelerators is limited by the available rf-power while the maximum energy of proton or ion accelerators and low energy electron accelerators is more likely limited by the maximum achievable magnetic fields in bending magnets.

What is a nuisance for researchers in one field can provide tremendous opportunities for others. Synchrotron radiation is emitted tangentially from the particle orbit and within a highly collimated angle of $\pm 1/\gamma$. The spectrum reaches from the far infrared up to hard x-rays, the radiation is polarized and the intensities greatly exceed other sources specifically in the vacuum ultra violet to x-ray region. With these properties synchrotron radiation was soon recognized to be a powerful research tool for material sciences, crystallography, surface physics, chemistry, biophysics, and medicine to name only a few areas of research. While in the past most of this research was done parasitically on accelerators built and optimized for high-energy physics the usefulness of synchrotron radiation for research has become important in its own right to justify the construction and operation of dedicated synchrotron radiation sources all over the world.

## 3.1 Radiation Power

Integrating the Poynting vector (2.14) over a closed surface enclosing the radiating charge we get with (2.13) and $\mathbf{n}^* \mathrm{d}\mathbf{A}^* = R^2 \sin \Theta^* \mathrm{d}\Theta^* \mathrm{d}\Phi^*$ the radiation power

$$P^* = \int \boldsymbol{S}^* \, \mathrm{d}\boldsymbol{A}^* = \tfrac{2}{3} r_\mathrm{c} \frac{mc^2}{c^3} \, a^{*2} \,, \tag{3.1}$$

where we have set $q^2 = [4\pi\epsilon_0] \, r_\mathrm{c} mc^2$. From the discussion of 4-vectors in Chapter C, we know that the square of the 4-acceleration is invariant to Lorentz transformations and get from (C.25) finally for the radiation power in the laboratory system

$$P = \tfrac{2}{3} r_\mathrm{c} mc \gamma^6 \left[ \dot{\boldsymbol{\beta}}^2 - \left( \boldsymbol{\beta} \times \dot{\boldsymbol{\beta}} \right)^2 \right] \,. \tag{3.2}$$

Equation (3.2) expresses the radiation power in a simple way and allows us to calculate other radiation characteristics based on beam parameters in the laboratory system. The radiation power is greatly determined by the geometric path of the particle trajectory through the quantities $\boldsymbol{\beta}$ and $\dot{\boldsymbol{\beta}}$. Specifically, if this path has strong oscillatory components we expect that motion to be reflected in the synchrotron radiation power spectrum. This aspect will be discussed later in more detail. Here we distinguish only between acceleration parallel $\dot{\boldsymbol{\beta}}_\parallel$ or perpendicular $\dot{\boldsymbol{\beta}}_\perp$ to the propagation $\boldsymbol{\beta}$ of the charge and set therefore

$$\dot{\boldsymbol{\beta}} = \dot{\boldsymbol{\beta}}_\parallel + \dot{\boldsymbol{\beta}}_\perp \,. \tag{3.3}$$

Insertion into (3.2) shows the total radiation power to be composed of separate contributions from parallel and orthogonal acceleration. Separating both contributions we get the synchrotron radiation power for both parallel and transverse acceleration respectively

$$P_{\parallel} = \tfrac{2}{3} r_c m c \gamma^6 \dot{\boldsymbol{\beta}}_{\parallel}^2 , \tag{3.4}$$

$$P_{\perp} = \tfrac{2}{3} r_c m c \gamma^4 \dot{\boldsymbol{\beta}}_{\perp}^2 . \tag{3.5}$$

Expressions have been derived that define the radiation power for parallel acceleration like in a linear accelerator or orthogonal acceleration found in circular accelerators or deflecting systems. We note a similarity for both contributions except for the energy dependence. At highly relativistic energies the same acceleration force leads to much less radiation if the acceleration is parallel to the motion of the particle compared to orthogonal acceleration. Parallel acceleration is related to the accelerating force by $m\dot{\boldsymbol{v}}_{\parallel} = \tfrac{1}{\gamma^3} d\boldsymbol{p}_{\parallel}/dt$ and after insertion into (3.4) the radiation power due to parallel acceleration becomes

$$P_{\parallel} = \frac{2}{3} \frac{r_c}{mc} \left( \frac{d\boldsymbol{p}_{\parallel}}{dt} \right)^2 . \tag{3.6}$$

The radiation power for acceleration along the propagation of the charged particle is therefore independent of the energy of the particle and depends only on the accelerating force or with $d\boldsymbol{p}_{\parallel}/dt = d\boldsymbol{E}/dz$ on the energy increase per unit length of accelerator. Different from circular electron accelerators we encounter therefore no practical energy limit in a linear accelerator at very high energies. In contrast very different radiation characteristics exist for transverse acceleration as it happens, for example, during the transverse deflection of a charged particle in a magnetic field. The transverse acceleration $\dot{\boldsymbol{v}}_{\perp}$ is expressed by the Lorentz force

$$\frac{d\boldsymbol{p}_{\perp}}{dt} = \gamma m \dot{\boldsymbol{v}}_{\perp} = e \frac{[c]}{c} [\boldsymbol{v} \times \boldsymbol{B}] \tag{3.7}$$

and after insertion into (3.5) the radiation power from transversely accelerated particles becomes

$$P_{\perp} = \frac{2}{3} \frac{r_c}{mc} \gamma^2 \left( \frac{d\boldsymbol{p}_{\perp}}{dt} \right)^2 . \tag{3.8}$$

From (3.6, 3.8) we find that the same accelerating force leads to a much higher radiation power by a factor $\gamma^2$ for transverse acceleration compared to longitudinal acceleration. For all practical purposes, technical limitations prevent the occurrence of sufficient longitudinal acceleration to generate noticeable radiation. From here on we will stop considering longitudinal acceleration unless specifically mentioned and eliminate, therefore, the index $\perp$ setting for the radiation power $P_{\perp} = P_{\gamma}$. We also restrict from now on the discussion to singly charged particles and set $q = e$ ignoring extremely high energies, where multiple charged ions may start to radiate. Replacing the force in (3.8) by the Lorentz force (3.7) we get

$$P_\gamma = \left[\frac{4\pi}{\mu_0}\right] \frac{2\,r_c^2\,c}{3\,(mc^2)^2} B^2\,E^2 = C_B B^2\,E^2 \,, \tag{3.9}$$

where

$$C_B = \left[\frac{4\pi}{\mu_0}\right] \frac{2\,r_c^2\,c}{3\,(mc^2)^2} = 6.077\,9 \times 10^{-8}\,\frac{\text{W}}{\text{T}^2\text{GeV}^2} = 379.35\,\frac{1}{\text{T}^2\text{GeV s}}\,. \tag{3.10}$$

The synchrotron radiation power scales like the square of the magnetic field and the square of the particle energy. Replacing the deflecting magnetic field $B$ by the bending radius $\rho$ (6.7) the instantaneous synchrotron radiation power becomes

$$P_\gamma = \tfrac{2}{3}\,r_c m c^3 \frac{\beta^4 \gamma^4}{\rho^2} \tag{3.11}$$

or in more practical units,

$$P_\gamma = \frac{c\,C_\gamma}{2\pi}\,\frac{E^4}{\rho^2}\,. \tag{3.12}$$

Here we use the definition of Sand's radiation constant for electrons [14]

$$C_\gamma = \frac{4\pi}{3}\,\frac{r_c}{(mc^2)^3} = 1.41733 \times 10^{-14}\,\frac{\text{msW}}{\text{GeV}^4} = 8.8460 \times 10^{-5}\,\frac{\text{m}}{\text{GeV}^3}\,. \tag{3.13}$$

The electromagnetic radiation of charged particles in transverse magnetic fields is proportional to the fourth power of the particle momentum $\beta\gamma$ and inversely proportional to the square of the bending radius $\rho$. The synchrotron radiation power increases very fast for high-energy particles and provides the most severe limitation to the maximum energy achievable in circular accelerators. We note, however, also a strong dependence on the kind of particles involved in the process of radiation. Because of the much heavier mass of protons compared to the lighter electrons we find appreciable synchrotron radiation only in electron accelerators.

In storage rings with different magnets and including insertion devices it is important to formulate the average radiation power of an electron during the course of one turn. In this case we calculate the average

$$\langle P_\gamma \rangle = \frac{c}{2\pi}C_\gamma E^4 \left\langle \frac{1}{\rho^2} \right\rangle = C_\gamma E^4 \frac{f_{\text{rev}}}{2\pi} \oint \frac{ds}{\rho^2}\,. \tag{3.14}$$

The radiation power of protons actually is smaller compared to that for electrons by the fourth power of the mass ratio or by the factor

$$\frac{P_{\mathrm{e}}}{P_{\mathrm{p}}} = 1836^4 = 1.1367 \times 10^{13} . \tag{3.15}$$

In spite of this enormous difference measurable synchrotron radiation has been predicted by Coisson [15] and was indeed detected at the 400 GeV proton synchrotron, SPS (Super Proton Synchrotron), at CERN in Geneva [16, 17]. Substantial synchrotron radiation is expected in multi-TeV proton colliders like the LHC (Large Hadron Collider) at CERN [18].

Knowledge of the synchrotron radiation power allows us now to calculate the energy loss per turn of a particle in a circular accelerator by integrating the radiation power along the circumference of the circular accelerator

$$U_0 = \oint P_\gamma \mathrm{d}t = \tfrac{2}{3} r_{\mathrm{e}} m c^2 \beta^3 \gamma^4 \oint \frac{\mathrm{d}s}{\rho^2} . \tag{3.16}$$

In an isomagnetic lattice, where the bending radius is the same for all bending magnets $\rho$ =const., the integration around a circular accelerator can be performed and the energy loss per turn due to synchrotron radiation is

$$U_0 = P_\gamma \frac{2\pi\rho}{\beta c} = \tfrac{4\pi}{3} r_{\mathrm{e}} m c^2 \beta^3 \frac{\gamma^4}{\rho} . \tag{3.17}$$

The integration obviously is to be performed only along those parts of the circular accelerator, where synchrotron radiation occurs, or along bending magnets only. In more practical units, the energy loss of relativistic electrons per revolution in a circular accelerator with an isomagnetic lattice and a bending radius $\rho$ is given by

$$U_{0,\mathrm{iso}}\,(\mathrm{GeV}) = C_\gamma \frac{E^4 (\mathrm{GeV}^4)}{\rho(m)} . \tag{3.18}$$

For a beam of $N_{\mathrm{e}}$ particles or a circulating beam current $I = e f_{\mathrm{rev}} N_{\mathrm{e}}$ the total average radiation power is

$$\langle P_{\mathrm{s}} \rangle = U_0 \frac{I}{e} , \tag{3.19}$$

or in more practical units

$$\langle P_{\mathrm{s}}\,(\mathrm{MW}) \rangle_{\mathrm{iso}} = 0.088463 \frac{E^4\,(\mathrm{GeV})}{\rho\,(\mathrm{m})} I\,(\mathrm{A}) . \tag{3.20}$$

The total synchrotron radiation power scales like the fourth power of the particle energy and is inversely proportional to the bending radius. The strong dependence of the radiation on the particle energy causes severe practical limitations on the maximum achievable energy in a circular accelerator.

## 3.2 Spectrum

Synchrotron radiation from relativistic charged particles is emitted over a wide spectrum of photon energies. The basic characteristics of this spectrum can be derived from simple principles as suggested in [19]. For an observer synchrotron light has the appearance similar to the light coming from a lighthouse. Although the light is emitted continuously an observer sees only a periodic flash of light as the aperture mechanism rotates in the lighthouse. Similarly, synchrotron light emitted from relativistic particles will appear to an observer as a single flash if it comes from a bending magnet in a transport line passed through by a particle only once or as a series of equidistant light flashes as bunches of particles orbit in a circular accelerator.

Since the duration of the light flashes is very short the observer notes a broad spectrum of frequencies as his eyes or instruments Fourier analyze the pulse of electromagnetic energy. The spectrum of synchrotron light from a circular accelerator is composed of a large number of harmonics with fundamental frequency equal to the revolution frequency of the particle in the circular accelerator. These harmonics reach a cutoff, where the period of the radiation becomes comparable to the duration of the light pulse. Even though the aperture of the observers eyes or instruments are assumed to be infinitely narrow we still note a finite duration of the light flash. This is a consequence of the finite opening angle of the radiation as illustrated in Fig. 3.1. Synchrotron light emitted by a particle travelling along the orbit cannot reach the observer before it has reached the point $P_0$ when those photons emitted on one edge of the radiation cone at an angle $-1/\gamma$ aim directly toward the observer. Similarly, the last photons to reach the observer are emitted from point $P_1$ at an angle of $+1/\gamma$. Between point $P_0$ and point $P_1$ we have therefore a deflection angle of $2/\gamma$. The duration of the light flash for the observer is not the time it takes the particle to travel from point $P_0$ to point $P_1$ but must be corrected for the finite time of flight for the photon emitted at $P_0$. If particle and photon would travel toward the observer with exactly the same velocity the light pulse would be infinitely short. However, particles move

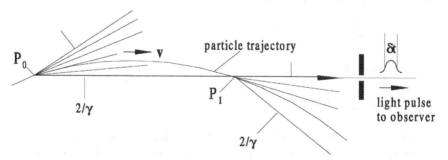

**Fig. 3.1.** Temporal pulse formation of synchrotron radiation

slower following a slight detour and therefore the duration of the light pulse equals the time difference between the first photons from point $P_0$ arriving at the observer and the last photons being emitted by the particles at point $P_1$. Although the particle reaches point $P_0$ at time $t = 0$ the first photon can be observed at point $P_1$ only after a time

$$t_\gamma = \frac{2\rho \sin \frac{1}{\gamma}}{c} . \tag{3.21}$$

The last photon to reach the observer is emitted when the particle arrives at point $P_1$ at the time

$$t_c = \frac{2\rho}{\beta c \gamma} . \tag{3.22}$$

The duration of the light pulse $\delta t$ is therefore given by the difference of both travel times (3.21, 3.22)

$$\delta t = t_c - t_\gamma = \frac{2\rho}{\beta c \gamma} - \frac{2\rho \sin \frac{1}{\gamma}}{c} . \tag{3.23}$$

The sinc-function can be expanded for small angles keeping linear and third order terms only and the duration of the light pulse at the location of the observer is after some manipulation

$$\delta t = \frac{4\rho}{3c\gamma^3} . \tag{3.24}$$

The total duration of the electromagnetic pulse is very short scaling inversely proportional to the third power of $\gamma$. This short pulse translates into a broad spectrum. Using only half the pulse length for the effective pulse duration the spectrum reaches up to a maximum frequency of about

$$\omega_c \approx \frac{1}{\frac{1}{2}\delta t} \approx \frac{3}{2} c \frac{\gamma^3}{\rho} , \tag{3.25}$$

which is called the critical photon frequency of synchrotron radiation. The critical photon energy $\varepsilon_c = \hbar \omega_c$ is then given by

$$\varepsilon_c = C_c \frac{E^3}{\rho} , \tag{3.26}$$

with

$$C_c = \frac{3\hbar c}{2 \left(mc^2\right)^3} . \tag{3.27}$$

For electrons, numerical expressions are

$$\varepsilon_c \, (\text{keV}) = 2.2183 \frac{E^3 \, (\text{GeV}^3)}{\rho \, (\text{m})} = 0.66503 \, E^2 \, (\text{GeV}^2) \, B \, (\text{T}) \ . \tag{3.28}$$

The synchrotron radiation spectrum from relativistic particles in a circular accelerator is made up of harmonics of the particle revolution frequency $\omega_0$ with values up to and beyond the critical frequency (3.28). Generally, a real synchrotron radiation beam from say a storage ring will not display this harmonic structure. The distance between harmonics is extremely small compared to the extracted photon frequencies in the VUV and x-ray regime while the line width is finite due to the energy spread and beam emittance.

For a single pass of particles through a bending magnet in a beam transport line we observe the same spectrum. Specifically, the maximum frequency is the same assuming similar parameters. Synchrotron radiation is emitted in a particular spatial and spectral distribution, both of which will be derived in Chapter 9, and we will use here only some of these results. A useful parameter to characterize the photon intensity is the photon flux per unit solid angle into a frequency bin $\Delta\omega/\omega$ and from a circulating beam current $I$ defined by

$$\frac{d^2 \dot{N}_{\text{ph}}}{d\theta d\psi} = C_\Omega E^2 I \frac{\Delta\omega}{\omega} \left( \frac{\omega}{\omega_c} \right)^2 K_{2/3}^2 (\xi) \, F (\xi, \theta) \ , \tag{3.29}$$

where $\psi$ is the angle in the deflecting plane and $\theta$ the angle normal to the deflecting plane,

$$C_\Omega = \frac{3\alpha}{4\pi^2 e \, (mc^2)^2} = 1.3255 \times 10^{16} \, \frac{\text{photons}}{\text{s mrad}^2 \, \text{GeV}^2 \, \text{A} \, 100\%\text{BW}} \ , \tag{3.30}$$

$\alpha$ the fine structure constant and

$$F (\xi, \theta) = \left( 1 + \gamma^2 \theta^2 \right)^2 \left( 1 + \frac{\gamma^2 \theta^2}{1 + \gamma^2 \theta^2} \frac{K_{1/3}^2 (\xi)}{K_{2/3}^2 (\xi)} \right) \ . \tag{3.31}$$

The functions $K_{1/3} (\xi)$ and $K_{2/3} (\xi)$, displayed in Fig. 3.2, are modified Bessel's functions with the argument

$$\xi = \tfrac{1}{2} \frac{\omega}{\omega_c} \left( 1 + \gamma^2 \theta^2 \right)^{3/2} \ . \tag{3.32}$$

Synchrotron radiation is highly polarized in the plane normal ($\sigma$-mode), and parallel ($\pi$-mode), to the deflecting magnetic field. The relative flux in both polarization directions is given by the two components in the second bracket of function $F (\xi, \theta)$ in (3.31). The first component is equal to unity and determines the photon flux for the polarization normal to the magnetic field or $\sigma$-mode, while the second term relates to the polarization parallel to the magnetic field which is also called the $\pi$-mode. Equation (3.29) expresses both the spectral and spatial photon flux for both the $\sigma$-mode radiation in

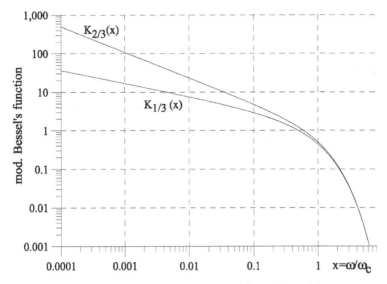

**Fig. 3.2.** Modified Bessel's functions $K_{1/3}(\xi)$ and $K_{2/3}(\xi)$

the forward direction within an angle of about $\pm 1/\gamma$ and for the $\pi$-mode off axis.

For highly relativistic particles the synchrotron radiation is collimated very much in the forward direction and we may assume that all radiation in the nondeflecting plane is accepted by the experimental beam line. In this case we are interested in the photon flux integrated over all angles $\theta$. This integration will be performed in Chap. 9 with the result (9.158)

$$\frac{\mathrm{d}\dot{N}_{\mathrm{ph}}}{\mathrm{d}\psi} = \frac{4\alpha}{9}\,\gamma\,\frac{I}{e}\frac{\Delta\omega}{\omega}\,S\left(\frac{\omega}{\omega_{\mathrm{c}}}\right)\,, \qquad (3.33)$$

where $\psi$ is the deflection angle in the bending magnet, $\alpha$ the fine structure constant and the function $S(x)$ is defined by

$$S\left(\frac{\omega}{\omega_{\mathrm{c}}}\right) = \frac{9\sqrt{3}}{8\pi}\frac{\omega}{\omega_{\mathrm{c}}}\int_{\omega/\omega_{\mathrm{c}}}^{\infty} K_{5/3}(\bar{x})\,\mathrm{d}\bar{x} \qquad (3.34)$$

with $K_{5/3}(x)$ a modified Bessel's function. The function $S(\omega/\omega_{\mathrm{c}})$ is known as the universal function of synchrotron radiation and is shown in Fig. 3.3. In practical units, the angle integrated photon flux is

$$\frac{\mathrm{d}\dot{N}_{\mathrm{ph}}}{\mathrm{d}\psi} = C_{\psi}\,E\,I\,\frac{\Delta\omega}{\omega}\,S\left(\frac{\omega}{\omega_{\mathrm{c}}}\right) \qquad (3.35)$$

with $C_{\psi}$ defined by

$$C_\psi = \frac{4\alpha}{9e\,mc^2} = 3.9614 \times 10^{19} \; \frac{\text{photons}}{\text{s rad A GeV}} \,. \tag{3.36}$$

The spectral distribution depends only on the particle energy, the critical frequency $\omega_c$ and a purely mathematical function. This result has been derived originally by Ivanenko and Sokolov [20] and independently by Schwinger [21]. Specifically it should be noted that the spectral distribution, if normalized to the critical frequency, does not depend on the particle energy and can therefore be represented by a universal distribution shown in Fig. 3.3.

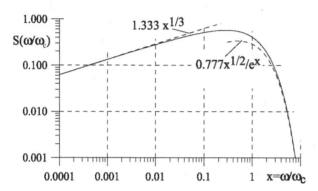

**Fig. 3.3.** Universal function of the synchrotron radiation spectrum, $S(\omega/\omega_c)$

The energy dependence is contained in the cubic dependence of the critical frequency acting as a scaling factor for the actual spectral distribution. The synchrotron radiation spectrum in Fig. 3.3 is rather uniform up to the critical frequency beyond which the intensity falls off rapidly. This synchrotron radiation spectrum has been verified experimentally soon after such radiation sources became available [22, 23].

Equation (3.33) is not well suited for quick calculation of the radiation intensity at a particular frequency. We may, however, express (3.33) in much simpler form for very low and very large frequencies making use of limiting expressions of Bessel's functions for large and small arguments. For small arguments $x = \frac{\omega}{\omega_c} \ll 1$ an asymptotic approximation [24] for the modified Bessel's function may be used to give instead of (3.35)

$$\frac{\mathrm{d}\dot{N}_{\mathrm{ph}}}{\mathrm{d}\psi} \approx C_\psi \, E\, I \, \frac{\Delta\omega}{\omega} \, 1.333 \left(\frac{\omega}{\omega_c}\right)^{1/3} . \tag{3.37}$$

Similarly, for high photon frequencies $x = \frac{\omega}{\omega_c} \gg 1$ we get

$$\frac{\mathrm{d}\dot{N}_{\mathrm{ph}}}{\mathrm{d}\psi} \approx C_\psi \, E\, I \, \frac{\Delta\omega}{\omega} \, 0.77736 \frac{\sqrt{x}}{e^x} , \tag{3.38}$$

where $x = \frac{\omega}{\omega_c}$. Both approximations are included in Fig. 3.3 and display actually a rather good representation of the real spectral radiation distribution over all but the central portion of the spectrum. Specifically, we note the slow increase in the radiation intensity at low frequencies and the exponential drop off above the critical frequency.

## 3.3 Spatial Photon Distribution

The expressions for the photon fluxes (3.29, 3.33) provide the opportunity to calculate the spectral distribution of the photon beam divergence. Photons are emitted into a narrow angle and we may represent this narrow angular distribution by a Gaussian distribution. The effective width of a Gaussian distribution is $\sqrt{2\pi}\sigma_\theta$ and we set

$$\frac{\mathrm{d}\dot{N}_{\mathrm{ph}}}{\mathrm{d}\psi} \approx \frac{\mathrm{d}^2\dot{N}_{\mathrm{ph}}}{\mathrm{d}\theta\,\mathrm{d}\psi}\sqrt{2\pi}\sigma_\theta\,. \tag{3.39}$$

With (3.29, 3.35) the angular divergence of the forward lobe of the photon beam or for a beam polarized in the $\sigma$-mode is

$$\sigma_\theta\,(\mathrm{mrad}) = \frac{C_\psi}{\sqrt{2\pi}C_\Omega}\frac{1}{E}\frac{S\,(x)}{x^2 K_{2/3}^2\left(\frac{1}{2}x\right)} = \frac{f\,(x)}{E\,(\mathrm{GeV})}\,, \tag{3.40}$$

where $x = \omega/\omega_c$. For the forward direction $\theta \approx 0$ the function $f(x) = \sigma_\theta\,(\mathrm{mrad})\,E\,(\mathrm{GeV})$ is shown in Fig. 3.4 for easy numerical calculations.

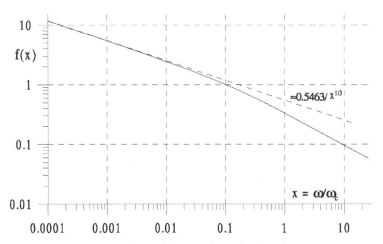

**Fig. 3.4.** Scaling function $f(x) = \sigma_\theta(\mathrm{mrad})\,E(\mathrm{GeV})$ for the photon beam divergence in (3.40)

For wavelengths $\omega \ll \omega_c$, (3.40) can be greatly simplified to become in more practical units

$$\sigma_\theta \,(\text{mrad}) \approx \frac{0.54626}{E\,(\text{GeV})} \left(\frac{\omega}{\omega_c}\right)^{1/3} = \frac{7.124}{\left[\rho\,(\text{m})\,\epsilon_{\text{ph}}\,(\text{eV})\right]^{1/3}}, \qquad (3.41)$$

where $\rho$ is the bending radius and $\epsilon_{\text{ph}}$ the photon energy. The photon beam divergence for low photon energies compared to the critical photon energy is independent of the particle energy and scales inversely proportional to the third root of the bending radius and photon energy.

## 3.4 Fraunhofer Diffraction

Synchrotron radiation is emitted from a rather small area equal to the cross section of the electron beam. In the extreme and depending on the photon wavelength the radiation may be spatially coherent because the beam cross section in phase space is smaller than the wavelength. This possibility to create spatially coherent radiation is important for many experiments specifically for holography and we will discuss in more detail the conditions for the particle beam to emit such radiation.

Reducing the particle beam cross section in phase space by diminishing the particle beam emittance reduces also the source size of the photon beam. This process of reducing the beam emittance is, however, effective only to some point. Further reduction of the particle beam emittance would have no effect on the photon beam emittance because of diffraction effects. A point like photon source appears in an optical instrument as a disk with concentric illuminated rings. For synchrotron radiation sources it is of great interest to maximize the photon beam brightness which is the photon density in phase space. On the other hand designing a lattice for a very small beam emittance can cause beam stability problems. It is therefore prudent not to push the particle beam emittance to values much less than the diffraction limited photon beam emittance. In the following we will therefore define diffraction limited photon beam emittance as a guide for low emittance lattice design.

For highly collimated synchrotron radiation it is appropriate to assume Fraunhofer diffraction. Radiation from an extended light source appears diffracted in the image plane with a radiation pattern which is characteristic for the particular source size and radiation distribution as well as for the geometry of the apertures involved. For simplicity, we will use the case of a round aperture being the boundaries of the beam itself although in most cases the beam cross section is more elliptical. In spite of this simplification, however, we will obtain all basic physical properties of diffraction which are of interest to us. We consider a circular light source with diameter $2a$.

The radiation field at point $P$ in the image plane is then determined by the Fraunhofer diffraction integral [25]

$$U(P) = C \int_0^a \int_0^{2\pi} \mathrm{e}^{-ik\rho w \cos(\Theta-\psi)} \mathrm{d}\Theta\, \rho \mathrm{d}\rho. \qquad (3.42)$$

Here $k$ is the wave number of the radiation and $w$ is the sine of the angle between the light ray and the optical axis as shown in Fig. 3.5.

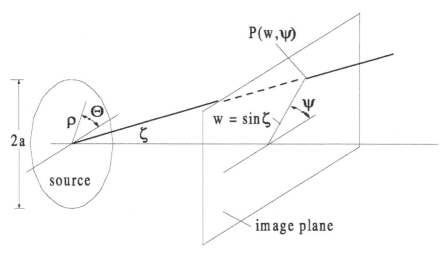

**Fig. 3.5.** Diffraction geometry

With $\alpha = \Theta - \psi$ and the definition of the lowest order Bessel's function $J_0(x) = \frac{1}{2\pi} \int_0^{2\pi} \mathrm{e}^{-ix \cos\alpha} \mathrm{d}\alpha$, (3.42) can be expressed by the integral

$$U(P) = 2\pi C \int_0^a J_0(k\rho w)\, \rho \mathrm{d}\rho. \qquad (3.43)$$

This integral can be solved analytically as well with the identity $\int_0^x J_0(y)\, y\, \mathrm{d}y = x J_1(x)$. The radiation intensity is proportional to the square of the radiation field and we get finally for the radiation intensity in the image plane at the point $P$

$$I(P) = I_0 \frac{4 J_1^2(kaw)}{(kaw)^2}, \qquad (3.44)$$

where $I(P) = |U(P)|^2$ and $I_0 = I(w \to 0)$ is the radiation intensity at the image center. This result has been derived first by Airy [26]. The radiation intensity from a light source of small circular cross section is distributed in the image plane due to diffraction into a central circle and concentric rings illuminated as shown in Fig. 3.6.

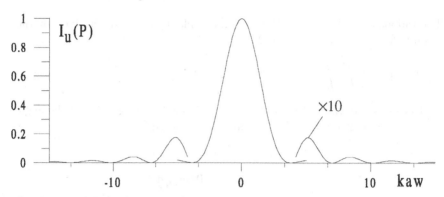

**Fig. 3.6.** Fraunhofer diffraction for a cicular uniform light source

Tacitly, we have assumed that the distribution of emission at the source is uniform which is generally not correct for a particle beam. A Gaussian distribution is more realistic resembling the distribution of independently radiating particles. We must be careful in the choice of the scaling parameter. The relevant quantity for the Fraunhofer integral is not the actual particle beam size at the source point but rather the apparent beam size and distribution. By folding the particle density distribution with the argument of the Fraunhofer diffraction integral we get the radiation field from a round, Gaussian particle beam,

$$U_G(P) = \text{const.} \int_0^\infty \exp\left(-\frac{\rho^2}{2\sigma_r^2}\right) J_0\left(k\rho w\right) \rho \mathrm{d}\rho, \tag{3.45}$$

where $\sigma_r$ is the apparent standard source radius. Introducing the variable $x = \rho/\sqrt{2}\sigma_r$ and replacing $k\rho w = \sqrt{2}xk\sigma_r w = 2x\sqrt{z}$ we get from (3.45)

$$U_G(P) = \text{const.} \int_0^\infty \mathrm{e}^{-x^2} x J_0\left(2x\sqrt{z}\right) \mathrm{d}x \tag{3.46}$$

and after integration

$$U_G(P) = \text{const. } \exp\left[-\tfrac{1}{2}\left(k\sigma_r w\right)^2\right]. \tag{3.47}$$

The diffraction pattern from a Gaussian light source (Fig. 3.7) does not exhibit the ring structure of a uniform source. The radiation field assumes rather the form of a Gaussian distribution in the emission angles $w$ with a standard width of $\sigma_{r'}^2 = \langle w^2 \rangle$ or

$$\sigma_{r'} = \frac{1}{k\,\sigma_r}. \tag{3.48}$$

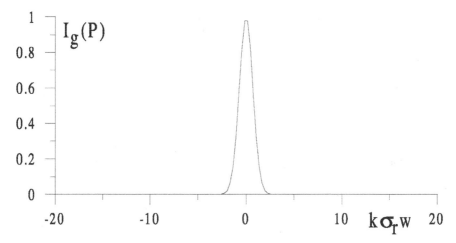

**Fig. 3.7.** Fraunhofer diffraction for a Gaussian luminescence at the light source

## 3.5 Spatial Coherence

Synchrotron radiation is emitted into a broad spectrum with the lowest frequency equal to the revolution frequency and the highest frequency not far above the critical photon energy. Detailed observation of the whole radiation spectrum, however, may reveal significant differences to these theoretical spectra at the low frequency end. At low photon frequencies we may observe an enhancement of the synchrotron radiation beyond intensities predicted by the theory of synchrotron radiation as discussed so far. We note from the definition of the Poynting vector that the radiation power is a quadratic effect with respect to the electric charge. For photon wavelengths equal and longer than the bunch length, we expect therefore all particles within a bunch to radiate coherently and the intensity to be proportional to the square of the number $N_e$ of particles rather than linearly proportional as is the case for high frequencies. This quadratic effect can greatly enhance the radiation since the bunch population can be from $10^8 - 10^{11}$ electrons.

Generally such radiation is not emitted from a storage ring beam because radiation with wavelengths longer than the vacuum chamber dimensions are greatly damped and will not propagate along a metallic beam pipe [27] . This radiation shielding is fortunate for storage ring operation since it eliminates an otherwise significant energy loss mechanism. Actually, since this shielding affects all radiation of sufficient wavelength both the ordinary synchrotron radiation and the coherent radiation is suppressed. New developments in storage ring physics, however, may make it possible to reduce the bunch length by as much as an order of magnitude below presently achieved short bunches of the order of 10 mm. Such bunches would then be much shorter than vacuum chamber dimensions and the emission of coherent radiation in some limited frequency range would be possible. Much shorter electron bunches of the or-

der of 1    2 mm and associated coherent radiation can be produced in linear
accelerators [28] [29], and specifically with bunch compression [30] a significant fraction of synchrotron radiation is emitted spontaneously as coherent
radiation [31].

In this section we will discuss the physics of spontaneous coherent synchrotron radiation while distinguishing two kinds of coherence in synchrotron
radiation, the temporal coherence  and the spatial coherence. Temporal coherence occurs when all radiating electrons are located within a short bunch
of the order of the wavelength of the radiation. In this case the radiation from
all electrons is emitted with about the same phase. For spatial coherence the
electrons may be contained in a long bunch but the transverse beam emittance must be smaller than the radiation wavelength. In either case there is
a smooth transition from incoherent radiation to coherent radiation as determined by a formfactor which depends on the bunch length or transverse
emittance.

Similar to the particle beam characterization through its emittance we
may do the same for the photon beam and doing so for the horizontal or
vertical plane we have with $\sigma_{x,y} = \sigma_r/\sqrt{2}$ and $\sigma_{x',y'} = \sigma_{r'}/\sqrt{2}$ the photon
beam emittance

$$\epsilon_{\mathrm{ph},x,y} = \tfrac{1}{2}\sigma_r\sigma_{r'} = \frac{\lambda}{4\pi} . \tag{3.49}$$

This is the diffraction limited photon emittance and reducing the electron
beam emittance below this value would not lead to an additional reduction
in the photon beam emittance. To produce a spatially coherent or diffraction
limited radiation source the particle beam emittance must be less than the
diffraction limited photon emittance

$$\epsilon_{x,y} \leq \frac{\lambda}{4\pi} . \tag{3.50}$$

Obviously, this condition is easier to achieve for long wavelengths. For
visible light, for example, the electron beam emittance must be smaller than
about $5 \times 10^{-8}$ rad-m to be a spatially coherent radiation source. After having
determined the diffraction limited photon emittance we may also determine
the apparent photon beam size and divergence. The photon source extends
over some finite length $L$ along the particle path which could be either the
path length required for a deflection angle of $2/\gamma$ or a much longer length in
the case of an undulator radiation source to be discussed in the next section.
With $\sigma_{r'}$ the diffraction limited beam divergence the photons seem to come
from a disc with diameter (Fig. 3.8)

$$D = \sigma_{r'}L . \tag{3.51}$$

On the other hand, we know from interference theory the correlation

$$D\sin\sigma_{r'} \approx D\sigma_{r'} = \lambda \tag{3.52}$$

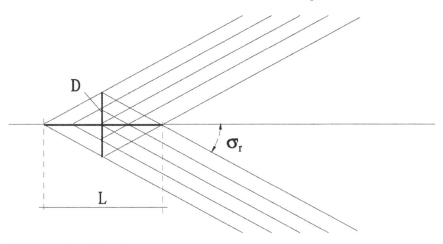

**Fig. 3.8.** Apparent photon source size

and eliminating $D$ from both equations gives the diffraction limited photon beam divergence

$$\sigma_{r'} = \sqrt{\frac{\lambda}{L}} \, . \tag{3.53}$$

With this we get finally from (3.48) also the diffraction limited source size

$$\sigma_r = \frac{1}{2\pi} \sqrt{\lambda L} \, . \tag{3.54}$$

The apparent diffraction limited, radial photon beam size and divergence depend both on the photon wavelength of interest and the length of the source.

## 3.6 Temporal Coherence

To discuss the appearance of temporal coherent synchrotron radiation, we consider the radiation emitted from each particle within a bunch. The radiation field at a frequency $\omega$ from a single electron is

$$\mathcal{E}_j \propto e^{i\left(\omega t + \varphi_j\right)} \, , \tag{3.55}$$

where $\varphi_j$ describes the position of the $j$-th electron with respect to the bunch center. With $z_j$ the distance from the bunch center, the phase is

$$\varphi_j = \frac{2\pi}{\lambda} z_j \, . \tag{3.56}$$

Here we assume that the cross section of the particle beam is small compared to the distance to the observer such that the path length differences from any point of the beam cross section to observer are small compared to the shortest wavelength involved. The radiation power is proportional to the square of the radiation field and summing over all electrons we get

$$P\left(\omega\right) \propto \sum_{j,l}^{N_e} \mathcal{E}_j \mathcal{E}_l^* \propto \sum_{j,l}^{N_e} e^{i\left(\omega t + \varphi_j\right)} e^{-i\left(\omega t + \varphi_l\right)}$$

$$= \sum_{j,l}^{N_e} \exp i(\varphi_j - \varphi_l) = N_e + \sum_{j \neq l}^{N_e} \exp i\left(\varphi_j - \varphi_l\right) . \tag{3.57}$$

The first term $N_e$ on the r.h.s. of (3.57) represents the ordinary incoherent synchrotron radiation with a power proportional to the number of radiating particles. The second term averages to zero for all but long wavelengths. The actual coherent radiation power spectrum depends on the particular particle distribution in the bunch. For a storage ring bunch it is safe to assume a Gaussian particle distribution and we use therefore the density distribution

$$\Psi_G\left(z\right) = \frac{N_e}{\sqrt{2\pi}\sigma} \exp\left(-\frac{z^2}{2\sigma^2}\right) , \tag{3.58}$$

where $\sigma$ is the standard value of the Gaussian bunch length. Instead of summing over all electrons we integrate over all phases and folding the density distribution (3.58) with the radiation power (3.57) we get with (3.56)

$$P\left(\omega\right) \propto N_e + N_e \frac{N_e - 1}{2\pi\sigma^2} I_1 I_2 , \tag{3.59}$$

where the integrals $I_1$ and $I_2$ are defined by

$$I_1 = \int_{-\infty}^{+\infty} \exp\left(-\frac{z^2}{2\sigma^2} + i\,2\pi\frac{z}{\lambda}\right) dz , \tag{3.60a}$$

$$I_2 = \int_{-\infty}^{+\infty} \exp\left(-\frac{w^2}{2\sigma^2} + i\,2\pi\frac{w}{\lambda}\right) dw , \tag{3.60b}$$

and $z = \frac{1}{2}\pi\lambda\varphi_j$ and $w = \frac{1}{2}\pi\lambda\varphi_l$. The factor $N_e - 1$ reflects the fact that we integrate only over different particles. Both integrals are equal to the Fourier transform for a Gaussian particle distribution. With

$$\int_{-\infty}^{+\infty} \exp\left(-\frac{z^2}{2\sigma^2} + i\,2\pi\frac{z}{\lambda}\right) dz = \sqrt{2\pi}\sigma \exp\left[-2\pi^2\left(\frac{\sigma}{\lambda}\right)^2\right] \tag{3.61}$$

we get from (3.59) for the total radiation power at the frequency $\omega = 2\pi c/\lambda$

$$P\left(\omega\right) = p\left(\omega\right) N_e \left[1 + (N_e - 1) \; g^2\left(\sigma, \lambda\right)\right] , \tag{3.62}$$

where $p\left(\omega\right)$ is the radiation power from one electron and the Fourier transform

$$g^2\left(\sigma, \lambda\right) = \exp\left[-2\pi^2\left(\frac{\sigma}{\lambda}\right)^2\right] \tag{3.63}$$

is called the formfactor. With the effective bunch length

$$\ell = \sqrt{2\pi}\sigma \tag{3.64}$$

this formfactor becomes finally

$$g^2\left(\ell, \lambda\right) = \exp\left[-\pi\frac{\ell^2}{\lambda^2}\right] . \tag{3.65}$$

The coherent radiation power falls off rapidly for wavelengths as short or even shorter than the effective bunch length $\ell$. In Fig. 3.9 the relative coherent radiation power is shown as a function of the effective bunch length in units of the radiation wavelength. The fast drop off is evident and for an effective bunch length of about $\ell \approx 0.6\,\lambda$ the radiation power is reduced to only about 10% of the maximum power for very short bunches, when $\ell \ll \lambda$. Particle beams from a linear accelerator have often a more compressed particle distribution of a form between a Gaussian and a rectangular distribution. If we take the extreme of a rectangular distribution

$$\Psi_{\rm r}\left(z\right) = \begin{cases} 1 & \text{for } -\frac{1}{2}\ell < z < \frac{1}{2}\ell \\ 0 & \text{otherwise} \end{cases}, \tag{3.66}$$

we expect to extend the radiation spectrum since the corners and sharp changes of the particle density require a broader spectrum in the Fourier transform. Following the procedure for the Gaussian beam we get for a rectangular particle distribution the Fourier transform

$$g\left(\ell\right) = \frac{\sin x}{x}, \tag{3.67}$$

where $x = \pi\ell/\lambda$. Fig. 3.9 also shows the relative coherent radiation power for this distribution and we note a significant but scalloping extension to

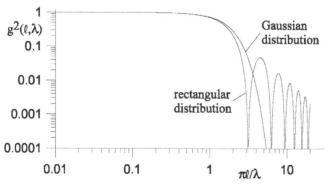

**Fig. 3.9.** Formfactor $g^2(\ell, \lambda)$ for a Gaussian and rectangular particle distribution

higher radiation frequencies. Experiments have been performed with picosecond electron bunches from linear accelerators both at Tohoku University [28] and at Cornell University [29] which confirm the appearance of this coherent part of synchrotron radiation.

## 3.7 Spectral Brightness

The optical quality of a photon beam is characterized by the spectral brightness defined as the six-dimensional volume occupied by the photon beam in phase space

$$\mathcal{B} = \frac{\dot{N}_{\mathrm{ph}}}{4\pi^2 \sigma_x \, \sigma_{x'} \, \sigma_y \, \sigma_{y'} \, \frac{d\omega}{\omega}}, \qquad (3.68)$$

where $\dot{N}_{\mathrm{ph}}$ is the photon flux defined in (3.35). For bending magnet radiation there is a uniform angular distribution in the deflecting plane and we must therefore replace the Gaussian divergence $\sigma_{x'}$ by the total acceptance angle $\Delta\psi$ of the photon beam line or experiment. The particle beam emittance must be minimized to achieve maximum spectral photon beam brightness. However, unlimited reduction of the particle beam emittance will, at some point, seize to further increase the brightness. Because of diffraction effects the electron beam emittance need not be reduced significantly below the limit (3.49) discussed in the previous section.

For a negligible particle beam emittance and deflection angle $\Delta\psi$ the maximum spectral brightness is therefore from (3.49, 3.68)

$$\mathcal{B}_{\mathrm{max}} = \frac{4}{\lambda^2 \frac{d\omega}{\omega}} \dot{N}_{\mathrm{ph}}. \qquad (3.69)$$

For a realistic synchrotron light source the finite beam emittance of the particle beam must be taken into account as well which is often even the dominant emittance being larger than the diffraction limited photon beam emittance. We may add both contributions in quadrature and have for the total source parameters

$$
\begin{aligned}
\sigma_{\mathrm{tot},x} &= \sqrt{\sigma_{\mathrm{b},x}^2 + \tfrac{1}{2}\sigma_r^2}, \\
\sigma_{\mathrm{tot},y} &= \sqrt{\sigma_{\mathrm{b},y}^2 + \tfrac{1}{2}\sigma_r^2}, \\
\sigma_{\mathrm{tot},x'} &= \sqrt{\sigma_{\mathrm{b},x'}^2 + \tfrac{1}{2}\sigma_{r'}^2}, \\
\sigma_{\mathrm{tot},y'} &= \sqrt{\sigma_{\mathrm{b},y'}^2 + \tfrac{1}{2}\sigma_{r'}^2},
\end{aligned}
\qquad (3.70)
$$

where $\sigma_{\mathrm{b}}$ refers to the respective particle beam parameters.

### 3.7.1 Matching

A finite particle beam emittance does reduce the photon beam brightness from it's ideal maximum. The amount of reduction, however, depends on the *matching to the photon beam*. The photon beam size and divergence are the result of folding the diffraction limited source emittance with the electron beam size and divergence. In cases where the electron beam emittance becomes comparable to the diffraction limited emittance the effective photon beam brightness can be greatly affected by the mutual orientation of both emittances. Matching both orientations will maximize the photon beam brightness.

This matching process is demonstrated in Fig. 3.10. The left side shows a situation of poor matching in 2-dimensional $x - x'$-phase space. In this case the electron beam width is very large compared to the diffraction limited source size while its divergence is small compared to the diffraction limit. The effective photon beam distribution in phase space is the folding of both electron beam parameters and diffraction limit and is much larger than either one of its components. The photon beam width is dominated by the electron beam width and the photon beam divergence is dominated by the diffraction limit. Consequently, the effective photon density in phase space and photon beam brightness is reduced.

To improve the situation one would focus the electron beam to a smaller beam size at the source point at the expense of beam divergence. The reduction of the electron beam width increases directly the photon beam brightness while the related increase of the electron beam divergence is ineffective because the diffraction limit is the dominant term. Applying more focusing may give a situation shown on the right side of Fig. 3.10 where the folded photon phase space distribution is reduced and the brightness correspondingly increased. Of course, if the electron beam is focused too much we have the opposite situation as discussed. There is an optimum focusing for optimum matching.

To find this optimum we use the particle beam parameters

$$\sigma_{b,x,y}^2 = \epsilon_{x,y}\beta_{x,y} \quad \text{and} \quad \sigma_{b,x',y'}^2 = \frac{\epsilon_{x,y}}{\beta_{x,y}}, \tag{3.71}$$

where $\beta_{x,y}$ are the betatron functions at the photon source location and $\epsilon_{x,y}$ the beam emittances, in the horizontal and vertical plane respectively. Including diffraction limits, the product

$$\sigma_{\text{tot},x}\sigma_{\text{tot},x'} = \sqrt{\epsilon_x\beta_x + \tfrac{1}{2}\sigma_r^2}\sqrt{\frac{\epsilon_x}{\beta_x} + \tfrac{1}{2}\sigma_{r'}^2} \tag{3.72}$$

has a minimum ($\frac{\mathrm{d}}{\mathrm{d}\beta_x}\sigma_{\text{tot},x}\sigma_{\text{tot},x'} = 0$) for

$$\beta_x = \frac{\sigma_r}{\sigma_{r'}} = \frac{L}{2\pi}. \tag{3.73}$$

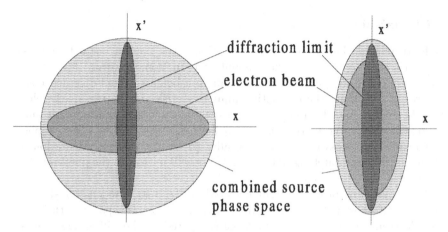

**Fig. 3.10.** Matching of the electron beam emittance to the diffraction limited emittance to gain maximum photon beam brightness

A similar optimum occurs for the vertical betatron function at the source point. The optimum value of the betatron functions at the source point depends only on the length of the undulator.

The values of the horizontal and vertical betatron functions should be adjusted according to (3.73) for optimum photon beam brightness. In case the particle beam emittance is much larger than the diffraction limited photon beam emittance, this minimum is very shallow and almost nonexistent in which case the importance of matching becomes irrelevant. As useful as matching may appear to be, it is not always possible to reach perfect matching because of limitations in the storage ring focusing system. Furthermore it is practically impossible to get a perfect matching for bending magnet radiation since the effective source length $L$ is very small, $L = 2\rho/\gamma$.

## Exercises *

**Exercise 3.1 (S).** Consider an electron storage ring at an energy of 800 MeV, a circulating current of 1 amp and a bending radius of $\rho = 1.784$ m. Calculate the energy loss per turn, and the total synchrotron radiation power from all bending magnets. What would the radiation power be if the particles were 800 MeV muons.

**Exercise 3.2 (S).** For the electron beam of exercise 3.1 calculate the critical energy and plot the radiation spectrum. What is the useful frequency range for experimentation assuming that the spectral intensity should be within

---

* The argument (S) indicates an exercise for which a solution is given in Appendix A.

1 % of the maximum value? Express the maximum useful photon energy in terms of the critical photon energy (only one significant digit!).

**Exercise 3.3 (S).** What beam energy would be required to produce x-rays from the storage ring of Exercise 3.1 at a critical photon energy of 10 keV? Is that energy feasible from a conventional magnet point of view or would the ring have to be larger? What would the new bending radius have to be?

**Exercise 3.4 (S).** The design of the European Large Hadron Collider [18], calls for a circular proton accelerator for energies up to 10 TeV. The circumference is 26.7 km and the bending radius $\rho = 2887$ m. Calculate the energy loss per turn due to synchrotron radiation and the critical photon energy. What is the synchrotron radiation power for a circulating beam of 164 ma?

**Exercise 3.5 (S).** Consider a 7 GeV electron ring with a circulating beam of 200 mA and a bending radius of $\rho = 20$ m. Your experiment requires a photon flux of $10^6$ photons/sec at a photon energy of 8 keV, within a band width of $10^{-4}$ onto a sample with a cross section of $10 \times 10$ $\mu m^2$ and your experiment is 15 m away from the source point. Can you do your experiment on a bending magnet beam line of this ring?

**Exercise 3.6 (S).** Bending magnet radiation ($\rho = 2$ m) from a 800 MeV, 500 mA storage ring includes a high intensity component of infrared radiation. Calculate the photon beam brightness for $\lambda = 10$ $\mu m$ radiation at the experimental station which is 5 m away from the source. The electron beam cross section is $\sigma_{b,x} \times \sigma_{b,y} = 1.1 \times 0.11$ mm and its divergence $\sigma_{b,x'} \times \sigma_{b,y'} = 0.11 \times 0.011$ mrad. What is the corresponding brightness for infrared radiation from a black body radiator at 2000 °K with a source size of $x \times y = 10 \times 2$ mm? (Hint: the source length $L = \rho 2\theta_{rad}$, where $\pm\theta_{rad}$ is the vertical opening angle of the radiation.)

**Exercise 3.7 (S).** How well are the electron beam parameters of Exercise 3.6 at the source matched to the photon beam? Show the phase space ellipses of both the electron and the photon beam in phase space and in $x$ and $y$.

**Exercise 3.8.** With the definition of the world time $\tau = \sqrt{-\tilde{s}^2}$ show that $\gamma d\tau = dt$ and express the 4-velocity and 4-acceleration in terms of laboratory coordinates.

**Exercise 3.9.** Verify the equality of (3.1) and (3.2)

**Exercise 3.10.** Verify the numerical validity of Eqs. (3.20, 3.28, 3.30, 3.36) and (3.40).

# 4. Radiation Sources

Deflection of a relativistic particle beam causes the emission of electromagnetic radiation which can be observed in the laboratory system as broadband radiation, highly collimated in the forward direction. The emission is related to the deflection of a charged particle beam and therefore sweeps like a search light across the detection apparatus of the observer. It is this shortness of the observable radiation pulse which implies that the radiation is detected as synchrotron radiation with a broad spectrum as shown in Fig. 3.3. The width of the spectrum is characterized by the critical photon energy (3.26) and depends only on the particle energy and the bending radius of the magnet. Generally, the radiation is produced in bending magnets of a storage ring, where an electron beam is circulating for hours.

In order to adjust the radiation characteristics to special experimental needs, other magnetic devices are being used as synchrotron radiation sources. Such magnets are known as insertion devices since they do not contribute to the overall deflection of the particle beam in the circular accelerator. Their effect is localized and the total deflection in an insertion device is zero. In this chapter, we give a short overview of all radiation sources and their characteristics and postpone more detailed discussions of insertion device radiation to Chap. 10.

## 4.1 Bending Magnet Radiation

The radiation from bending magnets is emitted tangentially from any point along the curved path like that of a searchlight and appears therefore as a swath of radiation around the storage ring as shown in Fig. 4.1. In the vertical, nondeflecting plane, however, the radiation is very much collimated with a typical opening angle of $\pm 1/\gamma$.

The temporal structure of synchrotron radiation reflects that of the electron beam. Electrons circulating in the storage ring are concentrated into equidistant bunches. The distance between bunches is equal to an integer multiple (usually equal to unity) of the rf-wavelength (60 cm for 500 MHz) while the bunch length itself is of the order of 1 to 3 cm or 30 to 100 ps depending on beam energy and rf-voltage. As a consequence, the photon beam

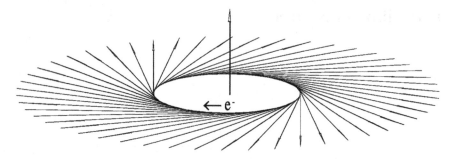

**Fig. 4.1.** Radiation swath from bending magnets in an electron storage ring

consists of a series of short 30 100 ps flashes every 2 ns (500 MHz) or integer multiples thereof.

Radiation is emitted in a broad spectrum (see Fig. 3.3) reaching, in principal, from microwaves up to the critically photon energy (3.26) and beyond with fast declining intensities. The long wavelength limit of the radiation spectrum is actually limited by the vacuum chamber, which causes the suppression of radiation at wavelength longer than its dimensions. The strength of bending magnets, being a part of the geometry of the storage ring cannot be freely varied to optimize for desired photon beam characteristics. This is specifically limiting in the choice of the critical photon energy. While the lower photon energy spectrum is well covered even for rather low energy storage rings, the x-ray region requires high beam energies and/or high magnetic fields. Often, the requirements for x-rays cannot be met with existing bending magnet and storage ring parameters.

## 4.2 Superbends

The critical photon energy from bending magnet radiation (3.28) is determined by the magnet field and the particle energy. The combination of both quantities may not be sufficient to extend the synchrotron radiation spectrum into the hard x-ray regime, especially in low energy storage rings. In this case, it is possible to replace some or all original bending magnets by much stronger but shorter magnets, called superbends. To be more specific, conventional bending magnets are replaced by high field, shorter superconducting magnets deflecting the electron beam by the same angle to preserve the storage ring geometry. Since conventional bending magnet fields rarely exceed 1.5 Tesla, but superconducting magnets can be operated at 5 to 6 Tesla or higher, one can gain a factor of 3 to 4 in the critical photon energy and extend the photon spectrum towards or even into the hard x-ray regime.

## 4.3 Wavelength Shifter

The installation of superbends is not always feasible or desirable. To still meet the need for harder x-ray radiation in a low energy storage ring, it is customary to use a wavelength shifter. Such a device may consist of three or five superconducting dipole magnets with alternating magnetic field directions. For this latter reason, a wavelength shifter is a true insertion device. The limitation to three or five poles is purely technical and may be eased as superconducting magnet and cryo-technology progresses. Figure 4.2 shows schematically a three-pole wavelength shifter.

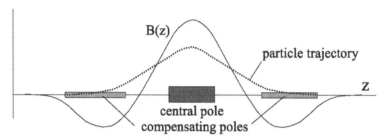

**Fig. 4.2.** Magnetic field distribution along the beam path for a wave length shifter

The particle beam passing through this wavelength shifter is deflected up and down or left and right in such a way that no net deflection remains. To meet this condition, the longitudinal field distribution of a horizontally deflecting wavelength shifter must obey the condition

$$\int_{-\infty}^{\infty} B_y\,(y = 0, z)\,\mathrm{d}\,z = 0\,. \tag{4.1}$$

A wavelength shifter with such field properties is neutral on the geometry of the particle beam path through a storage ring and therefore can be made in principle as strong as necessary or technically feasible.

Only the central high field pole is used as the radiation source, while the two side poles compensate the beam deflection from the central pole. In a five-pole wavelength shifter the three central poles would be used as radiators, while both end poles again act as compensators. Mostly, the end poles are longer than the central poles and operate at a lower field. As their name implies, the primary objective in wavelength shifters is to extend the photon spectrum while the enhancement of intensity through radiation accumulation from many poles, while desirable, is of secondary importance. To maximize the desired effect, wavelength shifters are often constructed as high field superconducting magnets to maximize the critical photon energy for the given particle beam energy. Some limitations apply for such devices as well as for any other insertion device. The end fields of magnets can introduce particle focusing and nonlinear field components may introduce aberrations and cause beam instability. Both effects must either be kept below a critical level or be compensated.

## 4.4 Wiggler Magnet Radiation

The principle of a wavelength shifter is extended in the case of a wiggler-magnet. Such a magnet consists of a series of equal dipole magnets with alternating magnetic field direction. Again, the end poles must be configured to make the total device neutral to the geometry of the particle beam path such that the condition (4.1) is met.

The main advantage of using many magnet poles is to increase the photon flux. Like a single bending magnet, each of the $N_m$ magnet poles produces a fan of radiation in the forward direction and the total photon flux is $N_m$-times larger than that from a single pole. Wiggler-magnets may be constructed as electromagnets with fields up to 2 T to function both as a flux enhancer and as a more modest wavelength shifter compared to the superconducting type. An example of an 8-pole, 1.8 T electromagnetic wiggler-magnet [32] is shown in Fig. 4.3.

**Fig. 4.3.** Electromagnetic wiggler magnet with eight 1.8 T poles

In this picture, the magnet gap is wide open, to display the flat vacuum chamber running through the magnet between the poles. The pole pieces in the lower row are visible surrounded by water cooled excitation coils. During operation, both rows of wiggler poles are closed to almost touch the flat vacuum chamber. When the magnet is closed, a maximum magnetic field of 1.8 T can be obtained. Strong fields can be obtained from electromagnets, but the space requirement for the excitation coils limits the number of poles that can be installed within a given length.

Progress in the manufacturing of high field permanent magnet material permits the installation of many more poles into the same space compared to an electromagnet. An example of a modern 26 pole, 2.0 T permanent magnet wiggler magnet is shown in Fig. 4.4 [33].

Figure 4.4 shows the wiggler magnet during magnetic measurement with the rail in front of the magnet holding and guiding the Hall probe. The increased number of poles and simplified design compared to the electromagnetic wiggler in Fig. 4.3 are clearly visible.

For short wiggler poles, we express the magnetic field by

$$B_y\left(x, y = 0, z\right) = B_0 \sin \frac{2\pi z}{\lambda_\mathrm{p}} \tag{4.2}$$

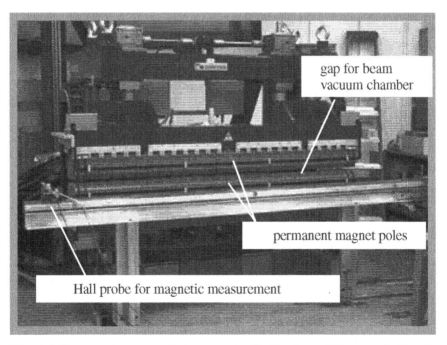

gap for beam
vacuum chamber

permanent magnet poles

Hall probe for magnetic measurement

**Fig. 4.4.** Permanent magnet wiggler magnet with 26 poles, a 175 mm period length and a maximum field of 2.0 T

and the maximum beam deflection from the axis is equal to the deflection angle per half pole (6.11)

$$\vartheta = \frac{B_0}{B\rho} \int_0^{\lambda_{\mathrm{p}}/4} \sin \frac{2\pi z}{\lambda_{\mathrm{p}}} \, \mathrm{d}\, z = \frac{B_0}{B\rho} \frac{\lambda_{\mathrm{p}}}{2\pi} , \qquad (4.3)$$

where $B\rho$ is the beam rigidity defined in (6.7). Multiplying this with the beam energy $\gamma$, we define the wiggler strength parameter

$$K = \gamma \vartheta = \frac{[c]\, e B_0}{mc^2} \frac{\lambda_{\mathrm{p}}}{2\pi} = 0.934 \, B_0 \, (\mathrm{T}) \, \lambda_{\mathrm{p}} \, (\mathrm{cm}) . \qquad (4.4)$$

This wiggler strength parameter is generally much larger than unity. Conversely, a series of alternating magnet poles is called a wiggler magnet if the strength parameter $K \gg 1$ and condition (4.1) is met. As we will see later, a weak wiggler magnet with $K \ll 1$ is called an undulator and produces radiation with significant different characteristics. The magnetic field strength can be varied in both electromagnetic wigglers as well as in permanent magnet wigglers . While this is obvious for electromagnets, the magnetic field strength in permanent magnets depends on the distance between magnet poles or on the gap height $g$. By varying mechanically the gap height of a permanent magnet wiggler, the magnetic field strength can be varied as well. The field strength also depends on the period length and on the design and magnet materials used. For a wiggler magnet constructed as a hybrid magnet with Vanadium Permendur poles, the field strength along the midplane axis scales approximately like [34]

$$B_y(\mathrm{T}) \approx 3.33 \exp\left[ -\frac{g}{\lambda_{\mathrm{p}}} \left( 5.47 - 1.8 \frac{g}{\lambda_{\mathrm{p}}} \right) \right] , \qquad \text{for } 0.1\,\lambda_{\mathrm{p}} \lesssim g \lesssim 10\,\lambda_{\mathrm{p}} ,$$

$$(4.5)$$

where $g$ is the gap aperture between magnet poles. This dependency is also shown in Fig. 4.5 and we note immediately that the field strength drops off dramatically for magnet gaps of the order of a period length or greater.

On the other hand, significant field strengths can be obtained for small gap apertures and it is therefore important to install the insertion device at a location, where the beam dimension normal to the deflection plane is small.

The total radiation power can be derived by integrating (3.9) through the wiggler magnet. The result of this integration is

$$\langle P_\gamma \rangle = \tfrac{1}{3} r_{\mathrm{e}} mc^2 \, c\, \gamma^2 K^2 \frac{4\pi^2}{\lambda_{\mathrm{p}}^2} , \qquad (4.6)$$

or in practical units

$$\langle P_\gamma \, (\mathrm{W}) \rangle = 632.7 \, E^2 \, B_0^2 \, I \, L_{\mathrm{u}} , \qquad (4.7)$$

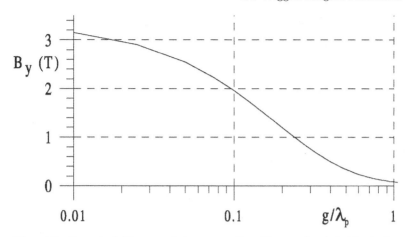

**Fig. 4.5.** On-axis field strength in a vanadium Permendur hybrid wiggler magnet as a function of gap aperture (4.5)

where $I$ is the ciculating beam current, and $L_u = N_p \lambda_p$ the length of the wiggler magnet.

For a sinusoidal field distribution $B_0 \sin \frac{2\pi}{\lambda_p} z$, the desired wavelength shifting property of a strong wiggler magnet can be obtained only in the forward direction. Radiation emitted at a finite angle with respect to the wiggler axis is softer because it is generated at a source point where the field is lower. The hardest radiation is emitted in the forward direction from the crest of the magnetic field. For a distance $\Delta z$ away from the crest, the emission angle in the deflection plane is $\psi = \frac{1}{\rho_0} \frac{\lambda_p}{2\pi} \sin \frac{2\pi}{\lambda_p} \Delta z$ and the curvature at the source point is $\frac{1}{\rho} = \frac{1}{\rho_0} \sqrt{1 - \left(\frac{\gamma\psi}{K}\right)^2}$, where we have made use of (4.4). Consequently, the critical photon energy for radiation in the direction $\psi$ with respect to the wiggler axis varies with the emission angle $\psi$ like

$$\varepsilon_c = \varepsilon_{c0} \sqrt{1 - \left(\frac{\gamma\psi}{K}\right)^2}. \tag{4.8}$$

At the maximum deflection angle $\psi_{max} = \theta = K/\gamma$ the critical photon energy has dropped to zero, reflecting a zero magnetic field at the source point.

This property is undesirable if more than one experimental station is supposed to receive hard radiation from the same wiggler magnet. The strength of the wiggler magnet sweeps the electron beam over a considerable angle, a feature which can be exploited to direct radiation not only to one experimental station along the axis but also to two or more side-stations on either side of the wiggler axis. However, these side beam lines at an angle $\psi \neq 0$ receive softer radiation than the main beam line. This can be avoided if the poles of the wiggler magnet are lengthened thus flattening the sinusoidal field crest.

As the flat part of the field crest is increased, hard radiation is emitted into an increasing angular cone.

## 4.5 Undulator Radiation

So far, we discussed insertion devices designed specifically to harden the radiation spectrum or to increase the radiation intensity. Equally common is the implementation of insertion devices to optimize photon beam quality by maximizing its brightness or to provide specific characteristics like elliptically polarized radiation. This is done with the use of undulator magnets, which are constructed similar to wiggler magnets, but are operated at a much reduced field strength.

Fundamentally, an undulator magnet causes particles to be only very weakly deflected with an angle of less than $\pm 1/\gamma$ and consequently the transverse motion of particles is nonrelativistic. In this picture, the electron motion viewed from far away along the beam axis appears as a purely sinusoidal transverse oscillation similar to the electron motion in a linear radio antenna driven by a transmitter and oscillating at the station's carrier frequency. The radiation emitted is therefore monochromatic with a period equal to the oscillation period.

To be more precise, viewed from far away the particle appears to be at rest or uniform motion as long as the electron has not yet reached the undulator magnet. Upon entering the magnet the electron performs sinusoidal transverse oscillations and returns to its original motion again after it exits the undulator. As a consequence of this motion and in light of earlier discussions, we observe emission of radiation at the frequency of the transverse oscillating beam motion. If $N_\mathrm{p}$ is the number of undulator periods, the electric field lines have been perturbed periodically $N_\mathrm{p}$-times and the radiation pulse is composed of $N_\mathrm{p}$ oscillations. In the particle rest frame $\mathcal{L}^*$ the undulator periodlength is Lorentz contracted to $\lambda_\gamma^* = \lambda_\mathrm{p}/\gamma$ which is the wavelength of the emitted radiation. Because the radiation includes only a finite number of $N_\mathrm{p}$ oscillations, the radiation is not quite monochromatic but rather quasi monochromatic with a band width of $1/N_\mathrm{p}$. This situation is illustrated in Fig. 4.6a.

In Fig. 4.6b the radiation lobe and spectrum is shown in the laboratory system. The monochromatic nature of the radiation is somewhat lost because radiation emitted at different angles experiences different Doppler shifts. Of course, the radiation is again quasi monochromatic even in the laboratory system when observed through a narrow pin hole along the axis. This monochromatic radiation is called the fundamental undulator radiation and has for $K \ll 1$ a wavelength of

$$\lambda_\gamma \approx \frac{\lambda_\mathrm{p}}{2\gamma^2}.$$

(4.9)

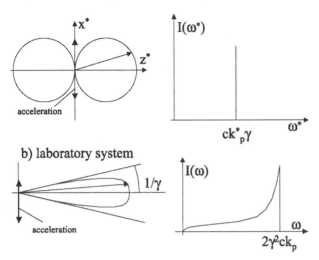

a) electron rest system

b) laboratory system

**Fig. 4.6.** Beam dynamics and radiation lobes in the particle rest system (**a**) and the laboratory system (**b**) for a weak undulator ($K \ll 1$).

The situation becomes more complicated as the undulator strength is increased. Two new phenomena appear, an oscillatory forward motion and a transverse relativistic effect. The first phenomenon that we need to discuss is the fact that the transverse motion becomes relativistic. As a consequence of this, the pure sinusoidal transverse motion becomes distorted. There is a periodic Lorentz contraction of the longitudinal coordinate, which is larger when the particle travels almost parallel to the axis in the vicinity of the oscillation crests and is smaller when in between crests. The cusps and valleys of the sinusoidal motion become Lorentz-contracted in the particle system thus perturbing the sinusoidal motion as shown in Fig. 4.7.

This perturbation is symmetric about the cusps and valleys causing the appearance of odd and only odd (3rd, 5th, 7th...) harmonics of the fundamental oscillation period. From an undulator of medium strength ($K \gtrsim 1$) we observe therefore along the axis a line spectrum of odd harmonics in addition to the fundamental undulator radiation.

The second phenomenon to be discussed is the periodic modulation of the longitudinal motion. The longitudinal component of the particle velocity is maximum when the particle travels close to the crest of the oscillations and at a minimum when it is close to the axis crossings. In a reference system which moves uniformly with the average longitudinal particle velocity along the axis, the particle performs periodic longitudinal oscillations in addition to the transverse oscillations. For each transverse period, the particle performs two longitudinal oscillations and its path looks therefore like a figure of "8". This situation is shown in Fig. 4.8.

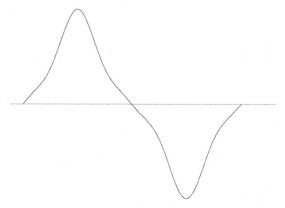

**Fig. 4.7.** Distortion of sinusoidal motion due to relativistic perturbation of transverse motion

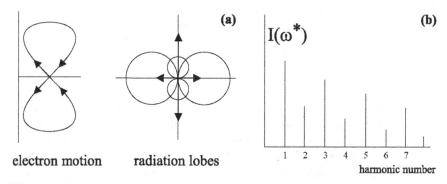

electron motion        radiation lobes

**Fig. 4.8.** Beam dynamics and radiation lobes in the particle rest system (**a**) and the laboratory system (**b**) for a stronger undulator ($K \gtrsim 1$).

We have now two orthogonal accelerations, one transverse and one longitudinal, and two radiation lobes as indicated in Fig. 4.8. Since the longitudinal motion occurs at twice the frequency of the transverse motion, we observe now radiation also at twice the fundamental frequency. Of course, the relativistic perturbation applies here too and we have therefore a line spectrum which includes two series, one with all odd harmonics and one with only even harmonics. Even and odd harmonic radiation is emitted in the particle system in orthogonal directions and therefore we find both radiation lobes in the laboratory system spatially separated as well. The odd harmonics all have their highest intensities along the undulator axis, while the even harmonic radiation is emitted preferentially into an angle $1/\gamma$ with respect to the axis and has zero intensity along the axis.

In another equally valid view of undulator radiation, the static and periodic magnetic undulator field appears in the rest frame of the electron as a Lorentz contracted electromagnetic field or as monochromatic photon of wavelength $\lambda^* = \lambda_p/\gamma$. The emission of photons can therefore be described as Thomson scattering of virtual photons by free electrons [35] resulting in monochromatic radiation in the direction of the particle path. Viewed from the laboratory system, the radiation is Doppler shifted and applying (2.20) the wavelength of the backscattered photons is

$$\lambda_{ph} = \frac{\lambda_p}{\gamma^2 \left(1 + \beta\, n_z^*\right)}. \tag{4.10}$$

Viewing the radiation parallel to the forward direction ($\vartheta \approx 0$), (2.23) becomes with $n_z = \cos \vartheta^* \approx 1 - \frac{1}{2}\vartheta^{*2}$, and $\beta \approx 1$

$$1 + \beta\, n_z^* = \frac{\beta + n_z^*}{n_z} \approx 2 - \frac{1}{2}\frac{\vartheta^{*2}}{n_z}. \tag{4.11}$$

Setting $n_z \approx 1$, the fundamental wavelength of the emitted radiation is

$$\lambda_1 = \frac{\lambda_p}{\gamma^2} \frac{1}{2 - \frac{1}{2}\frac{\vartheta^{*2}}{n_z}} \approx \frac{\lambda_p}{2\gamma^2}\left(1 + \tfrac{1}{4}\,\vartheta^{*2}\right). \tag{4.12}$$

With (2.24) the angle $\vartheta^*$ of the particle trajectory with respect to the direction of observation is transformed into the laboratory system for $\vartheta^* = 2\gamma\vartheta$. We distinguish two configurations. One where $\vartheta = K/\gamma =$const. describing the particle motion in a helical undulator, where the magnetic field being normal to the undulator axis rotates about this axis. The other more common case is that of a flat undulator , where the particle motion follows a sinusoidal path in which case $\vartheta = \vartheta_{und} + \vartheta_{obs}$ . Here $\vartheta_{und} = \frac{K}{\gamma}\sin k_p z$ is the observation angle due to the periodic motion of the electrons in the undulator and $\vartheta_{obs}$ is the actual observation angle. With these definitions and taking the average $\langle\vartheta_{und}^2\rangle$ we get $\gamma^2\vartheta^2 = \frac{1}{2}K^2 + \gamma^2\vartheta_{obs}^2$. Depending on the type of undulator, the wavelength of radiation from an undulator with a strength parameter $K$ is

$$\lambda_1 = \begin{cases} \dfrac{\lambda_p}{2\gamma^2}\left(1 + K^2 + \gamma^2\vartheta_{obs}^2\right) & \text{for a helical undulator} \\[2ex] \dfrac{\lambda_p}{2\gamma^2}\left(1 + \tfrac{1}{2}K^2 + \gamma^2\vartheta_{obs}^2\right) & \text{for a flat undulator.} \end{cases} \tag{4.13}$$

From now on only flat undulators will be considered in this text and readers interested in more detail of helical undulators are referred to [36]. No special assumptions have been made here which would prevent us to apply this derivation also to higher harmonic radiation and we get the general expression for the wavelength of the $k$-th harmonic

$$\lambda_k = \frac{\lambda_p}{2\gamma^2 k}\left(1 + \tfrac{1}{2}K^2 + \gamma^2\vartheta_{obs}^2\right). \tag{4.14}$$

The additional terms $\frac{1}{2}K^2 + \gamma^2\vartheta^2_{obs}$ compared to (4.9) comes from the correct application of the Doppler effect. Since the particles are deflected periodically in the undulator, we view even the on-axis radiation at a periodically varying angle which accounts for the $\frac{1}{2}K^2$-term. Of course, observation of the radiation at a finite angle $\vartheta_{obs}$ generates an additional red-shift expressed by the term $\gamma^2\vartheta^2_{obs}$.

In more practical units, the undulator wavelengths for the $k$-th harmonic are expressed from (4.14) by

$$\lambda_k \left(\text{Å}\right) = 13.056 \frac{\lambda_p \left(\text{cm}\right)}{k\,E^2 \left(\text{GeV}^2\right)} \left(1 + \tfrac{1}{2}\,K^2 + \gamma^2\vartheta^2_{obs}\right) \tag{4.15}$$

and the corresponding photon energies are

$$\epsilon_k \left(\text{eV}\right) = 950 \frac{k\,E^2 \left(\text{GeV}^2\right)}{\lambda_p \left(\text{cm}\right) \left(1 + \tfrac{1}{2}\,K^2 + \gamma^2\vartheta^2_{obs}\right)} . \tag{4.16}$$

Recollecting the discussion of undulator radiation, we found that the first harmonic or fundamental radiation is the only radiation emitted for $K \ll 1$. As the undulator parameter increases, however, the oscillatory motion of the particle in the undulator deviates from a pure sinusoidal oscillation. For $K > 1$ the transverse motion becomes relativistic, causing a deformation of the sinusoidal motion and the creation of higher harmonics. These harmonics appear at integral multiples of the fundamental radiation energy. Only odd harmonics are emitted along the axis ($\vartheta \approx 0$) while even harmonics are emitted into a small angle from the axis. As the undulator strength is further increased more and more harmonics appear, each of them having a finite width due to the finite number of undulator periods, and finally merging into the well-known broad spectrum of bending or wiggler magnet radiation (Fig. 4.9).

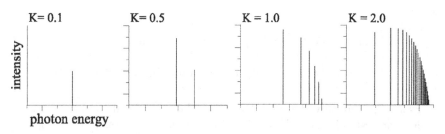

**Fig. 4.9.** Transition from quasi-monochromatic undulator radiation to broad band wiggler radiation

We find no fundamental difference between undulator and wiggler magnets, one being just a stronger version of the other. From a practical point of view, the radiation characteristics are very different and users of synchrotron radiation make use of this difference to optimize their experimental capabilities. In Chap. 10 we will discuss the features of undulator radiation in much more detail.

The electron motion through an undulator with $N_p$ periods includes that many oscillations and so does the radiation field. Applying a Fourier transformation to the field, we find the spectral width of the radiation to be

$$\frac{\Delta\lambda}{\lambda} = \frac{1}{N_p}. \tag{4.17}$$

In reality, this line width is increased due to the finite aperture of the radiation detection elements, and due to a finite energy spread and finite divergence of the electron beam. Typical experimental undulator spectra are shown in Fig. 4.10 for increasing undulator strength $K$ [37].

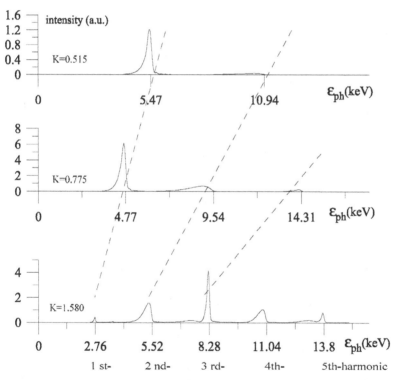

**Fig. 4.10.** Measured radiation spectrum from an undulator for different strength parameters $K$. The intensity at low photon energies are reduced by absorption in a Be-window

Although this radiation was measured through a pin hole and on-axis, we still recognize even harmonic radiation since the pin hole covers a finite solid angle and lets some even harmonic radiation through. Furthermore, the measured intensities of the line spectrum does not reflect the theoretical expectation for the lowest harmonics at higher values of $K$. This is an artifact of the experimental circumstances, where the x-rays have been extracted from the storage ring vacuum chamber through a Be-window. Such a window works very well for hard x-rays but absorbs heavily at photon energies below some 3 keV.

The concentration of all radiation into one or few spectral lines is very desirable for many experiments utilizing monochromatic photon beams since radiation is produced only in the vicinity of the desired wavelength at high brightness. Radiation at other wavelengths creating undesired heating effects on optical elements and samples is greatly eliminated.

## 4.6 Back Scattered Photons

The principle of Thomson backscattering or Compton scattering of the static undulator fields can be expanded to that of photon beams colliding head on with the particle beam. In the electron system of reference the electromagnetic field of this photon beam looks fundamentally no different than the electromagnetic field from the undulator magnet. We may therefore apply similar arguments to determine the wavelength of back scattered photons. The basic difference of both effects is that in the case of back scattered photons the photon beam moves with the velocity of light towards the electron beam and therefore the electron *sees* twice the Lorentz contracted photon frequency and we expect therefore a back scattered photon beam at twice Doppler shifted frequency. That extra factor of two does not apply for undulator radiation since the undulator field is static and the relative velocity with respect to the electron beam is $c$. If $\lambda_L$ is the wavelength of the incident radiation or incident laser, the wavelength of the backscattered photons is

$$\lambda_\gamma = \frac{\lambda_L}{4\gamma^2} \left(1 + \gamma^2 \vartheta_{\mathrm{obs}}^2\right) , \tag{4.18}$$

where $\vartheta_{\mathrm{obs}}$ is the angle between the direction of observation and the particle beam axis. Scattering, for example, a high intensity laser beam from high-energy electrons produces a monochromatic beam of hard x-rays which is highly collimated within an angle of $\pm 1/\gamma$. If the laser wavelength is, for example, $\lambda_L = 10$ μm and the particle energy is 100 MeV the wavelength of the backscattered x-rays would be 1.3 Å or the photon energy would be 9.5 keV which is well within the hard x-ray regime.

### 4.6.1 Photon Flux

The intensity of the backscattered photons can be calculated in a simple way utilizing the Thomson scattering cross section [35]

$$\sigma_{\mathrm{Th}} = \tfrac{8\pi}{3} r_{\mathrm{c}}^2 = 6.65 \times 10^{-25} \ \mathrm{cm}^2. \tag{4.19}$$

The total scattering event rate or the number of back scattered photons per unit time is then

$$N_{\mathrm{sc}} = \sigma_{\mathrm{Th}} \mathcal{L} , \tag{4.20}$$

where $\mathcal{L}$ is called the luminosity. The value of the luminosity is independent of the nature of the physical reaction and depends only on the intensities and geometrical dimensions of the colliding beams. The definition of the luminosity is the product of the target density of one beam by the "particle"-flux of the other beam onto this target. Therefore the luminosity can be determined by folding the particle density in one beam with the incident "particles" per unit time of the other beam. Obviously, only those parts of the beam cross sections count which overlap with the cross section of the other beam. For simplicity, we assume a Gaussian distribution in both beams and assume that both beam cross sections are the same. In a real setup one would focus the electron beam and the photon beam to the same optimum cross section given by the Rayleigh length (11.58). We further consider the particle beam as the target for the photon beam.

With $N_{\mathrm{e}}$ electrons in each bunch of the particle beam within a cross section of $2\pi\sigma_x\sigma_y$ the particle density is $N_{\mathrm{e}} / 2\pi\sigma_x\sigma_y$. We consider now a photon beam with the same time structure as the electron beam. If this is not the case only that part of the photon beam which actually collides with the particle beam within the collision zone may be considered. For an effective photon flux $\dot{N}_{\mathrm{ph}}$ the luminosity is

$$\mathcal{L} = \frac{N_{\mathrm{e}} \dot{N}_{\mathrm{ph}}}{2\pi\sigma_x\sigma_y} . \tag{4.21}$$

Although the Thomson cross-section and therefore the photon yield is very small, this technique can be used to produce photon beams with very specific characteristics. By analyzing the scattering distribution this procedure can also be used to determine the degree of polarization of an electron beam in a storage ring.

So far, it was assumed that the incident and scattered photon energies are much smaller than the particle energy in which case it was appropriate to use the classical case of Thomson scattering. However, we note from (4.18) that the backscattered photonenergy increases quadratically with the particle energy and therefore at some energy the photon energy becomes larger than the particle energy which is nonphysical. In case of large photon energies comparable with the particle energy, Compton corrections [38] [39] [40] must be included. The Compton cross-section for head-on collision is given by [41]

$$\sigma_{\mathrm{C}} = \frac{3\,\sigma_{\mathrm{Th}}}{4x} \left[ \left( 1 - \frac{4}{x} - \frac{8}{x^2} \right) \ln\left(1+x\right) + \tfrac{1}{2} + \frac{8}{x^2} - \frac{1}{2\left(1+x\right)^2} \right] , \tag{4.22}$$

where $x = \frac{4\gamma\hbar\omega_0}{mc^2}$, and $\hbar\omega_0$ the incident photon energy. The energy spectrum of the scattered photons is then [41]

$$\frac{d\sigma_C}{dy} = \frac{3\,\sigma_{Th}}{4x}\left[1 - y + \frac{1}{1-y} - \frac{4y}{x\,(1+y)} + \frac{4y^2}{x^2\,(1-y)^2}\right],\qquad (4.23)$$

where $y = \hbar\omega/E$ is the scattered photon energy in units of the particle energy.

# Exercises *

**Exercise 4.1 (S).** Assume a proton storage ring in space surrounding the earth at an average radius of 150 km. Further, assume a circulating current of 10 mA and 50 W of rf-power available to compensate for synchrotron radiation. What is the maximum proton energy that can be reached with permanent magnets producing a maximum field of 2 Tesla? Is the energy limited by the maximum magnetic field or synchrotron radiation losses? Calculate the energy loss per turn, critical photon energy, and the total radiation power.

**Exercise 4.2 (S).** Specify main parameters for a synchrotron radiation source for digital subtractive angiography. In this medical procedure, two hard x-ray beams are selected from monochromators, one just below and the other just above the K-edge of iodine. With each beam an x-ray picture of, for example, the human heart with peripheral arteries is taken, while the blood stream contains some iodine. Both pictures differ only where there is iodine because of the very different absorption coefficient for both x-ray beams. Displaying the difference of both pictures shows the blood vessels alone. Select parameters for beam energy, wiggler magnet field, number of poles and beam current to produce an x-ray beam at 33 keV of $2 \times 10^{14}$ photons/sec/0.1%BW into an opening angle of 25 mrad, while keeping the 66 keV and 99 keV contamination to less than 1% of the 33 keV radiation.

**Exercise 4.3 (S).** Consider a 30-pole wiggler magnet with 10 cm wide poles, a field distribution $B_y(T) = 2.0\sin\frac{2\pi}{\lambda_p}z$ and a period length of $\lambda_p = 7.0$ cm. Determine the magnetic force between the upper and lower row of poles. Is this force attractive or repulsive? why?

**Exercise 4.4 (S).** Derive an expression for the total synchrotron radiation power from a wiggler magnet.

**Exercise 4.5 (S).** In Chapter 2 we described undulator radiation as a result of Compton scattering of the undulator field by the electrons. Derive the fundamental undulator wavelength from the process of Compton scattering.

---

* The argument (S) indicates an exercise for which a solution is given in Appendix A.

**Exercise 4.6 (S).** An undulator is constructed from hybrid permanent magnet material with a period length of $\lambda_p = 5.0$ cm. What is the fundamental wavelength range in a 800 MeV storage ring and in a 7 GeV storage ring if the undulator gap is to be at least 10 mm?

**Exercise 4.7 (S).** Determine the tuning range for a hybrid magnet undulator in a 2.5 GeV storage ring with an adjustable gap $g \geqq 10$ mm. Plot the fundamental wavelength as a function of magnet gap for two different period lengths, $\lambda_p = 15$ mm and $\lambda_p = 75$ mm. Why are the tuning ranges so different?

**Exercise 4.8.** Consider an electron storage ring at an energy of 1 GeV, a circulating current of 200 mA and a bending radius of $\rho = 2.22$ m. Calculate the energy loss per turn, the critical energy and the total synchrotron radiation power. At what frequency in units of the critical frequency has the intensity dropped to 1% of the maximum? Plot the radiation spectrum and determine the frequency range available for experimentation.

**Exercise 4.9.** What beam energy would be required to produce x-rays from the storage ring of exercise 4.8 at a critical photon energy of 10 keV? Is that energy feasible from a conventional magnet point of view or would the ring have to be larger? What would the new beam energy and bending radius have to be?

**Exercise 4.10.** Consider a storage ring with an energy of 1 GeV and a bending radius of $\rho = 2.5$ m. Calculate the angular photon flux density $d\dot{N}/d\psi$ for a high photon energy $\hat{\varepsilon}$ where the intensity is still 1% of the maximum spectral intensity. What is this maximum photon energy? Installing a wavelength shifter with a field of $B = 6$ T allows the spectrum to be greatly extended. By how much does the spectral intensity increase at the photon energy $\hat{\varepsilon}$ and what is the new photon energy limit for the wavelength shifter?

**Exercise 4.11.** Derive an expression for the average velocity component $\bar{\beta}_z = \bar{v}_z/c$ of a particle traveling through an undulator magnet of strength $K$.

**Exercise 4.12.** Consider an electromagnetic wavelength shifter in a 1 GeV storage ring with a central pole length of 30 cm and a maximum field of 6 T. The side poles are 60 cm long and for simplicity, assume that the field in all poles has a sinusoidal distribution along the axis. Determine the focal length due to edge focusing for the total wavelength shifter. To be negligible, the focal length should typically be longer than about 30 m. Is this the case for this wavelength shifter?

**Exercise 4.13.** Use the tuning graphs of the two undulators of problem 4.7 and add the tuning ranges for the 3rd and 5th harmonic to it. Is it possible in both cases to produce radiation over the whole spectral range between 1st and 5th harmonic?

**Exercise 4.14.** Consider a 26-pole wiggler magnet with a field $B_y(\text{T}) = 2.0\sin\frac{2\pi}{\lambda_p}z$ and a period length of $\lambda_p = 15.0$ cm as the radiation source for a straight through photon beam line and two side stations at an angle $\vartheta = \pm 4$ mr in a storage ring with a beam energy of 2.0 GeV. What is the critical photon energy for the photon beam in the straight ahead beam line and in the two side stations?

**Exercise 4.15.** Collide a 25 MeV electron beam with a 1 kW $CO_2$ laser beam ($\lambda = 10.0$ $\mu$m). What is the energy of the backscattered photons? Assume a diffraction limited interaction length of twice the Rayleigh length and an electron beam cross section matching the photon beam. Calculate the x-ray photon flux for an electron beam from a 3 GHz linear accelerator with a pulse length of 1 $\mu$m and a pulse current of 100 mA.

# 5. Accelerator Physics

A beam of charged particles can emit synchrotron radiation whenever it is deflected by a magnetic or electric field. The intensity and spectrum of the radiation depends greatly on the relativistic factor $\gamma$ of the charged particle. For this reason, only electron or positron beams are considered as potential synchrotron radiation sources and we concentrate on accelerator systems which can produce highly relativistic electron or positron beams. In the future we will not distinguish anymore between electrons and positrons. Some synchrotron radiation facilities operate with a positron beam to avoid sometimes detrimental effects of positive ion-clouds surrounding an electron beam.

The energy of an electron is measured in units of an "electron Volt". This is equal to the kinetic energy gained by an electron while being accelerated in the field between two electrodes with a potential difference of 1 Volt. Electrons become relativistic if their kinetic energy exceeds that of the restmass or about 511000 eV. Most synchrotron radiation sources are based on electron beams with kinetic energies of several hundred million electron volts and higher. We use for such high energies the units MeV ($10^6$ eV) or GeV($10^9$ eV). The photon energy of synchrotron radiation is also measured in eV. Photon energies of general interest reach up to about 20 keV, where 1 keV=1000 eV. For scaling it is useful to remember that a photon wavelength of 1Å is equivalent to an energy of 12398 eV or 12.4 keV.

In this brief overview on accelerator physics, we consider only magnetic fields for relativistic electron beam guidance since technically feasible magnetic fields are much more effective than equally feasible electric fields (1 Tesla of magnetic field corresponds to 3.0 MV/cm of electric field!). For application and research one would like to have a continuous emission of photons which can be accomplished in an electron storage ring.

A storage ring is a circular accelerator which is widely used as a synchrotron radiation source. After injection, electrons circulate in this ring for many hours at constant energy serving as the source of continuous synchrotron radiation. A storage ring is therefore not a true accelerator although a beam can be accelerated very slowly if required (e.g. if the injection energy is lower than the operating energy). While the electrons circulate in the storage ring they emit electromagnetic radiation whenever they pass through a magnetic field. This radiation can be extracted from the ring through long pipes, called photon beam lines leading to experimental stations and be used for basic and applied science.

The intensity of synchrotron radiation is proportional to the number of electrons circulating in the storage ring. We define a circulating beam current by

$$I_b = e N_e f_{rev} \,,$$

where $N_e$ is the total number of electrons circulating in the storage ring, $e$ the unit of electric charge, and $f_{rev}$ is the revolution frequency.

The storage ring operating at energies above several hundred MeV is enclosed in a concrete tunnel or behind a concrete shielding wall to shield people from ionizing radiation. The photon beam escapes this radiation environment through small holes in the shielding wall to the experimental stations. To establish and sustain an electron beam in a storage ring, many technical components are required. The nature and functioning of the major ones will be discussed in more detail.

Every circular accelerator is composed of technical components, like magnets, ultra-high vacuum system, rf-system, injector system, beam monitoring, control system etc. Basically, all main components are installed along a closed loop defining the orbit along which the electrons travel. In the schematic Fig. 5.1 the principle layout of the main components is displayed.

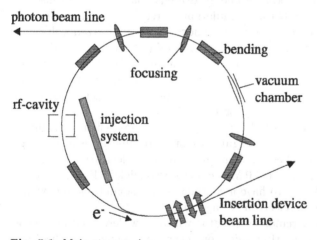

**Fig. 5.1.** Main storage ring components

- **Bending magnets** are used to deflect the electron beam. Placing bending magnets in a well ordered arrangement such as to form a closed ring forces the beam to follow a closed path along the circular accelerator. The location and deflection angle of bending magnets defines the geometry of the storage ring. Although we call this a circular accelerator, the shape is actually not circular. A series of arc sections (bending magnets) is interrupted by straight sections to make space for other components. Bending magnets also serve as sources of synchrotron radiation.

- **Quadrupole magnets** are placed in straight sections between bending magnets. Quadrupoles act much like glass lenses in light optics by providing a restoring or focusing force to particles which deviate too much from the orbital path or ideal orbit, thus assuring the survival of the particle beam for many revolutions. We will borrow many terms and techniques from light optics since the functions are very similar.

- **Sextupole magnets** are used to correct for chromatic aberrations caused by focusing errors on particles with different energies.

- **Vacuum system:** The electron beam must be enclosed in a vacuum chamber where the air pressure is reduced to some $10^{-9}$ Torr or lower to prevent particle losses due to scattering on residual gas atoms. Electrons, once injected into a storage ring, are expected to circulate and produce synchrotron radiation for many hours with a minimum rate of loss. This low pressure is achieved by placing many vacuum pumps along the circular path. Due to gas desorption caused by radiation hitting the vacuum chamber surface continuous pumping is required.

- **Rf-system:** Electrons are expected to circulate for many hours at constant energy in a storage ring to produce synchrotron radiation. Although the particle energy is kept constant, energy loss into synchrotron radiation occurs and must be compensated by equivalent acceleration. Special accelerating cavities are installed along the orbit generating an accelerating electric field in synchronism with the arrival of electrons. The acceleration exactly compensates for the energy loss to radiation. The electric fields oscillate at frequencies of the order of 500 MHz and proper acceleration occurs only when electrons pass through the cavity at a specific time which is the reason for the bunched character of the circulating electron beam. The circulating beam is composed of one or more electron clusters, called bunches, where the distance between bunches is an integer multiple of the rf-wavelength. For the same reason, the circumference also must be an integer multiple of the rf-wavelength.

- **Beam controls:** A number of beam controls are included in the design of a storage ring. Beam monitors are used to measure the circulating beam current, beam lifetime and transverse beam position. Due to field and alignment errors of main magnets, the particle beam follows a distorted closed loop. These distortions must be corrected as much as possible by steering magnets. Generally, a storage ring is controlled by a computer, setting and recording component parameters as well as monitoring beam current and safety equipment.

- **Injection system:** Electrons are generated in an injector system consisting of an electron source, a low energy accelerator (mostly a linear accelcrator) and a booster synchrotron to accelerate the electrons from the low linac energy to the operating energy of the storage ring. After acceleration in the booster, the electrons are transferred to the storage ring. To reach high beam intensities in the storage ring many booster pulses are injected.

**Insertion devices.** Synchrotron radiation emitted from bending magnets do not always meet all requirements of the users. In order to provide the desired radiation characteristics (photon energy, broad band, narrow band etc.) insertion devices are placed in magnetfree sections along the orbit. Such magnets are composed of more than one pole with opposing polarities such that the total beam deflection in the insertion device is zero.

- **Wiggler** magnets are used to produce high intensity broad band radiation, up to photon energies which can greatly exceed those available from bending magnets. In addition, a wiggler is composed of many poles thus increasing the total photon flux by a factor equal to the number of wiggler poles.
- **Wavelength shifters** are generally 3-pole wiggler magnets with a superhigh field in the central pole to reach hard x-rays in low energy storage rings. The lateral poles are of opposite and much lower field strength to compensate the deflection of the central pole.
- **Undulator magnets** are essentially weak field wiggler magnets and produce high brightness, quasi monochromatic radiation.
- **Other,** specially designed magnets, may produce circularly polarized radiation.

## Exercise *

**Exercise 5.1 (S).** An electron bunch of $\tau = 30$ ps duration and an instantaneous current of current of $I_b = 100$ mA is injected into a storage ring with a circumference of 300 m. Calculate the circulating beam current per bunch. How many electrons are injected into the storage ring during each pulse containing only one bunch. How many such pulses must be injected to reach a circulating beam current of 200 mA? What is the storage ring filling time, if the injection system can operate at 10 Hz? How many bunches must be injected per pulse to keep the injection time at 5 min or less?

---

* The argument (S) indicates an exercise for which a solution is given in Appendix A.

# 6. Particle Beam Optics

Bending and focusing of high energy, relativistic particles are effected by the Lorentz force

$$\boldsymbol{F} = e\boldsymbol{E} + e\frac{[c]}{c}[\boldsymbol{v} \times \boldsymbol{B}]. \tag{6.1}$$

A magnetic field of 1 Tesla exerts the same force on a relativistic electron as does an electric field of 3.0 MV/cm. A magnetic field of 1 Tesla is rather easy to produce while the corresponding electric field is beyond technical feasibility. For the manipulation of relativistic particles we use therefore magnetic fields.

## 6.1 Deflection in Bending Magnets

Charged particle beams are deflected in the uniform field of bending magnets. A transverse magnetic field being constant and homogeneous in space at least in the vicinity of the particle beam, is the lowest order field in beam guidance or beam transport systems. Such a field is called a dipole field and can be generated, for example, between the poles of an electromagnetic bending magnet with a cross section as shown schematically in Fig. 6.1.

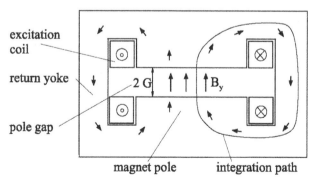

**Fig. 6.1.** Cross section of bending magnet

The magnetic field $\boldsymbol{B}$ is generated by an electrical current $I$ in current carrying coils surrounding the magnet poles. A ferromagnetic return yoke provides an efficient return path for the magnetic flux. The magnetic field is determined by one of Maxwell's equations

$$\nabla \times \frac{\boldsymbol{B}}{\mu_{\mathrm{r}}} = \left[\frac{\mu_0\, c}{4\pi}\right] \frac{4\pi}{c} \boldsymbol{j}\,, \tag{6.2}$$

where $\mu_{\mathrm{r}}$ is the relative permeability of the ferromagnetic material and $\boldsymbol{j}$ is the current density in the coils. We integrate (6.2) over an area enclosed by the integration path as shown in Fig. 6.1 and apply Stoke's theorem (B.29) to get

$$\frac{1}{\mu_{\mathrm{r}}} \int \nabla \times \boldsymbol{B} \,\mathrm{d}\boldsymbol{A} = \frac{1}{\mu_{\mathrm{r}}} \oint \boldsymbol{B} \,\mathrm{d}\boldsymbol{s} = \frac{4\pi}{c} \left[\frac{\mu_0 c}{4\pi}\right] \int \boldsymbol{j}\mathrm{d}\boldsymbol{A}\,. \tag{6.3}$$

The integration on the r.h.s. is just the total current in both excitation coils $2I_{\mathrm{tot}} = \int \boldsymbol{j}\mathrm{d}\boldsymbol{A}$. The l.h.s. must be evaluated along an integration path surrounding the excitation coils. We choose an integration path which is convenient for analytical evaluation. Starting in the middle of the lower magnet pole and integrating straight to the middle of the upper pole, we know from symmetry that the magnetic field along this path has only a vertical nonvanishing component $B_y \neq 0$, which is actually the desired field in the magnet gap and $\mu_{\mathrm{r}} = 1$. Within the iron the contribution to the integral vanishes since we assume no saturation effects and set $\mu_{\mathrm{r}} = \infty$. The total path integral becomes therefore

$$G\,B_y \;=\; \left[\frac{\mu_0\, c}{4\pi}\right] \frac{4\pi}{c} I_{\mathrm{tot}}\,, \tag{6.4}$$

where $I_{\mathrm{tot}}$ is the total current through one coil. Solving (6.4) for the total excitation current in each coil we get in more practical units

$$I_{\mathrm{tot}}\,(\mathrm{Amp}) \;=\; \frac{1}{\mu_0} B_y\,[\mathrm{T}]\;G[\mathrm{m}] \;=\; 795774\,B_y\,[\mathrm{T}]\;G[\mathrm{m}]\,. \tag{6.5}$$

The total required excitation current in each magnet coil is proportional to the desired magnetic field and proportional to the gap between the magnet poles.

As a practical example, we consider a magnetic field of 1 Tesla in a dipole magnet with a gap of $2G = 10$ cm. From (6.5) we find a total electrical current of about 40,000 A is required in each of two excitation coils to generate this field. Since the coil in general is composed of many turns, the actual electrical current is usually much smaller by a factor equal to the number of turns and the total coil current $I_{\mathrm{tot}}$, therefore, is often measured in units of Ampereturns. For example two coils each composed of 40 windings with sufficient cross section to carry an electrical current of 1000 A would provide the total required current of 40,000 A-turns each to produce a magnetic field of 1 Tesla.

Beam deflection in a magnetic field is derived from the equilibrium of the centrifugal force and Lorentz force

$$\frac{\gamma m v^2}{\rho} = [c] \frac{e}{c} v B \,, \tag{6.6}$$

where we assumed that the direction of the particle velocity $\boldsymbol{v}$ is orthogonal to the magnetic field: $\boldsymbol{v} \perp \boldsymbol{B}$. A pure dipole field deflects a charged particle beam onto a circular path with a bending radius $\rho$ given by

$$\frac{1}{\rho} = [c] \frac{eB}{\beta E} = \frac{B}{B\rho} \,, \tag{6.7}$$

where $\beta = v/c$, $E$ the particle energy, and $B\rho$ is defined as the beam rigidity.

In practical units

$$\frac{1}{\rho} (\mathrm{m}^{-1}) = C_\rho \frac{B(\mathrm{T})}{cp \,(\mathrm{GeV})} \,, \tag{6.8}$$

with

$$C_\rho = [c]\, e = 0.299792 \, \frac{\mathrm{GeV}}{\mathrm{m\ T}} \,. \tag{6.9}$$

The beam rigidity is

$$B\rho \,(\mathrm{Tm}) = 3.3356 \, cp \,(\mathrm{GeV}) \,. \tag{6.10}$$

A magnet of length $\ell_\mathrm{b}$ deflects a particle beam by the angle

$$\psi = \ell_\mathrm{b}/\rho \,. \tag{6.11}$$

Distributing a set of magnets bending the electron beam by appropriate deflection angles along a closed loop establishes the geometric shape of the storage ring.

## 6.2 Beam Focusing

A ring consisting only of bending magnets would not work since any particle beam has the tendency to spread out. Similar to a light beam, we require focusing elements to confine the particle beam to the vicinity of the orbit defined by the location of the bending magnets.

## 6.2.1 Principle of Focusing

We borrow much from light optics to describe the focusing of particle beams. To learn how to focus charged particles, we recall the principle of focusing in light optic (Fig. 6.2). The deflection of a light ray parallel to the optical axis by a focusing lens is proportional to the distance of the ray from the optical axis. The distance from the lens to the focal point, where all parallel rays are focused to a point, is called the focal length.

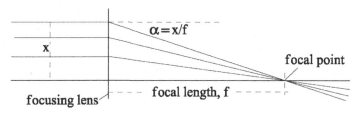

**Fig. 6.2.** Principle of focusing

Applying this to the focusing of particle beams, we note from the discussion on bending magnets that in a focusing magnet the deflection angle $\alpha$ must increase linearly with $x$

$$\alpha = \frac{\ell}{\rho} = [c]\,\frac{eB\ell}{cp} \propto x\,. \tag{6.12}$$

To accomplish this, we consider now a field expressed by $B = B_0 + gx$ which gives two contributions to the deflection, one a constant deflection due to the uniform dipole field $B_0$ and the other is an $x$-dependent deflection

$$\alpha = [c]\frac{eg\ell}{cp}\,x = kl\,x\,, \tag{6.13}$$

as desired for focusing. The $x$-dependent field component $gx$ can be created by a quadrupole magnet, which functions as the focusing element for charged particle beams. The quantity $g$ is the field gradient , and $k$ is defined as the quadrupole strength

$$k(\mathrm{m}^{-2}) = C_\rho\,\frac{g(\mathrm{T/m})}{cp\,(\mathrm{GeV})}\,. \tag{6.14}$$

The focal length of the quadrupole is $1/f = k\ell_\mathrm{q}$ where $\ell_\mathrm{q}$ is the length of the quadrupole.

## 6.2.2 Quadrupol Magnet

How do we produce the desired field gradient or a field: $B_y = gx$ in a quadrupole magnet? Static magnetic fields can be derived from a magnetic potential and a field $B_y = gx$ can be derived from the potential $V = -gxy$ by simple differentiation giving the field-components

$$B_y = -\frac{\partial V}{\partial y} = gx, \quad \text{and} \quad B_x = -\frac{\partial V}{\partial x} = gy. \tag{6.15}$$

Because ferromagnetic surfaces are equipotential surfaces (just like metallic surfaces are equipotential surfaces for electric fields) we use magnetic poles shaped in the form of a hyperbola (Fig. 6.3) given by

$$x \cdot y = \pm\tfrac{1}{2}R^2, \tag{6.16}$$

where $R$ is the aperture radius between the four hyperbolas. Along the $z$-axis the magnet cross section is assumed to be the same. A quadrupole is made of four hyperbolic poles with alternating magnetization producing the desired focusing field gradient.

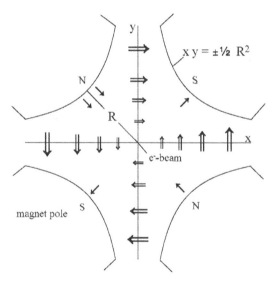

**Fig. 6.3.** Cross section of a quadrupole (schematic)

Each of the four poles is excited by electrically powered coils wound around it. Although quadrupoles function as focusing elements just like glass lenses function as focusing elements in light optics there is a fundamental difference. Quadrupoles focus in one plane but defocus in the other depending on the sign of the excitation current. An actual particle beam, however, requires focusing in both planes. To solve this problem, we borrow again from light optics the characteristics of focusing in a lens-doublet. The focal length of two lenses is $1/f^* = 1/f_1 + 1/f_2 - d/f_1f_2$ where $f_1$ and $f_2$ are the focal lengths, and $d$ is the distance between both lenses. If we choose, for example, $f_1 = -f_2 = f$ we get $1/f^* = d/f^2 > 0$ which is focusing in both planes. By using combinations of focusing and defocusing quadrupoles, one can create overall focusing systems. The quadrupoles in a storage ring or a general beam transport line are therefore polarized alternately as focusing or defocusing quadrupoles. It is common in this regard to consider a quadrupole to be focusing if it focuses a particle beam in the horizontal plane. Tacitly, we know that this same quadrupole is defocusing in the vertical plane.

## 6.3 Equation of Motion

Bending magnets and quadrupoles are the main components to guide and focus a charged particle beam. Sextupole magnets are used to correct chromatic aberrations introduced by quadrupole focusing. In the following discussion we will formulate mathematical equations to describe the path of individual particles along a beam transport line including any of theses magnet types.

The equation of motion in the presence of dipole ($B_0$), quadrupole ($g$) and sextupole ($g'$) fields can be derived from the general expression for the curvature $1/\rho$ of paraxial beams. Here, we define the curvature very general to include all fields (dipole, quadrupole, sextupole...) although in any one magnet only one field type may be present. There are exceptions to this separation of fields in some special cases. Specifically, in some new synchrotron light sources, we find a combination of a dipole and gradient field in the same bending magnet. For paraxial beams ($x' \ll 1$) the curvature $1/\rho$ is

$$\frac{1}{\rho} \approx x'' = [c] \, \frac{eB}{cp} \,, \tag{6.17}$$

where

$$B = B_0 + gx + \tfrac{1}{2} g' x^2 + \dots \qquad \text{or} \tag{6.18}$$

$$[c] \frac{eB}{cp} = \frac{1}{\rho_0} + kx + \tfrac{1}{2} mx^2 + \dots \,.$$

Before we proceed to derive the equation of motion, we notice that the field term $\frac{1}{\rho_0}$ describes the ideal beam guidance through the bending magnets. The solution of the equations of motion would be a rather complicated expression describing the alternating straight and curved segments of the ideal orbit as defined by the location and strength of the bending magnets. We are not interested in this solution, because we already know from the placement of bending magnets where the beam should be. We rather concern ourselves with deviations of particle trajectories from this ideal orbit and redefine (6.17) by a transformation eliminating the geometric expression of the ideal orbit and set

$$\frac{1}{\rho} - \frac{1}{\rho_0} \approx x'' = [c] \, e \left( \frac{B}{cp} - \frac{B_0}{cp_0} \right). \tag{6.19}$$

With the expansions (6.18) inserted into (6.19) the equation of motion becomes

$$x'' = \quad -kx \dots\dots\dots\dots \text{focusing term}$$

$$+ \frac{1}{\rho_0}\delta \dots\dots\dots\text{dispersion}$$

$$+ kx\delta \dots\dots\dots\text{chromatic aberration} \qquad (6.20)$$

$$- \tfrac{1}{2}mx^2 \dots\dots\dots\text{chromatic and geometric aberration}$$

$$+ \mathcal{O}(3) \dots\dots\dots\text{higher order terms,}$$

where $\delta = \Delta E/E_0$ is the relative energy deviation of a particle from the ideal energy $E_0$ and $m$ the sextupole strength defined in (6.18).

Each term in the equation of motion makes its specific (sometimes unwanted) contribution to beam dynamics. Deflection of the beam in a bending magnet depends on the particle energy. Particles with a slightly different energy are deflected differently and this difference cannot be neglected and is covered by the term $\frac{1}{\rho_0}\delta$. The same is true for focusing giving rise to the chromatic aberration term, $kx\delta$. Finally, to avoid beam instability, we must correct one of the most serious chromatic effects, the chromaticity, by the installation of sextupole magnets. Unfortunately, such sextupole magnets introduce nonlinear terms into the otherwise linear beam dynamics causing significant stability problems for particles at large amplitudes $x$. A similar equation exists for the vertical plane keeping in mind that the magnet parameters must change signs ($k \to -k$), etc.

In the approximation of linear beam optics, we keep only linear terms in (6.18) and get the differential equation of motion

$$x'' + k(s)x = \frac{1}{\rho_o}\delta. \qquad (6.21)$$

This equation is similar to that of a perturbed harmonic oscillator although in this case we have a $s$-dependent rather than a constant restoring force.

The solution of this equation is composed of the solution of the homogeneous differential equation and a particular solution of the inhomogeneous differential equation. The physical significance of both solutions is the following. Solutions for the homogeneous equation represent oscillations about an equilibrium orbit. Such oscillations are called betatron oscillations. In this case, the equilibrium orbit is $x \equiv 0$ because it is the path we defined by the placement of the magnets and $x$ represents only the deviation from this ideal orbit. For off-energy particles, we must consider the perturbation term on the right hand side of the equation. This perturbation is on average a constant generating a shift of the particle trajectory from the ideal orbit. For example, particles with a higher energy, $\delta > 0$, would oscillate about a path which is mostly outside ($x > 0$) of the ideal path. The solution of the inhomogeneous equation of motion therefore defines the reference orbit for particles with energy $E = E_0(1 + \delta)$. Such particles perform oscillations about this reference orbit. In the following we discuss both solutions in more detail.

### 6.3.1 Solutions of the Equations of Motion

First, we set $\delta = 0$ and solve the homogeneous equation: $x'' + k(s)x = 0$. We cannot solve this equation in general since $k = k(s)$ is a function of $s$ describing the distribution of quadrupoles along the beam transport line. Within each individual quadrupole, however, the solution for $k = \text{const} > 0$ is simply

$$x(s) = a\,\cos(\sqrt{k}s) + b\,\sin(\sqrt{k}s)\,, \tag{6.22a}$$
$$x'(s) = -a\sqrt{k}\,\sin(\sqrt{k}s) + b\sqrt{k}\,\cos(\sqrt{k}s)\,. \tag{6.22b}$$

The integration constants $a, b$ are determined by initial conditions. With $x = x_0$ and $x' = x_0'$ at $s = 0$, we get at the location $s$ the particle coordinates

$$x = x_0\,\cos(\sqrt{k}s) + x_0'\,\frac{1}{\sqrt{k}}\sin(\sqrt{k}s)\,, \tag{6.23a}$$
$$x' = -x_0\sqrt{k}\,\sin(\sqrt{k}s) + x_0'\,\cos(\sqrt{k}s)\,. \tag{6.23b}$$

For a particle with ideal energy, these two equations express the position and slope at point $s$ as a function of initial particle coordinates at $s = 0$.

### 6.3.2 Matrix Formalism

Both equations can be expressed in matrix formulation

$$\begin{pmatrix} x \\ x' \end{pmatrix} = \begin{pmatrix} \cos(\sqrt{k}s) & \frac{1}{\sqrt{k}}\sin(\sqrt{k}s) \\ -\sqrt{k}\,\sin(\sqrt{k}s) & \cos(\sqrt{k}s) \end{pmatrix} \begin{pmatrix} x_0 \\ x_0' \end{pmatrix}\,. \tag{6.24}$$

For the case of a defocusing quadrupole $k < 0$ we derive a similar transformation matrix

$$\begin{pmatrix} x \\ x' \end{pmatrix} = \begin{pmatrix} \cosh(\sqrt{|k|}s) & \frac{1}{\sqrt{|k|}}\sinh(\sqrt{|k|}s) \\ \sqrt{|k|}\,\sinh(\sqrt{|k|}s) & \cosh(\sqrt{|k|}s) \end{pmatrix} \begin{pmatrix} x_0 \\ x_0' \end{pmatrix}\,. \tag{6.25}$$

A special case appears for $k \to 0$ describing a drift space of length $s$ for which the transformation matrix is

$$\begin{pmatrix} x \\ x' \end{pmatrix} = \begin{pmatrix} 1 & s \\ 0 & 1 \end{pmatrix} \begin{pmatrix} x_0 \\ x_0 \end{pmatrix}\,. \tag{6.26}$$

The transformations are expressed for quadrupoles of finite length. Sometimes it is desirable to perform quick calculations in which case we use the thin lens approximation just like in light optics by setting $s \rightarrow 0$ while $1/f = ks = $ const and get a thin lens transformation matrix for a quadrupole

$$\begin{pmatrix} x \\ x' \end{pmatrix} = \begin{pmatrix} 1 & 0 \\ -\frac{1}{f} & 1 \end{pmatrix} \begin{pmatrix} x_0 \\ x_0 \end{pmatrix} . \tag{6.27}$$

Of course, in this case half the length of the actual quadrupole must be assigned to the drift space on either side.

Transformation of particle trajectories through complicated, multi-magnet arrangements, called lattice, with transformation matrices $M_1, M_2, \ldots$ are derived by simple matrix multiplication

$$M = M_n \cdot M_{n-1} \cdots M_2 \cdot M_1 . \tag{6.28}$$

This matrix formalism is quite powerful and well matched to the capabilities of computers.

### 6.3.3 FODO Lattice

As an example of a beam line appropriate to demonstrate the usefulness of the matrix formalism we use what is called the FODO structure. This magnet structure consist of an alternating series of focusing and defocusing quadrupoles, thence the name FODO lattice as shown in Fig. 6.4. This lattice has been used to construct large high energy physics storage rings by filling the space between the quadrupoles with bending magnets. The FODO lattice is a very stable magnet configuration and its simplicity lends itself to "back-of-an-envelope" calculations.

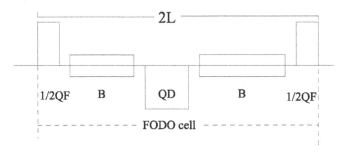

**Fig. 6.4.** FODO lattice

To formulate particle dynamics in a FODO-lattice we keep the formalism simple by starting in the middle of a quadrupole. The transformation through a half cell, 1/2 QF DRIFT 1/2 QD, in thin lens approximation becomes

$$\begin{pmatrix} 1 & 0 \\ -\frac{1}{f_d} & 1 \end{pmatrix} \begin{pmatrix} 1 & L \\ 0 & 1 \end{pmatrix} \begin{pmatrix} 1 & 0 \\ -\frac{1}{f_f} & 1 \end{pmatrix} = \begin{pmatrix} 1 - \frac{L}{f_f} & L \\ -\frac{1}{f^*} & 1 - \frac{L}{f_d} \end{pmatrix} , \tag{6.29}$$

where $2L$ is the length of the FODO-cell, $1/f^* = 1/f_{\mathrm{f}} + 1/f_{\mathrm{d}} - L/(f_{\mathrm{f}} f_{\mathrm{d}})$. Bending magnets are ignored in (6.29) because they do not contribute to focusing in first approximation used here. For the second half cell we get the transformation matrix by replacing $f_{\mathrm{f}} \leftrightarrows f_{\mathrm{d}}$ and vice versa. For simplicity, we consider a symmetric FODO lattice by setting $f_{\mathrm{f}} = -f_{\mathrm{d}} = f$ and get from both half cell transformation matrices the transformation matrix of a full FODO-cell starting in the center of a QF and ending in the middle of the next QF

$$M_{\mathrm{cell,qf}} = \begin{pmatrix} 1 - \frac{2L^2}{f^2} & 2L\left(1 - \frac{L}{f}\right) \\ -\frac{2L}{f^2}\left(1 - \frac{L}{f}\right) & 1 - \frac{2L^2}{f^2} \end{pmatrix}. \tag{6.30}$$

This transformation matrix is valid for the horizontal plane. In the vertical plane, we have the same matrix except that the sign of the focal length must be changed, $f \to -f$.

We may combine the transformation in both planes into one $4 \times 4$ transformation matrix which we can use to transform a particle trajectory with initial coordinates $x_0, x_0', y_0, y_0'$ through a full FODO-cell

$$\begin{pmatrix} x \\ x' \\ y \\ y' \end{pmatrix} = \begin{pmatrix} 1 - \frac{2L^2}{f^2} & 2L\left(1 + \frac{2L^2}{f}\right) & 0 & 0 \\ -\frac{2L}{f^2}\left(1 - \frac{L}{f}\right) & 1 - \frac{2L^2}{f^2} & 0 & 0 \\ 0 & 0 & 1 - \frac{2L^2}{f^2} & 2L - \frac{2L^2}{f} \\ 0 & 0 & -\frac{2L}{f^2}\left(1 + \frac{L}{f}\right) & 1 - \frac{2L^2}{f^2} \end{pmatrix} \begin{pmatrix} x_0 \\ x_0' \\ y_0 \\ y_0' \end{pmatrix}. \tag{6.31}$$

For practical reasons, however, mostly only $2 \times 2$ matrices are used to describe beam dynamics in one plane only.

The matrix formalism is a very powerful tool to calculate the trajectories of single particles. Yet, in a storage ring there are of the order of $10^{11}$ or more particles circulating and it would be prohibitive to have to recalculate the trajectories of all particles whenever a quadrupole strength is changed. A more simple formalism has been developed which allows us to determine the overall behavior of a multi-particle beam.

## 6.4 Betatron Function

Although the quadrupole strength is a function of $s$, $k = k(s)$, the homogeneous part of the equation of motion( 6.21) looks very much like that of a harmonic oscillator $x'' + k(s)x = 0$. As an analytical solution of the equation of motion we try the solution of a harmonic oscillator with variable amplitude and phase

$$x(s) = a_i \sqrt{\beta(s)} \cos [\psi(s) + \varphi_i] \,, \tag{6.32}$$

where $a_i$ and $\varphi_i$ are the integration constants for particle $i$ and $\beta(s)$, $\psi(s)$ are so far unidentified functions of $s$. We insert this ansatz into the differential equation and get from the coefficients of both the sine-and cosine-terms two conditions for $\beta(s)$ and $\psi(s)$

$$\frac{\mathrm{d}^2 \sqrt{\beta}}{\mathrm{d}s^2} + k(s)\sqrt{\beta} + \beta^{-\frac{3}{2}} = 0 \,, \tag{6.33}$$

$$\psi'(s) = \frac{1}{\beta(s)} \,, \tag{6.34}$$

where the primes $'$ are derivations with respect to $s$.

Equation (6.32) describes the oscillatory motion of particles about the ideal orbit leading through the center of all magnets. These oscillations are defined separately in both the horizontal and vertical plane and are called betatron oscillations. The function $\beta(s)$ is called the betatron function and is defined by the placement and strength of the quadrupole magnets. It is a periodic function of $s$ resembling the periodic distribution of quadrupoles along the ring circumference. The periodicity is the circumference of the machine or shorter if quadrupoles are arranged around the ring in a higher periodicity.

Because of the nonlinearity of the differential equation for the betatron function (6.33), there is only one periodic solution in each plane for a given lattice configuration. Matrix formalism is used to determine this one betatron function in each plane utilizing special computer programs. For each lattice configuration a tabulated list of the values of betatron functions exists which can be used to determine beam sizes. The betatron function is, however, of much higher importance in beam dynamics beyond the ability to calculate single particle trajectories as we will discuss in the next section.

### 6.4.1 Betatron Phase and Tune

The second equation (6.34) can be integrated immediately for

$$\psi(s) = \int_{s_0}^{s} \frac{\mathrm{d}\sigma}{\beta(\sigma)} \tag{6.35}$$

defining the phase of the betatron oscillation at point $s$ and is measured from the starting point $s = s_0$. Integrating along the full orbit produces the betatron tune $\nu_{x,y}$ of the machine which is equal to the number of betatron oscillations per revolution. Again, tunes are defined separately in both planes.

$$\nu_{x,y} = \frac{\psi_{x,y}(C)}{2\pi} = \frac{1}{2\pi} \oint \frac{\mathrm{d}\sigma}{\beta_{x,y}(\sigma)} \,. \tag{6.36}$$

The significance of the tune is that it may not be an integer or a half integer value. If one or the other assumes such a value, beam dynamics becomes instantly unstable leading to beam loss. This becomes obvious for an integer resonance ($\nu_{x,y} = n$) when considering a small dipole field perturbation at say one point along the orbit. This dipole field gives the beam a transverse kick at the same phase of its betatron oscillations after every turn building up larger and larger oscillation amplitudes until the beam gets lost on the vacuum chamber walls.

### 6.4.2 Beam Envelope

Solution (6.32) describes individual particle trajectories with different amplitudes $a_i$ and phases $\varphi_i$. If we choose only particles with the largest amplitude $a_i = \hat{a}$ within a beam and further look among these particles for the one for which $\cos[\psi(s) + \varphi_i] = 1$, we have defined the beam envelope at point $s$ by

$$E_{x,y}(s) = \pm \hat{a}_{x,y} \sqrt{\beta_{x,y}(s)} . \tag{6.37}$$

No other particle will have a greater amplitude at this point. Knowledge of the betatron functions in both the $x$ and $y$-plane and knowledge of the quantity $\hat{a}$ in both planes will allow us to define the beam dimensions at any point along the ring orbit.

As important the knowledge of the beam width and height at some point $s$ is, we do not yet have the tools to calculate either the numerical value of the betatron functions nor that of the beam emittances in both the horizontal and vertical plane.

## 6.5 Phase Ellipse

Particles perform oscillatory motion, called betatron oscillations, about the ideal reference orbit and its amplitude and slope is given by

$$x(s) = a\sqrt{\beta(s)} \cos[\psi(s) + \varphi] , \tag{6.38}$$

$$x'(s) = -a \frac{\alpha}{\sqrt{\beta(s)}} \cos[\psi(s) + \varphi] - a\sqrt{\beta(s)} \sin[\psi(s) + \varphi] \cdot \psi'(s), \tag{6.39}$$

where $\alpha = -\frac{1}{2}\beta'$ and $\gamma = \frac{1+\alpha^2}{\beta}$. All functions $\alpha(s), \beta(s)$ and $\gamma(s)$ are defined separately in $x$ and $y$ and are known as betatron functions or lattice functions. Eliminating the phase $\psi(s) + \varphi$ from both equations results in a constant of motion or the Courant Snyder invariant[42].

$$\gamma x^2 + 2\alpha xx' + \beta x'^2 = a^2 . \tag{6.40}$$

This equation describes an ellipse with an area $\pi a$. While travelling around the storage ring and performing betatron oscillations, individual particles move in phase space along the contour of an ellipse. Each particle $i$ has a different amplitude $a_i$ and travels therefore along a different phase ellipse. Again, there exist two different sets of phase ellipses for each particle, one in $(x, x')$-and the other in $(y, y')$-phase space.

## 6.6 Beam Emittance

Liouville's theorem states with respect to particle dynamics that in the presence of only external macroscopic fields the particle density in phase space is a constant of motion. That means, no particle can cross the phase ellipse of any other particle. We may therefore look for particles with the maximum amplitude $\hat{a}$ travelling along an ellipse which encloses all other particles. This phase ellipse then becomes representative for the whole beam (Fig. 6.5) because it encloses all other particles and due to Liouville no particle can cross this maximum phase ellipse. We have thereby succeeded in describing the dynamics of a many-particle beam by the dynamics of a single particle. Due to the variation of the betatron functions along the orbit, the phase ellipses also change their form and orientation but the area of the phase ellipses stay constant. At a particular point $s$ along the closed orbital path in a storage ring the phase ellipse has always the same shape/orientation while the actual particle appears at different points on the ellipse after each revolution. The constant $\pi \hat{a}^2$ is the area of the largest phase ellipse in a beam and we use this area to define the beam emittance. Because of synchrotron radiation, an electron beam in a storage ring has a Gaussian distribution and it is customary to define the beam emittance for a Gaussian particle distribution by the amplitude of the one-sigma particle

$$\epsilon_x = \frac{\langle x^2 \rangle}{\beta_x(s)} = \frac{\sigma_x^2}{\beta_x(s)}.$$

(6.41)

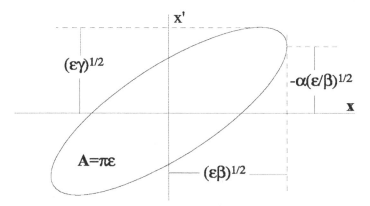

**Fig. 6.5.** Phase space ellipse with ellipse area $A$

From (6.38) we find $\langle x^2 \rangle / \beta_x(s) = \frac{1}{2}\hat{a}_x^2$ or $\hat{a}_x^2 = 2\epsilon_x$. The standard beam size is then defined by

$$\sigma_x = \sqrt{\epsilon_x \beta_x} \qquad \text{and} \qquad \sigma_y = \sqrt{\epsilon_y \beta_y} \qquad (6.42)$$

and the beam divergence by

$$\sigma_{x'} = \sqrt{\epsilon_x \gamma_x} \qquad \text{and} \qquad \sigma_{y'} = \sqrt{\epsilon_y \gamma_y} . \qquad (6.43)$$

Ignoring for the moment effects due to the finite energy spread in the beam and diffraction, the photon source parameters in transverse phase space are equal to those of the electron beam:

electron beam parameter $\equiv$ photon beam parameter .

### 6.6.1 Variation of the Phase Ellipse

While the area of the phase ellipse is a constant of motion, the shape of the ellipse is not. The orientation and aspect ratio continuously change through the action of quadrupole focusing and even along a fieldfree drift space. In Fig. 6.6 the variation of the phase ellipse is shown for a beam in a drift space while converging to a minimum followed by divergence.

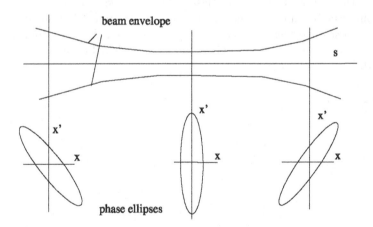

**Fig. 6.6.** Evolution of the phase ellipse along a drift space

The phase ellipse for a converging beam is tilted to the left while that for a diverging beam is tilted to the right. Of special interest is the upright ellipse which occurs at any symmetry point. At such a point $\alpha_{x,y} = 0$ and $\gamma_{x,y} = 1/\beta_{x,y}$ and the beam emittance is simply $\epsilon = \hat{x} \cdot \hat{x}'$ or for a Gaussian beam $\epsilon = \sigma_x \cdot \sigma_{x'}$ .

Fig. 6.7 shows the variation of the phase ellipse as the beam travels through a focusing quadrupole. We note the divergent nature of the beam before the quadrupole. Focusing turns the phase ellipse around resembling a convergent beam.

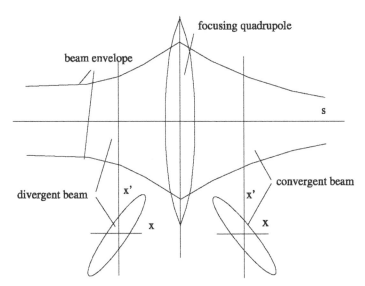

**Fig. 6.7.** Evolution of phase ellipse through a focusing quadrupole

### 6.6.2 Transformation of Phase Ellipse

The transformation matrix for a single particle can be used to determine the transformation of a phase ellipse from one point to another. Liouville's theorem requires that

$$\gamma x^2 + 2\alpha xx' + \beta x'^2 = \gamma_0 x_0^2 + 2\alpha_0 x_0 x_0' + \beta_0 x_0'^2 = a^2, \tag{6.44}$$

where the particle coordinates $(x_0, x_0')$ and $(x, x')$ are related by the transformation matrix $M = \begin{pmatrix} a & b \\ c & d \end{pmatrix}$. Replacing coordinates $(x_0, x_0')$ by $(x, x')$ and collecting all coefficients for $x^2$, $x'^2$ and $xx'$, we obtain a relation of the betatron functions from one point to another. Noting that the coordinates are independent variables we expect all three coefficients to be equal to zero independently. These three conditions are sufficient to determine the three lattice functions $\beta, \alpha$ and $\gamma$. In matrix formulation the transformation of betatron functions is then given by

$$\begin{pmatrix} \beta \\ \alpha \\ \gamma \end{pmatrix} = \begin{pmatrix} a^2 & -2ab & b^2 \\ -ac & ad+bc & -bd \\ c^2 & -2cd & d^2 \end{pmatrix} \begin{pmatrix} \beta_0 \\ \alpha_0 \\ \gamma_0 \end{pmatrix} = M_\beta \cdot \begin{pmatrix} \beta_0 \\ \alpha_0 \\ \gamma_0 \end{pmatrix}. \tag{6.45}$$

The transformation of betatron functions can be expressed by the elements of single particle transformation matrices. We apply this transformation to a drift space to formulate the evolution of the betatron function and find from (6.45) for a drift space

$$\beta(s) = \beta_0 - 2\alpha_0 s + \gamma_0 s^2. \tag{6.46}$$

Specifically, we look at a beam waist where $\alpha_0 = 0$ and get the betatron function a distance $s$ away from the waist

$$\beta(s) = \beta_0 + \frac{s^2}{\beta_0}. \tag{6.47}$$

The betatron function increases quadratically with distance from the waist and the beam size evolves like

$$\sigma(s) = \sigma_0 \sqrt{1 + \frac{\epsilon^2 s^2}{\sigma_0^4}}, \tag{6.48}$$

where $\epsilon$ is the beam emittance.

## 6.7 Dispersion Function

So far, we have treated particle dynamics for a monochromatic beam only and the solutions for particle trajectories cover only those of the homogeneous differential equation of motion. This is not correct for a real beam and we must consider corrections due to effects related to energy errors. Chromatic effects are described by a particular solution of the inhomogeneous differential equation

$$x'' + \left(k + \frac{1}{\rho_0^2}\right) x = \frac{1}{\rho_0} \delta, \tag{6.49}$$

where $\delta = \Delta E/E_0$. On the left hand side, we have added the term $1/\rho_0^2$ which takes care of a second order focusing effect from a bending magnet. So far we have neglected this weak focusing term, but we need to include it now because we are about to determine a chromatic aberration due to the right hand term which is of the same order of magnitude. The general solution is $x = x_\beta + x_\delta$, where the betatron oscillation $x_\beta$ is the solution of the homogeneous differential equation and $x_\delta$ the offset for off-momentum particles. Assuming a pure dipole field ($k = 0$) and $\delta = 1$ a special solution to (6.49), called the dispersion function $D(s) = x/\delta$, is

$$D(s) = \rho_0 \left( 1 - \cos \frac{s}{\rho_0} \right) . \tag{6.50}$$

This solution is nonvanishing only within a bending magnet of length $L_b$ and for $0 \le s \le L_b$. The $2 \times 2$-matrix formulation can be expanded to include off-energy particles by defining $3 \times 3$-transformation matrices

$$\begin{pmatrix} x \\ y \\ \delta \end{pmatrix} = \begin{pmatrix} a & b & D \\ c & d & D' \\ 0 & 0 & 1 \end{pmatrix} \begin{pmatrix} x_0 \\ y_0 \\ \delta \end{pmatrix} . \tag{6.51}$$

The matrix elements $(D, D')$ are nonzero only for bending magnets. For drift spaces and quadrupoles $D \equiv 0$, and $D' \equiv 0$. This is not to say that the dispersion in drift spaces and quadrupoles is zero if there was a bending magnet upstream. The chromatic contribution to the particle trajectory starting at the first bending magnet will transform through the beam line just like a regular trajectory, but getting specially modified within each bending magnet.

Knowledge of the dispersion function allows the calculation of chromatic offsets for any value of the energy deviation $\delta$. The dispersion function $\delta D(s)$ defines the reference path for particles with an energy deviation $\delta$ just like the ideal path $(x \equiv 0)$ is the reference path for $\delta = 0$. Particles with energy $E_0(1 + \delta)$ perform betatron oscillations about their respective reference paths defined by $\delta D(s)$.

## 6.8 Periodic Lattice Functions

In circular accelerators the betatron functions at any point $s$ are the same from turn to turn and therefore $\beta = M_\beta \beta_0 = \beta_0$. This periodic solution of the betatron functions can be derived from the eigenvalues of the eigenfunction equation

$$(M_\beta - I)\,\beta = 0 , \tag{6.52}$$

where $M_\beta$ is the transformation matrix from point $s$ through a whole orbit to point $s + C$, $C$ the ring circumference and $I$ is the unit matrix.

Generally, storage rings are composed of a number of equal sections with equal magnet distributions. In this case, each section, called either cell or unit, is representative for all cells and we need to find the periodic solution for one cell only. This solution then repeats from cell to cell as we progress along the orbit.

### 6.8.1 Periodic Betatron Function in a FODO Lattice

As an example, we look for the periodic solution of the betatron function in a FODO lattice. With $\alpha_0 = 0$, and $\gamma_0 = 1/\beta_0$ in the middle of the QF, we solve

(6.52) with (6.30) for the periodic betatron function and get from (6.45) with
(6.30) the periodic solution $\beta = \beta_0 = (1 - 2L^2/f^2)^2\beta_0 + 4L^2(1 + L/f)^2/\beta_0$.
Similarly, we can go from the middle of the QD to the middle of the next QD
and get the value of the periodic betatron function in the middle of the QD.
Since a QF is a QD in the vertical plane and vice versa, the two solutions
just described present the horizontal and vertical betatron functions in the
middle of the QF and interchangeably those in the middle of the QD. With
the FODO parameter $\kappa = f/L$, the betatron functions in the middle of the
QF are

$$\beta_{x,0} = L\frac{\kappa(\kappa+1)}{\sqrt{\kappa^2-1}} \quad \text{and} \quad \beta_{y,0} = L\frac{\kappa(\kappa-1)}{\sqrt{\kappa^2-1}}. \tag{6.53}$$

For values of the betatron functions in the middle of the QD we merely
interchange indices $x$ with $y$. Obviously, a solution and therefore beam stability exists only if $\kappa > 1$ or if $f > L$. Knowing the betatron functions at
one point is enough to allow the calculation of those functions at any other
point in the lattice by virtue of (6.45). In Fig. 6.8 the betatron functions are
shown for one cell of a FODO lattice. Note the similarity of the horizontal
and vertical betatron functions being only one quadrupole distance shifted
with respect to each other. The periodic nature of the solutions allows us to
construct a circular accelerator made up of a series of equal FODO cells with
periodic repetition of betatron functions and beam sizes.

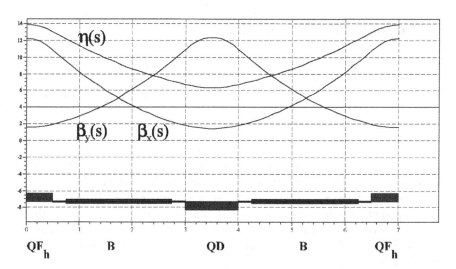

**Fig. 6.8.** Betatron functions in a FODO lattice

## 6.8.2 Periodic Dispersion or $\eta$-Function

In a circular accelerator there is also only one periodic solution for the dispersion function. Utilizing the $3 \times 3$ -transformation matrix for a cell or the whole circumference, the periodic dispersion function, called the $\eta$-function, is defined by

$$
\begin{pmatrix} \eta_0 \\ \eta_0' \\ 1 \end{pmatrix} = \begin{pmatrix} a & b & D \\ c & d & D' \\ 0 & 0 & 1 \end{pmatrix} \begin{pmatrix} \eta_0 \\ \eta_0' \\ 1 \end{pmatrix} . \tag{6.54}
$$

Again, knowledge of the $\eta$-function at one point permits the calculation of $\eta(s)$ at any point around the circular accelerator. The $\eta$-function defines the equilibrium orbit $\eta(s) \delta$ for off-energy particles about which they perform betatron oscillations. The ideal orbit, $x \equiv 0$, is the equilibrium orbit for particles with the design energy $E_0$ or $\delta = 0$. The $\eta$-function varies generally between zero and positive values. Particles with $\delta > 0$ follow therefore a path which is farther away from the ring center while lower energy particles follow a path closer to the ring center than the ideal orbit. The length of the closed path is also energy dependent which becomes of great significance later in connection with synchrotron oscillations as will be discussed in the next chapter.

**Periodic dispersion or $\eta$-function in a FODO lattice.** To illustrate the determination of the $\eta$-function, we take the FODO lattice as an example. A storage ring can be constructed from a series of FODO cells which include bending magnets between the quadrupoles. These bending magnets cause the appearance of a dispersion establishing a periodic $\eta$-function. Based on the transformation matrix through one FODO cell and assuming for simplicity that the bending magnets are as long as the thin-lens FODO cell we may calculate the $3 \times 3$ transformation matrix through one FODO cell. From that we get for the $\eta$-functions in the middle of the QF or QD

$$
\eta_{\mathrm{qf}} = \frac{L^2}{2\rho} \kappa (2\kappa + 1) \quad \text{and} \quad \eta_{\mathrm{qd}} = \frac{L^2}{2\rho} \kappa (2\kappa - 1) , \tag{6.55}
$$

respectively. Knowing the $\eta$-function at one point allows us to calculate its values at any other point around the storage ring. The result for a FODO lattice is shown in Fig. 6.8.

## 6.8.3 Beam Size

With the addition of these chromatic effects, the total beam sizes (6.42, 6.43) are modified to become

$$
\sigma_{x_{\mathrm{tot}}} = \sqrt{\epsilon_x \beta_x + \eta^2 \delta^2} \quad \text{and} \quad \sigma_y = \sqrt{\epsilon_y \beta_y} , \tag{6.56}
$$

where $\delta = \delta E / E_0$ is the relative energy spread in the beam. The total beam divergence is

$$\sigma_{x'_{tot}} = \sqrt{\epsilon_x \gamma_x + \eta'^2 \delta^2} \qquad \text{and} \qquad \sigma_{y'} = \sqrt{\epsilon_y \gamma_y} . \qquad (6.57)$$

There is no effect on the vertical beam parameters since we assume a storage ring with only horizontal bending magnets.

# Exercises *

**Exercise 6.1 (S).** Consider a uniform static magnetic field $B$ in the laboratory system. Determine the electromagnetic field in the restframe of a moving electron. Show that the Lorentz force in both reference systems point in the same direction. In the laboratory system, the static magnetic field causes a transverse deflection of the moving charge. Show, that the angular deflection is the same in both systems.

**Exercise 6.2 (S).** Consider a transverse electric field of 10 kV/cm and a transverse magnetic field of 1 Tesla. At what kinetic energy of an electron are the forces of both fields the same?

**Exercise 6.3 (S).** Construct a $E = 3$ GeV circular accelerator from a series of bending magnets. The magnet length shall be $\ell = 2$ m each and the magnetic field shall not exceed $B = 1.2$ T. Determine the number $n_b$ of magnets needed and the exact field strength to complete the ring. What is the bending radius and the deflection angle per magnet?

**Exercise 6.4 (S).** Derive equation (6.16).

**Exercise 6.5 (S).** Focus a parallel 1.5 GeV beam to a point 5 m downstream from a thin quadrupole ($l = 0.2$ m). Determine the quadrupole strength $k$ and field gradient $g$. For an aperture radius of $R = 5$ cm calculate the required excitation current to reach this field gradient $g$.

**Exercise 6.6 (S).** Consider a drift space of length $L$ and a symmetric electron beam cross section along this drift space. Derive an expression for the value of the betatron function at the beginning of the drift space that results in the minimum beam size anywhere along this drift space. By how much does the beam size vary between the waist and ends of the drift space?

**Exercise 6.7 (S).** Use the bending magnets of Exercise 6.3 and add quadrupoles to form a FODO lattice. Choose the optimum quadrupole strength for minimal beam sizes and calculate the value of the horizontal and vertical betatron function in the middle of the defocusing quadrupole? What is the focal length of the quadrupole?

---

* The argument (S) indicates an exercise for which a solution is given in Appendix A.

**Exercise 6.8 (S).** For a FODO lattice derive an expression (thin lens approximation) for one betatron function between quadrupoles and calculate the phase advance for the full FODO cell.

**Exercise 6.9.** Construct a bending magnet with a length of 1.2 m which deflects a 1.5 GeV beam by 20 degrees. What is the required magnetic field. For a gap height between poles of $g = 7$ cm calculate the required excitation current per coil to reach the design field and how many turns must each coil have if the power supply can deliver a current of about 500 A?

**Exercise 6.10.** Derive (6.33) and (6.34) form (6.32).

**Exercise 6.11.** Consider an electron beam entering a magnetfree straight section with $(\beta_0, \alpha_0) = (5.0 \text{ m}, 2.0)$. Plot the beam size $\sigma(s)$ for an emittance of $\epsilon = 10$ nm for 0 m$< s < 5.0$ m. What is the value of the betatron function at the symmetry point $\alpha = 0$ ?

**Exercise 6.12.** The horizontal beam size reaches a maximum value in the middle of the focusing quadrupole (QF) in a FODO lattice. Plot the beam size in the middle of the QF as a function of the FODO parameter $\kappa$. For which value of $\kappa$ becomes the beam size in a QF a minimum?

# 7. Radiation Effects

Particle dynamics is greatly influenced by the emission of radiation as well as by the restoration of the energy loss in accelerating cavities. These processes are both beneficial and perturbing while fundamentally determining the electron beam parameters and thereby the characteristics of the photon beam. We will briefly discuss these effects to illuminate the basic physics responsible for the photon beam characteristics in a synchrotron light source.

As electrons travel through magnetic fields, they experience a Lorentz force which deflects the beam orthogonal to the field and velocity vector. This force is the cause for the emission of electromagnetic radiation and the instantaneous radiation power is given by (3.12) or (3.20). This loss of energy into synchrotron radiation during each revolution, while small compared to the electron energy, is sufficiently strong to cause perturbations in beam dynamics which must be compensated.

## 7.1 Synchrotron Oscillations

One or more rf-cavities are located along the orbit of the ring. These cavities are excited by external microwave sources to generate electric fields parallel to the beam path providing the acceleration needed. Since the rf-fields oscillate very fast (order of 500 MHz) the particle arrival time at the cavity is very critical (Fig. 7.1). Ideally, particles should pass through the cavities exactly at a phase such that they gain the same energy from the accelerating field as they lost to synchrotron radiation. That phase or time is called the synchronous phase $\psi_s$ or synchronous time $t_s$. Not all particles follow that ideal timing.

Observing orbiting particles, we notice that particles with a higher than ideal energy follow a path which is mostly outside the ideal orbit while particle with lower energies follow a path mostly inside of the ideal path. All particles are highly relativistic, travel close to the speed of light and the going around travel time depends therefore on the length of the path around the ring and therefore on the particle energy. In spite of variations in the revolution time or arrival time at the cavity, a stable beam is ensured by virtue of the principle of phase focusing [43][44], which forces particles to arrive at the cavity, if not exactly, then at least close to the synchronous phase or time. The way this works can be explained with the help of Fig. 7.1. A particle with the ideal

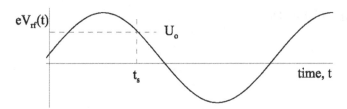

**Fig. 7.1.** Cavity acceleration voltage as a function of time

energy $E_0$ passing through the cavity at the synchronous time $t_s$ will arrive again at the synchronous time after one turn. This synchronous particle will gain from the cavity fields the energy

$$U_0 = eV_{\rm rf}\sin\psi_s \,, \tag{7.1}$$

where $U_0$ is the energy loss per turn of a particle with ideal energy $E_0$ to synchrotron radiation. A particle with a higher than ideal energy starting, for example, at time $t_s$ will take longer to orbit the storage ring and will arrive at the cavity after a time $t > t_s$ to gain an energy $\Delta E < U_0$ thus reducing the positive energy deviation of this particle. The next time this same particle will arrive closer to the synchronous time $t_s$. A similar process occurs for lower energy particles. The energy dependent revolution time together with the time varying rf-voltage provide a restoring force for particles with a wrong energy. By this process, nonideal particles are made to oscillate about the synchronous time $t_s$ similar to betatron oscillation due to the restoring forces from fields of quadrupole magnets. These oscillations are called synchrotron or phase oscillations.

To formulate this phase focusing, we consider the total energy change per turn $\mathrm{d}\varepsilon = eV_{\rm rf}(t) - U(E)$, where $eV_{\rm rf}(t)$ is the energy gained by a particle passing through a cavity at time $t$, and $U(E)$ the energy loss per turn at energy $E$. We have defined the ideal energy $E_0$ and synchronous time $t_s$ such that the ideal particle would gain an energy exactly equal to its lost energy $U_0 = U(E_0)$. Expanding at $t = t_s + \tau$ and $E = E_0 + \varepsilon$, keeping only linear terms and dividing by the revolution time $T_0$ we get the total energy gain per turn

$$\frac{\mathrm{d}\varepsilon}{\mathrm{d}t} = \frac{1}{T_0}\left[eV_{\rm rf}(t_s) + e\left.\frac{\partial V_{\rm rf}}{\partial t}\right|_{t_s}\tau - U(E_0) - \left.\frac{\mathrm{d}U}{\mathrm{d}E}\right|_{E_0}\varepsilon\right]$$

$$= \frac{1}{T_0}\left[e\left.\frac{\partial V_{\rm rf}}{\partial t}\right|_{t_s}\tau - \left.\frac{\mathrm{d}U}{\mathrm{d}E}\right|_{E_0}\varepsilon\right], \tag{7.2}$$

where we made use of the fact that $eV_{\rm rf}(t_s) = U(E_0)$. The revolution time depends on the particle velocity and momentum dependent path length ($\alpha_c$ momentum compaction factor ). The revolution time for a particle with energy deviation $\varepsilon$ is different from the ideal revolution time $T_0$ and we find with $\Delta T/T_0 = \mathrm{d}\tau/\mathrm{d}t$

$$\frac{\mathrm{d}\tau}{\mathrm{d}t} = -(\gamma^{-2} - \alpha_\mathrm{c})\frac{\varepsilon}{E_0} \,. \tag{7.3}$$

The $1/\gamma^2$-term is due to the velocity variation with energy and the momentum compaction factor $\alpha_\mathrm{c}$ is defined by the ratio of the relative change of the orbital path length to the relative energy error,

$$\frac{\Delta C}{C_0} = \alpha_\mathrm{c}\frac{\Delta E}{E_0} \tag{7.4}$$

and is

$$\alpha_\mathrm{c} = \left\langle \frac{\eta}{\rho} \right\rangle_\mathrm{s} \tag{7.5}$$

averaged over the circumference. Differentiation of (7.2) with respect to $t$ and replacing $\dot\tau$ by (7.3) results in the equation of motion for synchrotron oscillations

$$\frac{\mathrm{d}^2\varepsilon}{\mathrm{d}t^2} + 2\alpha_\mathrm{s}\frac{\mathrm{d}\varepsilon}{\mathrm{d}t} + \Omega^2\varepsilon = 0 \,, \tag{7.6}$$

where the damping decrement for synchrotron oscillations has been defined by

$$\alpha_\mathrm{s} = \frac{1}{2T_0}\frac{\mathrm{d}U}{\mathrm{d}E}\bigg|_{E_0} = \frac{1}{2}\frac{\mathrm{d}\langle P_\gamma \rangle}{\mathrm{d}E} = 2\frac{\langle P_\gamma \rangle}{E_0} \,, \tag{7.7}$$

and $\langle P_\gamma \rangle$ has been defined in (3.14). The synchrotron oscillation frequency $\Omega$ has been introduced with $\omega_0 = 2\pi/T_0$ by the definition

$$\Omega^2 = \frac{(\gamma^{-2} - \alpha_\mathrm{c})}{E_0 T_0}\frac{e\,\partial V_\mathrm{rf}(t)}{\partial t} = \omega_0^2\frac{h\,(\gamma^{-2} - \alpha_\mathrm{c})\,eV_\mathrm{rf}\cos\psi_\mathrm{s}}{2\pi E_0} \,. \tag{7.8}$$

A more detailed derivation of the damping decrement reveals a correction, which is necessary to add for specific bending magnet types. The corrected expression reads like (see for example [45])

$$\alpha_\mathrm{s} = \frac{1}{2}\frac{\mathrm{d}\langle P_\gamma \rangle}{\mathrm{d}E} = J_s\frac{\langle P_\gamma \rangle}{E_0} \,, \tag{7.9}$$

where

$$J_s = 2 + \vartheta \tag{7.10}$$

is the synchrotron partition number with

$$\vartheta = \frac{\oint \frac{\eta}{\rho^3} \left(1 + 2k\rho^2\right) \, ds}{\oint \frac{1}{\rho^2} \, ds}. \tag{7.11}$$

The term $\vartheta$ is a correction to the damping resulting from specific properties in some bending magnets. The first term under the integral in the nominator occurs only in sector magnets and is zero in rectangular magnets. The second term becomes nonzero only in gradient magnets where $k \neq 0$ and $\rho \neq 0$.

The synchrotron radiation power $\langle P_\gamma \rangle = U(E_0)/T_0$ is the average radiation power along the circumference. We also made use of the fact that the integrated rf-field changes sinusoidally with time like $V_{rf} = V_0 \sin\left(\omega_{rf} t\right)$ and therefore $\dot{V} = \omega_{rf} V_0 \cos \psi_s = 2\pi h/T_0 V_0 \cos \psi_s$, where $h$ is the integer harmonic number defined by $h = C_0/\lambda_{rf}$ and $\psi_s = \omega_{rf} t_s$ the synchronous phase at which a particle with energy $E_0$ is accelerated to compensate exactly for the energy loss $U_0$ to radiation.

The differential equation of motion exhibits two significant terms. For one, we may expect stable synchrotron oscillations with frequency $\Omega$ only if the synchronous phase is chosen such that this frequency is real and not imaginary. Furthermore, the damping term tells us that any deviation $\varepsilon_0$ of a particle from the ideal parameters in longitudinal phase space is damped due to the emission of synchrotron radiation. Due to the fact that synchrotron radiation depends on the particle energy in such a way that higher/lower energy particles loose more/less energy, we observe an energy correcting effect of the emission of synchrotron radiation.

The solution to the differential equation of motion (7.6) is that of a damped harmonic oscillator

$$\varepsilon(t) = \varepsilon_0 \, e^{-\alpha_s t} \cos \Omega t. \tag{7.12}$$

Generally, the damping time $\tau_s = 1/\alpha_s$ is of the order of milliseconds, while the synchrotron oscillation time is much shorter of the order of 10 50 $\mu$s and we can safely assume that $\tau_s \gg 1/\Omega$. This different time scale allows us to treat synchrotron oscillations while ignoring damping.

Particles orbiting in the storage ring perform oscillations in energy about the ideal energy $E_0$. At the same time, there is also an oscillation about the synchronous time described by $\tau$, which from (7.3) is 90° out of phase and described by

$$\tau(t) = \tau_0 \, e^{-\alpha_s t} \sin \Omega t, \tag{7.13}$$

where from (7.3)

$$\tau_0 = \frac{\gamma^{-2} - \alpha_c}{\Omega} \frac{\varepsilon_0}{E_0}, \tag{7.14}$$

while ignoring damping. This longitudinal oscillation about the synchronous time $t_s$ leads to a longitudinal distribution of particles and defines the bunch length $\ell_b$, which is from (7.14)

$$\ell_{\mathrm{b}} = \beta c \tau_0 \,. \tag{7.15}$$

Later, we will quantify this bunch length in a storage ring in more detail. The energy spread in the beam is directly related to the bunch length and the bunch length or bunch duration eventually defines the pulse length of the photon pulse from each electron bunch.

### 7.1.1 Longitudinal Phase Space Motion

It is interesting at this point to discuss in some detail the particle motion in longitudinal phase space because it relates directly to the process of radiation production in a Free Electron Laser to be discussed in Chap. 11. We ignore damping and eliminate from (7.12, 7.13) the trigonometric functions to get the formulation for the phase ellipse

$$\frac{\varepsilon^2}{\varepsilon_0^2} + \frac{\tau^2}{\tau_0^2} = 1 \,. \tag{7.16}$$

While performing synchrotron oscillations, particles move along an ellipse in phase space. This is true, however, only for small deviations from the reference point because we used only the linear term in the expansion (7.2) of the rf-voltage in the vicinity of the synchronous time. For large synchrotron oscillation amplitudes, the actual sinusoidal voltage variation must be taken into account. In this case the typical phase space trajectories of particle motion under the influence of an electromagnetic field are shown in Fig. 7.2.

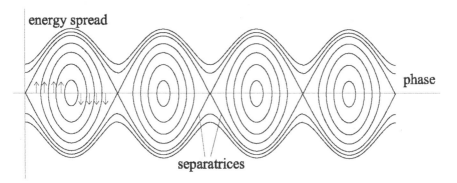

**Fig. 7.2.** Longitudinal phase space ellipses with separatrices

The phase space trajectories still look very much like ellipses for small amplitudes, which become somewhat distorted towards the shape of a lens as the oscillation amplitudes increase. The region of stable phase motion is enclosed by two intertwined separatrices, which separate the region of oscillatory motion with that of libration (see Fig. 7.2). This is similar to the dynamics of a swing which performs oscillations at small amplitudes or energies and keeps rotating over the top in one direction for large energies. For a stable beam, we require that all particles are contained well within the separatrices. Observing an individual particle, we notice that it gains and looses energy as it interacts with the electromagnetic field in the accelerating cavity and travels along its phase space trajectory in a clockwise direction. This motion will become specifically significant for a free electron laser as we will discuss in more detail in Chapter 11.

## 7.2 Damping

Due to synchrotron radiation, all particle motion in 6-dim phase space becomes damped. As we have seen above, energy damping occurs at a rate proportional to the average synchrotron radiation power $\langle P_\gamma \rangle$. Damping occurs also in the transverse plane due to a geometric effect. While particles perform betatron oscillations, they emit radiation in the direction of travel which is generally at a finite angle with the beam axis. During this emission of radiation the particles loose longitudinal as well as transverse momentum. Yet in the rf-cavity, acceleration occurs only in the longitudinal direction. As a consequence, the combined process of emission and acceleration results in a net loss of transverse momentum which is equivalent to a reduction in the betatron amplitude or transverse damping.

To be more quantitative, we note that the direction of the particle motion does not change with the emission of radiation since radiation is emitted in the forward direction within a negligibly small angle of $1/\gamma$. The momentum vector of the particle before emission is for small values of $x'_0$ in $x, z$-space $(cp_0 x'_0, cp_0)$. Emission of a photon caries away the momentum $(-c\Delta p x'_0, -c\Delta p)$ and the transverse particle momentum becomes $([cp_0 - c\Delta p] x'_0, cp_0 - c\Delta p)$. To compensate for this loss of momentum, the particle gains energy in an accelerating cavity and its associated momentum gain in the cavity is $(0, \beta P_{\mathrm{rf}} \, dt)$. Since the transverse momentum will not be changed by the acceleration, which for simplicity we assume to occur at the location of the radiation emission, we have

$$(cp_0 - c\Delta p) x'_0 = (cp_0 - c\Delta p + \beta \langle P_\gamma \rangle \, dt) x'_1 , \qquad (7.17)$$

where $x'_0, x'_1$ are the angles of the particle trajectories with respect to the beam axis before and after acceleration, respectively and the acceleration is equal to the energy loss, $P_{\mathrm{rf}} \, dt = \langle P_\gamma \rangle \, dt$. With $x' = \dot{x}/\beta c$ and the particle energy after emission of a photon but before acceleration $\beta E = cp$ the particle direction after acceleration is

$$\dot{x}_1 = \frac{E}{E + \langle P_\gamma \rangle \, dt} \dot{x}_0 \approx \left( 1 - \frac{\langle P_\gamma \rangle \, dt}{E} \right) \dot{x}_0 , \qquad (7.18)$$

where we made use of the fact that $\langle P_\gamma \rangle \, dt \ll E$ and $E_0 \approx E$. From this, the horizontal damping decrement becomes

$$\alpha_x = -\frac{1}{\dot{x}_0} \frac{d\dot{x}}{dt} = J_x \frac{\langle P_\gamma \rangle}{E_0} \,. \tag{7.19}$$

Similar to the synchrotron damping decrement, this expression also has to be corrected for second order effects in sector and gradient magnets, which is done by introducing the partition number

$$J_x = 1 - \vartheta \,, \tag{7.20}$$

where $\vartheta$ is defined in (7.11).

A similar expression is valid for the vertical plane:

$$\alpha_y = -\frac{1}{\dot{y}_0} \frac{d\dot{y}}{dt} = J_y \frac{\langle P_\gamma \rangle}{E_0} \,, \tag{7.21}$$

where the vertical partition number $J_y = 1$ for a flat storage ring without vertical bending. The amplitude of a betatron oscillation $x(t) = A(t)\sqrt{\beta}\cos\psi(t)$ scales then like

$$A(t) = A_0 \, e^{-\alpha_s t} \tag{7.22}$$

because of damping. Under certain circumstances, one or more damping decrements may be modified, e.g. when we have a field gradient in a bending magnet. However, due to very general principles, the sum of all damping decrements is a constant

$$\alpha_s + \alpha_x + \alpha_y = 4\frac{\langle P_\gamma \rangle}{E_0} \tag{7.23}$$

or

$$J_x + J_y + J_s = 4 \,, \tag{7.24}$$

also called the Robinson criterion. Whenever one decrement is modified another one will be modified in the opposite direction. From here on we will ignore such details and point to related discussions on this point to available literature (for example [46]).

## 7.3 Quantum Effects

Damping of 6-dim phase-space coordinates is counterbalanced in a storage ring by quantum effects [14]. We evaluate the emission of radiation in terms of photon emission and discuss the effect of a single photon emission process on particle dynamics. Consider, for example, a particle performing synchrotron oscillations, $A = A_0 e^{i\Omega(t-t_0)}$, where the last photon emission occurred at time $t_0$. A new photon of energy $\varepsilon$ be emitted at time $t_1$ causing an energy jump in phase space which alters the synchrotron oscillation like

$$
\begin{aligned}
A &= A_0 \exp\left[i\,\Omega(t - t_0)\right] - \varepsilon \exp\left[i\,\Omega(t - t_1)\right] \\
&= A_1 \exp\left[i\,\Omega(t - t_1)\right].
\end{aligned}
\tag{7.25}
$$

Solving for the new amplitude we get

$$
A_1^2 = A_0^2 + \varepsilon^2 - 2\,\varepsilon A_0 \cos[\Omega(t_1 - t_0)].
\tag{7.26}
$$

Radiation emission occurs many times during a synchrotron oscillation period and we may therefore average over all times to get the average change in the oscillation amplitude due to the emission of a photon with energy $\varepsilon$.

$$
\langle \delta A^2 \rangle = \langle A_1^2 - A_0^2 \rangle = \langle \varepsilon^2 \rangle.
\tag{7.27}
$$

Furthermore, we average now over all photon energies in the radiation spectrum and get the rate of change of the synchrotron oscillation amplitude

$$
\left\langle \frac{\mathrm{d}A^2}{\mathrm{d}t} \right\rangle_{s,\text{excitation}} = \int_0^\infty \varepsilon^2 \dot{n}(\varepsilon)\,\mathrm{d}\varepsilon = \left\langle \dot{N}_{\text{ph}} \langle \varepsilon^2 \rangle \right\rangle_s ,
\tag{7.28}
$$

where $\dot{n}(\varepsilon)$ is the number of photons of energy $\varepsilon$ emitted per unit time and unit energy bin $\mathrm{d}\varepsilon$, and $\dot{N}_{\text{ph}}$ is the total number of all photons emitted per unit time. The subscript $_s$ indicates that the integral be taken along the circumference of the ring. Since both photon energy and flux are positive we get from this effect a continuous increase of the oscillation amplitude.


## 7.4 Equilibrium Beam Parameters

Two radiation effects reflect on the particle beam, damping and quantum excitation which eventually determine the transverse beam sizes, beam divergencies, energy spread and bunch length. The geometric and temporal particle bunch parameters transform directly into those of the photon pulses. Equilibrium values for all of these quantities are determined by damping and quantum excitation.


### 7.4.1 Equilibrium Energy Spread

The equilibrium energy spread and bunch length can be derived by the requirement that the amplitudes do not change or with (7.28) that

$$
\left\langle \frac{\mathrm{d}A^2}{\mathrm{d}t} \right\rangle_s = \left\langle \frac{\mathrm{d}A^2}{\mathrm{d}t} \right\rangle_{s,\text{excitation}} + \left\langle \frac{\mathrm{d}A^2}{\mathrm{d}t} \right\rangle_{s,\text{damping}} = 0,
\tag{7.29}
$$

where $\left\langle \frac{dA^2}{dt} \right\rangle_s \Big|_{\text{damping}} = -2\alpha_s \langle A^2 \rangle$. With $\tau_s = 1/\alpha_s$ we get

$$\langle A^2 \rangle = \tfrac{1}{2}\tau_s \left\langle \dot{N}_{\text{ph}} \langle \varepsilon^2 \rangle \right\rangle_s , \tag{7.30}$$

where from the discussion in Section 9.7.1 (9.171)

$$\left\langle \dot{N}_{\text{ph}} \langle \varepsilon^2 \rangle \right\rangle_s = \frac{55}{24\sqrt{3}} r_c c\, mc^2\, \hbar c \gamma^7 \left\langle \frac{1}{\rho^3} \right\rangle_s . \tag{7.31}$$

The Gaussian energy spread is from $\varepsilon = A \sin \Omega t$ defined by $\sigma_\varepsilon^2 = \langle \varepsilon^2 \rangle = \tfrac{1}{2} \langle A^2 \rangle$ and the equilibrium energy spread becomes from (7.30) and (7.31)

$$\left( \frac{\sigma_\varepsilon}{E} \right)^2 = C_q \frac{\gamma^2 \langle 1/\rho^3 \rangle}{2\, J_s \, \langle 1/\rho^2 \rangle} , \tag{7.32}$$

where

$$C_q = \frac{55}{32\sqrt{3}} \frac{\hbar c}{mc^2} = 3.84\ 10^{-13}\ \text{m}. \tag{7.33}$$

It should be noted that the bending radius $\rho$ is always taken to be positive independent of the direction of deflection. The emission of synchrotron radiation does not depend on the sign of deflection. For an isomagnetic ring, where all bending magnets are of the same strength, the energy spread is

$$\left( \frac{\sigma_\varepsilon}{E} \right)^2 = C_q \frac{\gamma^2}{2 J_s \rho} \tag{7.34}$$

depending only on the beam energy and magnet fields, $B \propto \gamma/\rho$.

## 7.4.2 Bunch Length

The energy oscillation is correlated with a longitudinal oscillation about the bunch center and a beam with a Gaussian energy spread will also have a Gaussian longitudinal particle distribution. From (7.3), and $\tau = \tau_0 \sin \Omega t$ we find for the bunch length $\sigma_\ell = c \frac{|(\gamma^{-2} - \alpha_c)|}{\Omega} \frac{\sigma_\varepsilon}{E_0}$ noting that $\sigma_\ell = c\sigma_\tau$. Replacing the synchrotron oscillation frequency by its definition (7.8) we get finally for the equilibrium bunch length (7.3)

$$\sigma_\ell = \frac{\sqrt{2\pi} c}{\omega_{\text{rev}}} \sqrt{\frac{|(\gamma^{-2} - \alpha_c)| E_0}{h\, e V_{\text{rf}} \cos \psi_s}} \frac{\sigma_\varepsilon}{E_0} . \tag{7.35}$$

The bunch length is proportional to the energy spread and can be reduced by increasing the rf-voltage although the reduction scales only like $1/\sqrt{V_{\text{rf}}}$.

### 7.4.3 Horizontal Beam Emittance

Similar to the beam energy spread, we find an excitation effect also in transverse phase space. The emission of a photon occurs in a time short compared to the damping time and therefore causes a sudden change in the particle energy and consequently a sudden change of its reference orbit. Since the electron cannot jump to the new reference orbit, it must oscillate about it with a new betatron amplitude. The particle position is the sum of the betatron amplitude and the offset due to its energy deviation. Emission of a photon with energy $\varepsilon$ does not change the particle position $x = x_\beta + x_\eta$ directly but causes a variation of its components

$$\delta x = 0 = \delta x_\beta + \eta \frac{\varepsilon}{E_0} \rightarrow \delta x_\beta = -\eta \frac{\varepsilon}{E_0}\,, \tag{7.36a}$$

$$\delta x' = 0 = \delta x'_\beta + \eta' \frac{\varepsilon}{E_0} \rightarrow \delta x'_\beta = -\eta' \frac{\varepsilon}{E_0}\,. \tag{7.36b}$$

Particles orbiting the storage ring travel along their phase ellipses described by

$$\gamma x_\beta{}^2 + 2\alpha x_\beta x'_\beta + \beta x'^2_\beta = a^2 \tag{7.37}$$

and its perturbation due to the emission of a photon is

$$\gamma \delta(x_\beta{}^2) + 2\alpha \delta(x_\beta x'_\beta) + \beta \delta(x'^2_\beta) = \delta a^2\,. \tag{7.38}$$

Expressing these variations by (7.36), we get

$$\delta(x_\beta^2) = (x_{0,\beta} + \delta x_\beta)^2 - x_{0,\beta}^2 = 2x_{0,\beta}\delta x_\beta + \delta x_\beta^2\,,$$
$$\delta(x_\beta x'_\beta) = (x_{0,\beta} + \delta x_\beta)(x'_{0,\beta} + \delta x'_\beta) - x_{0,\beta}x'_{0,\beta}\,,$$
$$= x_{0,\beta}\delta x'_\beta + x'_{0,\beta}\delta x_\beta + \delta x_\beta \delta x'_\beta\,, \tag{7.39}$$
$$\delta(x'^2_\beta) = (x'_{0,\beta} + \delta x'_\beta)^2 - x'^2_{0,\beta} = 2x'_{0,\beta}\delta x'_\beta + \delta x'^2_\beta\,.$$

Emission of a photon can happen at any phase of the betatron oscillation and we therefore average over all betatron phases which causes the terms linear in $x_\beta$ and $x'_\beta$ to vanish. Replacing the variations in (7.38) by their expressions from (7.39) we get

$$\delta a^2 = \frac{\langle \varepsilon^2 \rangle}{E_0^2}\mathcal{H}(s)\,, \tag{7.40}$$

where we have averaged over all photon energies and have defined

$$\mathcal{H}(s) = \gamma \eta^2 + 2\alpha \eta \eta' + \beta \eta'^2\,. \tag{7.41}$$

This equation looks very similar to (7.27) and the rate of change of the betatron oscillation amplitude is analogous to (7.28)

$$\left\langle \frac{\mathrm{d}\,\langle a^2 \rangle}{\mathrm{d}t} \right\rangle_{s,\text{excitation}} = \frac{\langle \dot{N}_{\mathrm{ph}}\,\langle \varepsilon^2 \rangle\,\mathcal{H} \rangle_s}{E_0^2}. \qquad (7.42)$$

This excitation is again to be combined with damping to get an equilibrium beam emittance

$$\langle a^2 \rangle_s = \tfrac{1}{2}\left\langle \dot{N}_{\mathrm{ph}}\,\langle \varepsilon^2 \rangle \mathcal{H}(s) \right\rangle_s. \qquad (7.43)$$

The rms beam size is $\sigma_x^2 = \langle x^2 \rangle = \tfrac{1}{2}a^2 \beta_x$ and the equilibrium beam emittance

$$\epsilon_x = \frac{\sigma_x^2}{\beta_x} = C_{\mathrm{q}}\gamma^2 \frac{\langle \mathcal{H}(s)/\rho^3 \rangle}{J_x\,\langle 1/\rho^2 \rangle}. \qquad (7.44)$$

The equilibrium beam emittance is proportional to the square of the particle energy and further depends on lattice parameters like the strength of the bending magnets and the function $\mathcal{H}(s)$. Depending on the design goal, the magnet lattice can be optimized for a large beam emittance as is desired, for example, in colliding beam storage rings for high energy physics. Many such rings were and are in use now for the production of synchrotron radiation and are known as first generation radiation sources. Another lattice design approach is to minimize the beam emittance, which is the preferred goal for synchrotron radiation sources to maximize photon beam brightness. Synchrotron light sources with intermediate beam emittances and few or no magnet-free straight sections for insertion devices are classified as second generation storage rings. Third generation storage rings have been designed for as small a beam emittance as feasible and include many magnet free sections to install insertion devices. Generally, the focusing power must be increased to minimize the beam emittance leading to significant chromatic and geometric aberrations which limit beam stability in a storage ring.

### 7.4.4 Vertical Beam Emittance

In most storage rings, there is no dispersion in the vertical plane, $\eta_y \equiv 0$, and it seems therefore that $\mathcal{H}_y(s) = 0$ resulting in a vanishing vertical beam emittance $\epsilon_y = 0$. In this situation we must reconsider the approximations made so far which include the transverse recoil a particle may receive if a photon is emitted not exactly in the forward direction but at a finite angle within $\pm 1/\gamma$. Due to this recoil $\Delta p_\perp \neq 0$ and the vertical equilibrium beam emittance due to the transverse recoil turns out to be

$$\epsilon_y = \frac{\sigma_y^2}{\beta_y} = C_{\mathrm{q}}\gamma^2 \frac{\langle \beta_y \rangle\,\langle 1/\rho^3 \rangle}{2\,J_y\,\langle 1/\rho^2 \rangle} \approx 10^{-13}\ \mathrm{rad\,m}, \qquad (7.45)$$

indeed a very small emittance. This small value justifies the fact that we neglected this effect for the horizontal beam emittance. The smallest beam emittance achieved so far in any electron storage ring operated in the world is of the order of $10^{-9}$ rad m, much higher than this fundamental lower limit (7.45).

This result is actually so small that still other effects must be considered. Coupling of horizontal betatron oscillations into the vertical plane due to magnet misalignments (rotational errors of quadrupole alignment) contribute much more to the vertical beam emittance. Actually, in existing storage rings this coupling dominates the vertical beam emittance. For a well aligned storage ring

$$\epsilon_y \lesssim 0.01\epsilon_x \,. \tag{7.46}$$

## 7.5 Transverse Beam Parameters

Beam parameters like width, height, length, divergence, beam emittances and energy spread are not all fixed independent quantities, but rather depend on lattice and rf parameters. These dependencies on technical design parameters allow the storage ring designer the adjustment of beam parameters, within limits, to be optimum for the intended application. In this section we will discuss such dependencies. A particle beam at any point along a beam transport line may be represented by a few phase ellipses for different particle momenta as shown in Fig. 7.3. The phase ellipses for different momenta are shifted proportional to the dispersion function at that point and its derivative. Generally, the form and orientation of the ellipses are slightly different

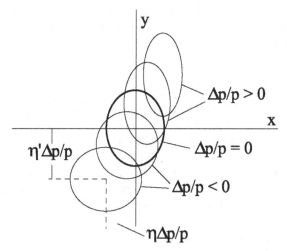

**Fig. 7.3.** Distribution of beam ellipses for a beam with finite emittance, dispersion and momentum spread (schematic). The variation in the shape of the phase ellipses for different energies reflects the effect of chromatic aberrations

too due to chromatic aberrations in the focusing properties of the beam line. For the definition of beam parameters we need therefore the knowledge of the lattice functions including chromatic aberrations and the beam emittance and momentum spread.

### 7.5.1 Beam Sizes

The particle beam width or beam height is determined by the beam emittance, the values of the betatron and dispersion functions, and the energy spread. The betatron and dispersion functions vary along a beam transport line and depend on the distribution of the beam focusing elements. The beam sizes are therefore also functions of the location along the beam line. From the magnet lattice these functions can be derived and the beam sizes be calculated.

The beam size of a particle beam is generally not well defined since the boundaries of a beam tends to become fuzzy. We may be interested in the beam size that defines all of a particle beam. In this case we look for that phase ellipse that encloses all particles and obtain the beam size in the form of the beam envelope. The beam half-width or half-height of this beam envelope is defined by

$$u_\beta(s) = \sqrt{\epsilon_u \beta_u(s)} \tag{7.47}$$

with $u = (x, y)$. If there is also a finite momentum spread the overall beam size or beam envelope is increased by the dispersion

$$u_\eta(s) = \eta_u(s) \frac{\Delta cp}{cp_0} \tag{7.48}$$

and the total beam size is

$$u_{\text{tot}}(s) = u_\beta(s) + u_\eta(s) = \sqrt{\epsilon_u \beta_u(s)} + \eta_u(s) \frac{\Delta cp}{cp_0}. \tag{7.49}$$

This definition of the beam size assumes a uniform particle distribution within the beam and is used mostly to determine the acceptance or the beam stay clear of a beam transport system. The acceptance of a beam transport system is defined as the maximum emittance a beam may have and still pass through the vacuum chambers of a beam line. In Fig. 7.3 this would be the area of that ellipse that encloses the whole beam including off momentum particles. In practice, however, we would choose a larger acceptance to allow for errors in the beam path.

Since the lattice functions vary along a beam line the required aperture, to let a beam with the maximum allowable emittance pass, is not the same every-where along the system. To characterize the aperture variation consistent with the acceptance, a beam stay clear area, BSC, is defined as the required material free aperture of the beam line.

For a more precise description of the actual beam size, the particle distribution must be considered. In a storage ring, most particle beams assume a Gaussian or near Gaussian density distribution in all six dimensions of phase space and therefore the contributions to the beam parameters from different sources add in quadrature. The beam parameters for Gaussian particle distributions are defined as the standard values of the Gaussian distributions

$$\sigma_x, \sigma_{x'}, \sigma_y, \sigma_{y'}, \sigma_\delta, \sigma_\ell , \tag{7.50}$$

where most designations have been defined and used in previous chapters and where $\sigma_\delta = \sigma_\varepsilon/cp_0$ and $\sigma_\ell$ the bunch length. The beam size for Gaussian beams is for $u = x$ or $u = y$

$$\sigma_{u,\text{tot}} = \sqrt{\epsilon_u \beta_u(s) + \eta_u^2(s)\sigma_\delta^2}. \tag{7.51}$$

Four parameters are required to determine the beam size in each plane although in most cases the vertical dispersion vanishes.

### 7.5.2 Beam Divergence

The angular distribution of particles within a beam depends on the rotation of the phase ellipse and we define analogous to the beam size an angular beam envelope by

$$\sigma_{u',\text{tot}} = \sqrt{\epsilon_u \gamma_u(s) + \eta_u'^2(s)\sigma_\delta^2}. \tag{7.52}$$

Again, there is a contribution from the betatron motion, from a finite momentum spread and associated chromatic aberration. The horizontal and vertical beam divergencies are also determined by four parameters in each plane.

## 7.6 Beam Emittance and Wiggler Magnets

In circular electron accelerators, the beam emittance is determined by the emission of synchrotron radiation and the resulting emittance is not always as desired. In such situations methods to alter the equilibrium emittance are desired and we will discuss in the next sections methods which may be used to either increase or decrease the beam emittance.

The beam emittance in an electron storage ring can be greatly modified by the use of wiggler magnets both to increase [47] or to decrease the beam emittance. Manipulation of the beam emittance in electron storage rings has become of great interest specifically, to obtain extremely small beam emittances, and we will therefore derive systematic scaling laws for the effect of wiggler magnets on the beam emittance as well as on the beam energy spread.

The particle beam emittance in a storage ring is the result of two competing effects, the quantum excitation caused by the quantized emission of photons and the damping effect. Both effects lead to an equilibrium beam emittance observed in electron storage rings. Independent of the value of the equilibrium beam emittance in a particular storage ring, it can be further reduced by increasing the damping without also increasing the quantum excitation. More damping can be established by causing additional synchrotron radiation through the installation of deflecting dipole magnets like strong wigglers magnets. In order to avoid quantum excitation of the beam emittance, however, the placement of wiggler magnets has to be chosen carefully. As discussed earlier, an increase of the beam emittance through quantum excitation is caused only when synchrotron radiation is emitted at a place in the storage ring where the dispersion function is finite.

Emittance reducing wiggler magnets must be placed in areas around the storage ring where the dispersion vanishes to minimize quantum excitation. To calculate the modified equilibrium beam emittance, we start from (7.42) and get with (7.31) an expression for the quantum excitation of the emittance which can be expanded to include wiggler magnets

$$\left. \frac{d\epsilon_x}{dt} \right|_{q0} = c C_Q E^5 \left\langle \frac{\mathcal{H}}{\rho^3} \right\rangle_0 , \tag{7.53}$$

where

$$C_Q = \frac{55}{24\sqrt{3}} \frac{r_c \hbar c}{(mc^2)^6} = 2.06 \, 10^{-11} \frac{m^2}{GeV^5} . \tag{7.54}$$

$E$ is the particle energy, and $\rho_0$ the bending radius of the regular ring magnets. The average $\langle \rangle$ is to be taken for the whole ring and the index $_0$ indicates that the average $\langle \mathcal{H}/\rho^3 \rangle_0$ be taken only for the ring magnets without wiggler magnets.

Since the contributions of different magnets, specifically, of regular storage ring magnets and wiggler magnets are independent of each other, we may use the results of the basic ring lattice and add to the regular quantum excitation and damping the appropriate additions due to the wiggler magnets (deflecting in the horizontal plane),

$$\left. \frac{d\epsilon_x}{dt} \right|_{qw} = c C_Q E^5 \left[ \left\langle \frac{\mathcal{H}}{\rho^3} \right\rangle_0 + \left\langle \frac{\mathcal{H}}{\rho^3} \right\rangle_w \right] . \tag{7.55}$$

Both, ring magnets and wiggler magnets produce synchrotron radiation and contribute to damping of the transverse particle oscillations. Again, we may consider both contributions separately and adding the averages the combined rate of emittance damping is

$$\left. \frac{d\epsilon_x}{dt} \right|_{dw} = -2\,\epsilon_{xw} C_d \frac{E^3}{J_x} \left[ \left\langle \frac{1}{\rho^2} \right\rangle_0 + \left\langle \frac{1}{\rho^2} \right\rangle_w \right] , \tag{7.56}$$

where $\epsilon_w$ is the beam emittance with wiggler magnets,

$$C_d = \frac{c}{3} \frac{r_c}{(mc^2)^3} = 2110 \frac{m^2}{GeV^3 s}, \tag{7.57}$$

and $J_x$ the horizontal damping partition number, assuming a flat storage ring in the horizontal plane. The equilibrium beam emittance is reached when the quantum excitation rate and the damping rates are of equal magnitude. We add therefore (7.55) and (7.56) and solve for the horizontal equilibrium beam emittance

$$\epsilon_{xw} = C_q \frac{\gamma^2}{J_u} \frac{\left\langle \frac{\mathcal{H}}{\rho^3} \right\rangle_0 + \left\langle \frac{\mathcal{H}}{\rho^3} \right\rangle_w}{\left\langle \frac{1}{\rho^2} \right\rangle_0 + \left\langle \frac{1}{\rho^2} \right\rangle_w}, \tag{7.58}$$

where $C_q$ is defined in (7.33). With $\epsilon_{x0}$ being the original beam emittance for $\rho_w \to \infty$, the relative emittance ratio due to the presence of wiggler magnets is

$$\frac{\epsilon_{xw}}{\epsilon_{x0}} = \frac{1 + \left\langle \frac{\mathcal{H}}{\rho^3} \right\rangle_w / \left\langle \frac{\mathcal{H}}{\rho^3} \right\rangle_0}{1 + \left\langle \frac{1}{\rho^2} \right\rangle_w / \left\langle \frac{1}{\rho^2} \right\rangle_0}. \tag{7.59}$$

Making use of the definition of average parameter values and the circumference of the storage ring $C$ we get

$$\left\langle \frac{\mathcal{H}}{\rho^3} \right\rangle_0 = \frac{1}{C} \oint \frac{\mathcal{H}}{\rho_0^3} ds, \tag{7.60a}$$

$$\left\langle \frac{\mathcal{H}}{\rho^3} \right\rangle_w = \frac{1}{C} \oint \frac{\mathcal{H}}{\rho_w^3} ds, \tag{7.60b}$$

$$\left\langle \frac{1}{\rho^2} \right\rangle_0 = \frac{1}{C} \oint \frac{1}{\rho_0^2} ds, \tag{7.60c}$$

$$\left\langle \frac{1}{\rho^2} \right\rangle_w = \frac{1}{C} \oint \frac{1}{\rho_w^2} ds. \tag{7.60d}$$

When evaluating these integrals, note, that the bending radii are always positive, $\rho_{0,w} > 0$. Evaluation of these integrals for a particular storage ring and wiggler magnet employed gives from (7.59) the relative change in the equilibrium beam emittance. The quantum excitation term scales like the cube while the damping scales only quadratically with the wiggler curvature. This feature leads to the effect that the beam emittance is always reduced for small wiggler fields and increases only when the third power terms become significant.

Concurrent with a change in the beam emittance a change in the momentum spread occurs due to the wiggler radiation which can be derived in a similar way for

$$\frac{\sigma_{\epsilon_w}^2}{\sigma_{\epsilon_0}^2} = \frac{1 + \langle 1/\rho^3 \rangle_w / \langle 1/\rho^3 \rangle_0}{1 + \langle 1/\rho^2 \rangle_w / \langle 1/\rho^2 \rangle_0} . \tag{7.61}$$

Closer inspection of (7.59,7.61) reveals basic rules and conditions for the manipulations of beam emittance and energy spread. If the ring dispersion function is finite in the wiggler section, we have $\langle \mathcal{H}_w \rangle \neq 0$ which can lead to strong quantum excitation depending on the magnitude of the wiggler magnet bending radius $\rho_w$. This situation is desired if the beam emittance must be increased [47]. If wiggler magnets are placed into a storage ring lattice were the ring dispersion function vanishes, only the small dispersion function created by the wiggler magnets themselves must be considered for the calculation of $\langle \mathcal{H}_w \rangle$ and therefore only little quantum excitation occurs. In this case the beam emittance can be reduced since the wiggler radiation contributes more strongly to damping and we call such magnets damping wigglers. Whenever wiggler magnets are used which are stronger than the ordinary ring magnets $\rho_w < \rho_0$ the momentum spread in the beam is increased. This is true for virtually all cases of interest.

Conceptual methods to reduce the beam emittance in a storage ring have been derived which are based on increased synchrotron radiation damping while avoiding quantum excitation effects. Optimum lattice parameters necessary to achieve this will be derived in the next section.

### 7.6.1 Damping Wigglers

General effects of wiggler magnet radiation on beam emittance has been discussed and we found that the beam emittance can be reduced if the wiggler is placed where $\eta = 0$ to eliminate quantum excitation $\langle \mathcal{H}_w \rangle = 0$. This latter assumption, however, is not quite correct. Even though we have chosen a place, where the storage ring dispersion function vanishes, we find the quantum excitation factor $\mathcal{H}_w$ to be not exactly zero once the wiggler magnets are turned on because they, being bending magnets, create their own dispersion function. To calculate this dispersion function we assume a sinusoidal wiggler field

$$B(z) = B_w \cos k_p z , \tag{7.62}$$

where $k_p = 2\pi/\lambda_p$ and $\lambda_p$ the wiggler period length as shown in Fig. 7.4. From (6.49) the differential equation for the dispersion function in a wiggler magnet

$$\eta'' = \frac{1}{\rho} = \frac{1}{\rho_w} \cos k_p z , \tag{7.63}$$

which can be solved by

$$\eta(z) = \frac{\theta_w}{k_p}(1 - \cos k_p z) , \tag{7.64}$$

$$\eta'(z) = \theta_w \sin k_p z , \tag{7.65}$$

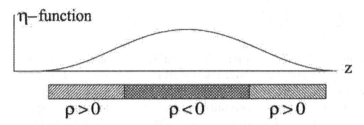

**Fig. 7.4.** Dispersion function in one period of a wiggler magnet

where we have assumed that the wiggler magnet is placed in a dispersion free location $\eta_0 = \eta_0' = 0$ and where from (4.3) $\theta_w = 1/(\rho_w k_p)$ is the deflection angle per wiggler halfpole. With this solution, the first two equations (7.60) can be evaluated. To simplify the formalism, we ignore the $z$-dependence of the lattice functions within the wiggler magnet and set $\alpha_x = 0$ and $\beta_x =$ const. Evaluating the integrals (7.60), we note that the absolute value of the bending radius $\rho$ must be used along the integration path because the synchrotron radiation does not depend on the sign of the deflection. With this in mind, we evaluate the integrals $\int (\eta^2/|\rho^3|)\,dz$ and $\int (\eta'^2/|\rho^3|)\,dz$. For each half period of the wiggler magnet, the contribution to the integral is with $\mathcal{H} = \eta^2/\beta_x + \eta'^2\beta_x$

$$\Delta \int_0^{\lambda_p/2} \frac{\mathcal{H}}{\rho^3}\,dz = \frac{36}{15}\frac{1}{\beta_x}\theta_w^5 + \frac{4}{15}\frac{\beta_x \theta_w^3}{\rho_w^2} \approx \frac{4}{15}\frac{\beta_x \theta_w^3}{\rho_w^2}\,, \qquad (7.66)$$

where the approximation $\lambda_p \ll \beta_x$ was used. For the whole wiggler magnet with $N_w$ periods, the total quantum excitation integral

$$\int_w \frac{\mathcal{H}}{\rho^3}\,dz \approx N_w \frac{8}{15}\frac{\beta_x}{\rho_w^2}\theta_w^3\,. \qquad (7.67)$$

Similarly, the damping integral for the total wiggler magnet is

$$\int_w \frac{1}{\rho^2}\,dz = \pi N_w \frac{\theta_w}{\rho_w}\,. \qquad (7.68)$$

Inserting expressions (7.60, 7.67, 7.68) into equation (7.59) we get for the emittance ratio

$$\frac{\epsilon_{xw}}{\epsilon_{x0}} = \frac{1 + \frac{8}{30\pi}\frac{\beta_x}{\langle \mathcal{H}_0\rangle_\rho}N_w \frac{\rho_0^2}{\rho_w^2}\theta_w^3}{1 + \frac{1}{2}N_w \frac{\rho_0}{\rho_w}\theta_w}\,, \qquad (7.69)$$

where $\langle \mathcal{H}_0\rangle_\rho$ is the average value of $\mathcal{H}$ in the ring bending magnets excluding the wiggler magnets. We note from (7.69) that the beam emittance indeed can be reduced by wiggler magnets if $\theta_w$ is kept small. For easier numerical calculation we replace $\langle \mathcal{H}_0 \rangle_\rho$ by the unperturbed beam emittance which from (7.58) in the limit $\rho_w \to \infty$ is

$$\langle \mathcal{H}_0 \rangle_\rho = \frac{J_x \, \rho_0 \, \epsilon_{x0}}{C_q \, \gamma^2} \tag{7.70}$$

and get instead of (7.69) with the wiggler strength parameter $K = \gamma \theta_w$

$$\frac{\epsilon_{xw}}{\epsilon_{x0}} = \frac{1 + \frac{8\,C_q}{30\pi J_x} \frac{\beta_x}{\epsilon_{x0}} \frac{K^2}{\rho_w} N_w \frac{\rho_0}{\rho_w} \theta_w}{1 + \frac{1}{2} N_w \frac{\rho_0}{\rho_w} \theta_w}. \tag{7.71}$$

The beam emittance is reduced by wiggler magnets whenever the second term in the nominator is smaller than the second term in the denominator or when the condition

$$\frac{8}{15\pi} \frac{C_q}{J_x} \frac{\beta_x}{\epsilon_{x0}} \frac{K^2}{\rho_w} \leq 1 \tag{7.72}$$

is fulfilled. For large numbers of wiggler periods, $N_w \to \infty$, the beam emittance reaches asymptotically a lower limit given by

$$\frac{\epsilon_{xw}}{\epsilon_{x0}} \to \frac{8\,C_q}{15\pi J_x} \frac{\beta_x}{\epsilon_{x0}} \frac{K^2}{\rho_w}. \tag{7.73}$$

For many wiggler periods the increase in momentum spread also reaches an asymptotic limit which is from (7.61)

$$\frac{\sigma_{\epsilon w}^2}{\sigma_{\epsilon 0}^2} \to \frac{\rho_0}{\rho_w} = \frac{B_w}{B_0}, \tag{7.74}$$

where $B_0$ is the magnetic field strength in the ring magnets. Beam stability and acceptance problems may occur if the beam momentum spread is allowed to increase too much and therefore inclusion of damping wigglers must be planned with some caution.

### 7.6.2 Variation of the Damping Distribution

Robinson's criterion (7.23) provides an expression for the overall damping in six-dimensional phase space without specifying the distribution of damping in the three degrees of freedom. In accelerators, we make an effort to decouple the particle motion in the three degrees of freedom as much as possible and as a result we try to optimize the beam parameters in each plane separately from the other planes for our application. Part of this optimization is the adjustment of damping and, as a consequence, of beam emittances to desired values. Robinson's criterion allows us to modify the damping in one plane at the expense of damping in another plane. This shifting of damping is done by varying damping partition numbers defined in Sect. 7.2.

From the definition of the $\vartheta$ parameter is is clear that damping partition numbers can be modified depending on whether the accelerator lattice is a combined function or a separated function lattice. By choosing a combination of gradient and separated function magnets we may generate virtually any distribution between partition numbers while staying within Robinson's criterion.

**Damping partition and rf-frequency.** Actually such "gradient magnets" can be introduced even in a separated function lattice. If the rf-frequency is varied, the beam will follow a path that meets the synchronicity condition. Increasing the rf-frequency corresponds to a shorter wavelength and therefore to a reduced orbit length to keep the harmonic number constant. As a consequence of the principle of phase stability, the beam energy is reduced and the beam follows a lower energy equilibrium orbit with the same harmonic number as the reference orbit for the reference energy. Decreasing the rf-frequency leads just to the opposite effect. Off momentum orbits pass systematically off center through quadrupoles which therefore function like combined function gradient magnets.

To quantify this effect, we use for $\vartheta$ only the second term in (7.11). The first term, coming from sector magnets, will stay unaffected. Displacement of the orbit in the quadrupoles will cause bending with a bending radius

$$\frac{1}{\rho_q} = k\,\delta x\,. \tag{7.75}$$

An rf-frequency shift causes a momentum change of

$$\frac{\Delta p}{p_0} = -\frac{1}{\alpha_c}\frac{\Delta f_{rf}}{f_{rf}}\,, \tag{7.76}$$

which in turn causes a shift in the equilibrium orbit of

$$\delta x = \eta\frac{\Delta p}{p_0} = -\frac{\eta}{\alpha_c}\frac{\Delta f_{rf}}{f_{rf}} \tag{7.77}$$

and the bending radius of the shifted orbit in quadrupoles is

$$\frac{1}{\rho_q} = k\delta x = k\eta\frac{\Delta p}{p_0} = -k\frac{\eta}{\alpha_c}\frac{\Delta f_{rf}}{f_{rf}}\,. \tag{7.78}$$

Inserted into the second term of (7.11), where $\rho_a$ is the actual bending radius of the ring bending magnets we get

$$\Delta\vartheta = -\frac{1}{\alpha_c}\frac{\oint 2\,k^2\,\eta^2\,ds}{\oint ds/\rho_a^2}\frac{\Delta f_{rf}}{f_{rf}}\,. \tag{7.79}$$

We see that all quantities in (7.79) are fixed properties of the lattice and changing the rf-frequency leads just to the effect we expected. Specifically, we note that all quadrupoles contribute additive irrespective of their polarity. We may apply this to a simple isomagnetic FODO lattice where all bending magnets and quadrupoles have the same absolute strength. Integration of the nominator in (7.79) leads to

$$\oint 2\,k^2\,\eta^2\,\mathrm{d}s \;=\; 2\,k^2(\eta_{\max}^2 + \eta_{\min}^2)\,l_{\mathrm{q}}2\,n_{\mathrm{c}}\,,$$

where $2l_{\mathrm{q}}$ is the quadrupole length in a FODO lattice, $\eta_{\max}$ and $\eta_{\min}$ the values of the $\eta$-function in the focusing QF and defocusing QD quadrupoles, respectively, and $n_{\mathrm{c}}$ the number of FODO cells in the ring. The denominator $\oint \mathrm{d}s/\rho^2 = 2\pi/\rho_{\mathrm{a}}$ and the variation of $\vartheta$ in a FODO lattice is

$$\Delta\vartheta \;=\; -\,n_{\mathrm{c}}\frac{2\,\rho_a}{\pi\,\alpha_{\mathrm{c}}\,l_{\mathrm{q}}}\frac{\eta_{\max}^2 + \eta_{\min}^2}{f^2}\frac{\Delta f_{\mathrm{rf}}}{f_{\mathrm{rf}}}\,, \tag{7.80}$$

where we have used the focal length $f^{-1} = k\,l_{\mathrm{q}}$. We replace in (7.80) the $\eta$ functions by the expressions (6.55) derived for a FODO lattice, recall the relation $f = \kappa\,L$ and get finally

$$\Delta\vartheta \;=\; -\frac{\rho_a}{\rho}\frac{1}{\alpha_{\mathrm{c}}}\frac{L}{l_{\mathrm{q}}}(4\,\kappa^2 + 1)\frac{\Delta f_{\mathrm{rf}}}{f_{\mathrm{rf}}}\,, \tag{7.81}$$

where $\rho$ is the average bending radius in the FODO cell as defined in Chap. 6. The variation of the $\vartheta$ parameter in a FODO lattice is the more sensitive to rf-frequency variations the longer the cell compared to the quadrupole length and the weaker the focusing. For other lattices the expressions may not be as simple as for the FODO lattice but can always be computed by numerical evaluation of (7.79).

By varying the rf-frequency and thereby the horizontal and longitudinal damping partition number, we have found a way to either increase or decrease the horizontal beam emittance. To decrease the horizontal beam emittance we would increase the horizontal partition number and at the same time the longitudinal partition number would be reduced. The adjustments, however, are limited. The limit is reached when the longitudinal motion becomes unstable or in practical cases when the partition number becomes less than about 0.5. Other more practical limits may occur before stability limits are reached if the momentum change becomes too large to fit in the vacuum chamber aperture or within the dynamic aperture, whichever is smaller.

### 7.6.3 Can we Eliminate the Beam Energy Spread?

To conclude the discussions on beam manipulation, we try to eliminate the energy spread in a particle beam. From beam dynamics, we know that the beam particles can be sorted according to their energy by introducing a dispersion function. The distance of a particle from the reference axis is proportional to its energy and given by

$$x_\delta = D\delta,\tag{7.82}$$

where $D$ is the value of the dispersion at the location under consideration and $\delta = \Delta E/E_0$ the energy error. For simplicity, we make no difference between energy and momentum during this discussion. We consider now a cavity excited in a mode ($\mathrm{TM}_{011}$-mode) such that the accelerating field is zero at the axis, but varies linearly with the distance from the axis. If now the accelerating field, or after integration through the cavity, the accelerating voltage off axis scales like

$$eV_{\mathrm{rf}}(x_\delta) = -\frac{x_\delta}{D}E_0,\tag{7.83}$$

we have just compensated the energy spread in the beam. The particle beam has become monochromatic, at least to the accuracy assumed here. In this process Liouville's theorem is violated because this scheme does not change the bunch length and the longitudinal emittance has been indeed reduced by application of macroscopic fields.

The problem is that we are by now used to consider transverse and longitudinal phase space separate. While this separation is desirable to manage the mathematics of beam dynamics, we must not forget, that ultimately beam dynamics occurs in 6-dimensional phase space. Since Liouville's theorem must be true, its apparent violation warns us to observe changes in other phase space dimensions. In the case of beam monochromatization, we notice that the transverse beam emittance has been increased by virtue of Maxwell's equations. The transverse variation of the longitudinal electric field causes the appearance of transverse magnetic fields which deflect the particle trajectories transversely thus increasing the transverse phase space at the expense of the longitudinal phase space.

This is a general feature of electromagnetic fields known as the Panofsky Wenzel Theorem [48] . The Lorentz force due to electromagnetic fields causes a change in the particle momentum which in the transverse direction is given by

$$c\boldsymbol{p}_\perp = \frac{e}{\beta}\int_0^d [\boldsymbol{E}_\perp + (\boldsymbol{\beta}\times\boldsymbol{B})_\perp]\,\mathrm{d}z.\tag{7.84}$$

Expressing the fields by the vector potential $\boldsymbol{E}_\perp = -\partial\boldsymbol{A}_\perp/\partial t$ and $\boldsymbol{B}_\perp = (\nabla\times\boldsymbol{A})_\perp$ the change in the transverse momentum can be expressed by

$$c\boldsymbol{p}_\perp = -e\nabla_\perp\int_0^d \boldsymbol{E}\mathrm{d}z,\tag{7.85}$$

where the integration is taken over the length $d$ of the cavity. This is the Panofsky Wenzel theorem which states in our case that transverse acceleration occurs whenever there is a transverse variation of the accelerating field. In conclusion, we find that indeed a particle beam can be monochromatized with the use of, for example, a $TM_{110}$ mode cavity, but only at the expense of transverse beam emittance.

## 7.7 Photon Source Parameters

With the knowledge of betatron functions, beam emittances and energy spread we are in a position to define the particle beam cross sections and photon source parameters. The total beam width or height is defined by the contribution of the betatron phase space $\sigma_{\beta,x,y}$ and the energy phase space $\sigma_{\eta,x,y}$ and is

$$\sigma_{\text{tot},x,y} = \sqrt{\sigma^2_{\beta,x,y} + \sigma^2_\eta} = \sqrt{\epsilon_{x,y}\beta, x, y + \eta_x^2 \left(\frac{\sigma_\varepsilon}{E_0}\right)^2} \qquad (7.86)$$

with $\sigma^2_{\beta,x,y} = \epsilon_{x,y}\beta_{x,y}$ and $\sigma_{\eta,x} = \eta_x \frac{\sigma_\varepsilon}{E_0}$, $\gamma_{x,y} = \frac{1+\alpha_{x,y}^2}{\beta_{x,y}}$ and $\alpha_{x,y} = -\frac{1}{2}\beta'_{x,y}$. Similarly, we get for the beam divergence

$$\sigma_{\text{tot},x',y'} = \sqrt{\sigma^2_{\beta,x',y'} + \sigma^2_{\eta'}} = \sqrt{\epsilon_{x,y}\gamma_{x,y} + \eta'^2\left(\frac{\sigma_\varepsilon}{E_0}\right)^2}. \qquad (7.87)$$

These beam parameters resemble in general the source parameters of the photon beam. Deviations occur when the beam emittance becomes very small, comparable to the photon wavelength of interest. In this case, the photon source parameters may be modified by diffraction effects which limit the apparent source size and divergence to some minimum values even if the electron beam cross section should be much smaller. For radiation at a wavelength $\lambda$, the diffraction limited, radial photon source parameters are[1]

$$\sigma_r = \frac{1}{2\pi}\sqrt{\lambda L} \qquad \text{and} \qquad \sigma'_r = \sqrt{\frac{\lambda}{L}}. \qquad (7.88)$$

Projection onto the horizontal or vertical plane gives $\sigma_{x,y} = \sigma_r/\sqrt{2}$ etc. Due to diffraction, it is not useful to push the electron beam emittance to values much smaller than

$$\epsilon_{x,y} = \frac{\lambda}{4\pi}. \qquad (7.89)$$

---

[1] Many authors use a different definition $\sigma_r = \sigma_r/\sqrt{2}$. The difference is mainly that the subscript $_r$ refers to radiation and the related beam parameters are already projected to the $x$ or $y$-plane. In this text we use the subscript $_r$ from the radial coordinate since we derive the diffraction effects from a round beam.

For an arbitrary electron beam cross section the photon source parameters are the quadratic sums of both contributions

$$\sigma^2_{\text{ph},x,y} = \sigma^2_{\text{tot},x,y} + \tfrac{1}{2}\sigma^2_r, \tag{7.90}$$

$$\sigma'^2_{\text{tot},x,y} = \sigma'^2_{\text{tot},x,y} + \tfrac{1}{2}\sigma'^2_r. \tag{7.91}$$

The contribution from diffraction can be ignored if

$$\epsilon_{x,y} \gg \frac{\lambda}{4\pi}. \tag{7.92}$$

# Exercises *

**Exercise 7.1 (S).** Derive the equation of motion for synchrotron oscillations with large amplitudes and for a sinusoidal variation of the rf voltage.

**Exercise 7.2 (S).** Calculate the synchrotron damping time for the storage ring in Exercise 6.3 or 4.1 and rectangular pure dipole magnets. What are the damping times in that ring? Calculate the equilibrium energy spread.

**Exercise 7.3 (S).** What is the probability for a 6 GeV electron to emit a photon with an energy of $\varepsilon = \sigma_\varepsilon$ per unit time travelling on a circle with radius $\rho = 25$ m. How likely is it that this particle emits another such photon within a damping time? In evaluating quantum excitation and equilibrium emittances, do we need to consider multiple photon emissions? (use isomagnetic ring)

**Exercise 7.4 (S).** How many photons are emitted by an electron of energy $E$ on average per turn.

**Exercise 7.5.** Consider an electron beam in a 6 GeV storage ring with a bending radius of $\rho = 20$ m . Calculate the rms energy spread $\sigma_\varepsilon/E_0$ and the damping time $\tau_s$.

**Exercise 7.6.** For the storage ring design in Exercise 6.3 estimate the average value $\langle\mathcal{H}\rangle$ and calculate the beam emittance.

**Exercise 7.7.** An electron beam circulating in a 1.5 GeV storage ring emits synchrotron radiation. The rms emission angle of photons is $1/\gamma$ about the forward direction of the particle trajectory. Determine the photon phase space distribution at the source point and at a distance of 10 m away while ignoring the finite particle beam emittance. Now assume a Gaussian particle distribution with a horizontal beam emittance of $\epsilon_x = 0.15 \times 10^{-6}$ rad m. Fold both the photon and particle distributions and determine the photon phase

---

* The argument (S) indicates an exercise for which a solution is given in
  Appendix A.

space distribution 10 m away from the source point if the electron beam size is $\sigma_x = 1.225$ mm, the electron beam divergence $\sigma_{x'} = 0.1225$ mrad and the source point is a symmetry point of the storage ring. Assume the dispersion function to vanish at the source point. For what minimum photon wavelength would the vertical electron beam size appear diffraction limited if the emittance coupling is 10%?

# 8. Storage Ring Design as a Synchrotron Light Source

Synchrotron radiation sources have undergone significant transitions and modifications over past years. Originally, most experiments with synchrotron radiation were performed parasitically on high energy physics colliding beam storage rings. Much larger photon fluxes could be obtained from such sources compared to any other source available. The community of synchrotron radiation users grew rapidly and so did the variety of applications and fields. By the time the usefulness of storage rings for high energy physics was exhausted some of these facilities were turned over to the synchrotron radiation community as fully dedicated radiation sources. Those are called first generation synchrotron radiation sources. They were not optimized for minimum beam emittance and maximum photon beam brightness. Actually, the optimization for high energy physics called for a maximum beam emittance to maximize collision rates for elementary particle events. The radiation sources were mostly bending magnets although the development and use of insertion devices started in these rings. Typically, the beam emittance is in the 100's of nm.

As the synchrotron radiation community further grew, funds became available to construct dedicated radiation facilities. Generally, these rings were designed as bending magnet sources but with reduced beam emittance ($\leq 100$ nm) to increase photon brightness. The design emittances were much smaller than those in first generation rings but still large by present day standards. The use of insertion devices did not significantly affect the storage ring designs yet. These rings are called second generation rings.

Third generation synchrotron radiation sources have been designed and constructed during the second half of the eighties and into the nineties. These rings were specifically designed for insertion device radiation and minimum beam emittance ($4 \leq \epsilon_x \leq 20$ nm) or maximum photon beam brightness. As such, they exhibit a large number of magnet-free insertion straight sections.

Finally, fourth generation synchrotron radiation sources are so far only under discussion. A consensus seems to emerge within the community that such sources may be based more on linear accelerators. For example, great efforts are underway in a number of laboratories to design x-ray lasers. Such a source would be based on the principle of a single pass FEL where a high energy and high quality electron beam passing through a long undulator produces coherent undulator radiation in the x-ray regime. A storage ring based alternative has been proposed which uses the ring structure only as a distributor of radiation to individual beam lines. An electron beam is injected continuously from a high performance electron linear accelerator. Such a linac beam can have a very low beam emittance which is preserved in the storage ring for some number of turns before quantum excitation takes over. To compensate for the high energy cost the spent electron beam is ejected from the storage ring again and its energy is recovered.

## 8.1 Storage Ring Lattices

To achieve a small particle beam emittance for maximum photon beam brightness a number of different magnet lattices for storage rings are available. All lattices can basically be used to achieve as small a beam emittance as desired, limited only by diffraction effects of the photon beams. Other, more practical considerations, however, limit the minimum beam emittance achievable in a particular lattice. A variety of magnet lattices have been used in the designs of existing storage ring based synchrotron radiation sources. In this section three basic types and some variations thereof will be discussed:

- the FODO lattice
- the double bend achromat lattice (dba)
- the triple bend achromat lattice (tba)

All lattice types can provide long magnet free sections for the installation of insertion devices, accelerating cavities and injection components. For insertion devices one would prefer to have dispersion free sections available which is easy to achieve in a dba-or tba-lattice but more complicated in a FODO lattice. On the other hand, a FODO lattice is very compact and is therefore mostly suitable for generating low emittance beams in so-called damping rings for applications in high energy accelerator systems. The dba-and tba-lattices are more open and provide easily magnetfree straight sections, a desired feature for high brightness synchrotron radiation sources. More recently, the demand for dispersionfree insertion straight sections has been relaxed in favor of an even lower beam emittance achievable this way. Quantum excitation is kept very low by using short period undulator and wiggler magnets.

### 8.1.1 FODO Lattice

We consider here briefly the FODO lattice because of its simplicity and its ability to give us a quick feeling for the scaling of beam emittance with lattice parameters. The beam emittance can be manipulated at design time by adjusting $\langle \mathcal{H} \rangle$ to the desired value. To calculate the average value $\langle \mathcal{H} \rangle$ in a FODO lattice is somewhat elaborate. Here, we are interested primarily in the scaling of the beam emittance with FODO lattice parameters. Recollecting the results for the symmetric solutions of the lattice functions in a FODO lattice (6.53,6.55) we notice the following scaling laws

$$\beta \propto L,\tag{8.1}$$
$$\beta' \propto L^0,\tag{8.2}$$
$$\eta \propto L^2/\rho,\tag{8.3}$$
$$\eta' \propto L/\rho,\tag{8.4}$$

where $L$ is the distance between the centers of adjacent quadrupoles. All three terms in the function $\mathcal{H}(s) = \gamma(s)\,\eta^2 + 2\alpha(s)\,\eta\eta' + \beta(s)\,\eta'^2$ scale in a similar fashion like

$$\{\mathcal{H}(s)\} = \left\{\frac{1}{L}\frac{L^4}{\rho};\ L^0\frac{L^2}{\rho}\frac{L}{\rho};\ L\frac{L^2}{\rho}\right\} \propto \frac{L^3}{\rho^2}\tag{8.5}$$

and the equilibrium emittance for a FODO lattice scales then like

$$\epsilon_x = C_q\gamma^2\frac{\langle\mathcal{H}/\rho\rangle}{\langle 1/\rho^2\rangle} \propto \gamma^2\frac{L^3}{\rho^3} \propto \gamma^2\varphi^3,\tag{8.6}$$

where $\varphi = \ell_b/\rho$ is the deflection angle in each bending magnet. The proportionality factor depends on the strengths of the quadrupoles and is large for very weak or very strong quadrupoles. A minimum can be reached for a focal length of $|f| \approx 1.06\,L$ in each half-quadrupole resulting in a minimum beam emittance achievable in a FODO lattice given in practical units by

$$\epsilon\,(\text{radm}) \approx 10^{-11}\,E^2(\text{GeV})\,\varphi^3\,(\text{deg}^3),\tag{8.7}$$

where $\varphi = 2\pi/N_M$, $N_M$ the number of bending magnets in the ring and $N_M/2$ the total number of FODO cells in the ring. This result is significant because it exhibits a general scaling law of the beam emittance proportional to the square of the beam energy and the cube of the deflecting angle in each bending magnet, which is valid for all lattice types. The coefficients, though, vary for different lattices. While the beam energy is primarily driven by the desired photon spectrum, we find that high brightness photon beams from low emittance electron beams require a storage ring design composed of many lattice cells with a small deflection angle per magnet. Of course, there are some limits on how far one can go with this concept due to other limitations, not the least being size and cost of the ring which both grow with the number of lattice cells.

## 8.2 Optimization of a Storage Ring Lattice

While the cubic dependence of the beam emittance on the bending angle is a significant design criterion we discuss here a more detailed optimization strategy. The emittance is determined by the beam energy, the bending radius and the $\mathcal{H}$-function. Generally, we have no choice on the beam energy which is mostly determined by the desired critical photon energy of bending magnet and insertion device radiation or cost. Similarly, the bending radius is defined by the ring geometry, desired spectrum etc. Interestingly, it is not the bending radius but rather the bending angle which influences the equilibrium beam emittance. The main process to minimize the beam emittance is to adjust the focusing such that the lattice functions in the bending magnets generate a minimum value for $\langle \mathcal{H} \rangle$.

### 8.2.1 Minimum Beam Emittance

The equilibrium beam emittance (7.44)

$$\epsilon_x = \frac{\sigma_x^2}{\beta_x} = C_q \gamma^2 \frac{\langle \mathcal{H}(s)/\rho^3 \rangle}{\langle 1/\rho^2 \rangle} \tag{8.8}$$

depends only on the lattice function $\mathcal{H}(s)$ inside bending magnets where $1/\rho \neq 0$. We may therefore, independent of any lattice type, consider this function only within bending magnets. For the purpose of this discussion we assume a regular periodic lattice, where all lattice functions within each bending magnet are the same, and concentrate therefore our discussion just on one bending magnet. The average value $\langle \mathcal{H}/\rho^3 \rangle$ for the whole ring will then be the same as that for one magnet.

The contribution of any individual bending magnet with bending radius $\rho$ to the beam emittance can be determined by calculation of the average

$$\langle \mathcal{H} \rangle = \frac{1}{\ell_b} \int_0^{\ell_b} \mathcal{H}(s)\, ds, \tag{8.9}$$

where $\ell_b$ is the length of the bending magnet and the bending radius is assumed to be constant within a magnet. From here on, we ignore the index $_x$ since we assume a flat storage ring in the horizontal plane. All lattice functions are therefore to be taken in the horizontal plane.

Since in first approximation there is no focusing within bending magnets we may treat such magnets as drift spaces. The lattice functions, starting at the entrance to the magnet (Fig. 8.1) with values $(\beta_0, \alpha_0, \gamma_0, \eta_0, \eta_0')$ vary within the bending magnet like

$$\begin{aligned}
\beta(s) &= \beta_0 - 2\alpha_0\, s + \gamma_0\, s^2, \\
\alpha(s) &= \alpha_0 - \gamma_0\, s, \\
\gamma(s) &= \gamma_0, \\
\eta(s) &= \eta_0 + \eta_0'\, s + \rho\,(1 - \cos\psi), \\
\eta'(s) &= \eta_0' + \sin\psi,
\end{aligned} \tag{8.10}$$

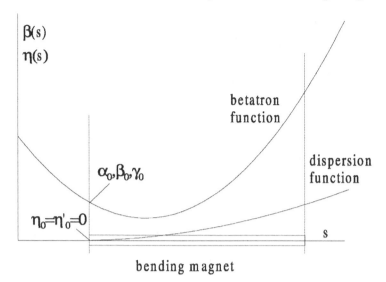

**Fig. 8.1.** Lattice functions in a bending magnet

where $0 \leq s \leq \ell_b$ and the deflection angle $\psi = \frac{s}{\rho}$. Before we use these equations, we assume lattices where $\eta_0 = \eta'_0 = 0$. The consequences of this assumption will be discussed later. Using (8.9) and (8.10) we get

$$\langle \mathcal{H} \rangle = \beta_0 B + \alpha_0 \rho A + \gamma_o \rho^2 C \,. \tag{8.11}$$

The coefficients $A, B,$ and $C$ are functions of the total deflection angle $\varphi = \ell_b / \rho$ defined by

$$B = \tfrac{1}{2} \left( 1 - \frac{\sin 2\varphi}{2\varphi} \right) , \tag{8.12}$$

$$A = 2 \frac{1 - \cos \varphi}{\varphi} - \frac{3 \sin^2 \varphi}{2\varphi} - \tfrac{1}{2}\varphi + \tfrac{1}{2}\sin 2\varphi \,, \tag{8.13}$$

$$C = \tfrac{3}{4} + 2\cos \varphi + \frac{5 \sin 2\varphi}{8\varphi} - 4 \frac{\sin \varphi}{\varphi} + \tfrac{1}{6}\varphi^2 \,. \tag{8.14}$$

For small bending angles and an isomagnetic ring, where all bending magnets are the same, we have:

$$B \approx \tfrac{1}{3}\varphi^2 \left( 1 - \tfrac{1}{5}\varphi^2 \right) , \tag{8.15}$$

$$A \approx -\tfrac{1}{4}\varphi^3 \left( 1 - \tfrac{5}{18}\varphi^2 \right) , \tag{8.16}$$

$$C \approx \tfrac{1}{20}\varphi^4 \left( 1 - \tfrac{5}{14}\varphi^2 \right) , \tag{8.17}$$

and the beam emittance (8.6) in the lowest order of approximation becomes

$$\epsilon = C_q \gamma^2 \varphi^3 \left[ \tfrac{1}{3}\frac{\beta_0}{\ell_b} - \tfrac{1}{4}\alpha_0 + \tfrac{1}{20}\gamma_0 \ell_b \right] , \tag{8.18}$$

where $C_q$ is defined in (7.33).

This equation shows clearly the cubic dependence of the beam emittance on the deflection angle $\varphi$ per bending magnet, which is a general property of all lattices since we have not yet made any assumption on a particular lattice. Equation (8.18) has a minimum for both $\alpha_0$ and $\beta_0$. From the derivative $\partial \langle \mathcal{H} \rangle / \partial \alpha_0 = 0$ we extract the optimum value for $\alpha_0$

$$\alpha_{0\,\mathrm{opt}} = -\frac{1}{2}\frac{A}{C}\frac{\beta_0}{\rho}. \tag{8.19}$$

Inserting this result into (8.18), and evaluating the derivative $\partial \langle \mathcal{H} \rangle / \partial \beta_0 = 0$ the optimum value for $\beta_0$ becomes

$$\beta_0^* = \frac{2\,C\,\rho}{\sqrt{4BC - A^2}}. \tag{8.20}$$

With $\beta_0^*$ the quantity $\alpha_0^*$ is

$$\alpha_0^* = \frac{-A}{\sqrt{4BC - A^2}} \tag{8.21}$$

and the minimum possible value for $\mathcal{H}$ is finally

$$\langle \mathcal{H} \rangle_{\mathrm{min}} = \sqrt{4BC - A^2}\,\rho. \tag{8.22}$$

For small deflection angles $\varphi \ll 1$, and neglecting second and higher order terms in $\varphi$, the optimum lattice functions at the entrance to the bending magnets are with $\eta_0 = \eta_0' = 0$

$$\alpha_0^* \approx \frac{(1 - \frac{5}{18}\varphi^2)\sqrt{15}}{\sqrt{1 - \frac{61}{105}\varphi^2}} \approx \sqrt{15}, \tag{8.23a}$$

$$\beta_0^* \approx \frac{\sqrt{12}\,(1 - \frac{5}{14}\varphi^2)\,\ell_b}{\sqrt{5}\sqrt{1 - \frac{61}{105}\varphi^2}} \approx \sqrt{\frac{12}{5}}\,\ell_b, \tag{8.23b}$$

$$\langle \mathcal{H} \rangle_{\mathrm{min}} \approx \frac{\varphi^3\,\rho}{4\sqrt{15}}\sqrt{1 - \frac{61}{105}\varphi^2} \approx \frac{\rho}{4\sqrt{15}}\,\varphi^3. \tag{8.23c}$$

With this, the minimum obtainable beam emittance in any lattice is from (7.44)

$$\epsilon_{\mathrm{dba,min}} = C_q\,\gamma^2\,\frac{\langle \mathcal{H}(s)/\rho^3 \rangle}{\langle 1/\rho^2 \rangle} \approx C_q\,\gamma^2\,\frac{\varphi^3}{4\sqrt{15}}. \tag{8.24}$$

The results are very simple for small deflection angles but for angles larger than about $33°$ per bending magnet the error for $\langle \mathcal{H} \rangle_{\mathrm{min}}$ exceeds 10% at which point higher order terms must be included. It is interesting to note that the next order correction due to larger bending angles gives a reduction in beam emittance compared to the lowest order approximation. Higher order terms, however, quickly stop and reverse this reduction.

For simplicity, we assumed that the dispersion functions $\eta_0 = 0$ and $\eta_0 = 0$. Numerical methods must be used to find the optimum solutions for finite dispersion functions. In the following we consider only very small values $\eta_0 \ll 1$ and $\eta_0' \ll 1$ to evaluate the impact of the correction for a finite dispersion on the beam emittance. Retaining only linear terms in $\eta_0$, $\eta_0'$, and $\varphi$, the expression for $\langle \mathcal{H} \rangle$ becomes

$$\langle \mathcal{H} \rangle_{\eta\,\min} = \langle \mathcal{H} \rangle_{\min} + \frac{1}{\sqrt{5}} \left( \frac{5}{3} \eta_0 + 6 \eta_0' \ell_b \right) \varphi. \tag{8.25}$$

Obviously, the beam emittance can be further reduced for negative values of $\eta_0$ and $\eta_0'$. This has been exploited in recent storage ring designs. Nonlinear terms, however, quickly cause an increase in the beam emittance again, thus limiting the gain.

In summary, it has been demonstrated that there are certain optimum conditions for lattice functions in bending magnets to minimize the equilibrium beam emittance. No assumption about a particular lattice has been made yet. Another observation is that the beam emittance is proportional to the third power of the magnet deflection angle suggesting to use small deflection angles in order to achieve a small beam emittance. Low emittance storage rings therefore are characterized by many short magnet and lattice periods.

### 8.2.2 The Double Bend Achromat (dba) Lattice

The dba-lattice is designed such as to make full use of the minimization possibilities for the beam emittance as just discussed and to provide dispersionfree insertion straight sections. Fig. 8.2 shows two renditions of the basic layout for a dba-lattice. Other slightly different modifications have been used but the basic design features are the same. Starting from the middle of an insertion straight section a set of two or more quadrupoles provide the proper focusing of the lattice functions into the bending magnet to achieve the minimum beam emittance. The insertions are kept dispersion free which is the main function of the focusing between the dipole magnets. The section between and including the bending magnets is called an achromat because the dispersion is zero outside of the achromat. The ideal minimum beam emittance in this lattice type is from (8.24)

$$\epsilon_{\text{dba,min}} = \frac{C_q}{4\sqrt{15}} \gamma^2 \varphi^3, \tag{8.26}$$

or in more practical units:

$$\epsilon_{\text{dba}} \, (\text{rad}\,\text{m}) = 5.036 \times 10^{-13} \, E^2(\text{GeV}^2) \, \varphi^3 \, (\text{deg}^3). \tag{8.27}$$

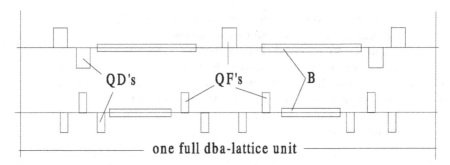

**Fig. 8.2.** dba Lattice

To achieve this minimum beam emittance, we must provide specific values for the lattice functions at the entrance to the bending magnets. Specifically, the initial horizontal betatron function must be strongly convergent reaching a minimum about one third through the bending magnet. At the end of the bending magnet the ideal betatron function, however, becomes quite large. Note, that the vertical lattice functions can be chosen freely since they do not affect the beam emittance as long as there is no vertical dispersion.

In an actual lattice design it appears difficult to achieve sufficient beam stability if the lattice parameters at the entrance to the bending magnets are set to the optimum values. A compromise between optimum lattice parameters and beam stability must be reached resulting in a somewhat increased beam emittance compared to the theoretical minimum. The source of the problem are the large value of the betatron function at the exit of the bending magnet causing strong chromatic aberrations which must be corrected by sextupole magnets. This correction, while essential for beam stability, also generates geometric aberrations and a compromise between correction of chromatic and generating geometric aberrations must be made. The result of this compromise in a well designed storage ring must be a sufficiently large aperture within which the beam can travel for many hours without losses. Outside of this aperture, called the dynamic aperture, particles are lost due to geometric aberrations. Generally, a sufficiently large dynamic aperture cannot be obtained for the ideal solution of minimum beam emittance. On the other hand, the dynamic aperture grows rapidly as the optimum conditions on the lattice functions are relaxed.

An example of an actual dba-lattice is shown in Fig. 8.3 for the 1.3GeV storage ring at the Laboratorio National de Luz Sincrotron (LNLS) in Campinas, Brazil. The central part of the lattice between the bending magnets may consist of one to four quadrupoles and its only function is to focus the dispersion function so that all insertions are dispersion free.

The choice of the optimum value for $\alpha_0^* = \sqrt{15}$ causes the betatron function to reach a sharp minimum at about one third into the bending magnet, $s_{\min} = \frac{\alpha_0^*}{\gamma_0^*} = \frac{3}{8}\,\ell_b$, and to increase from there on to large values causing prob-

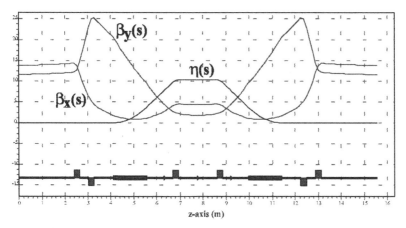

**Fig. 8.3.** dba Lattice of the Laboratorio National de Luz Sincrotron, LNLS in Campinas, Brazil

lems with nonlinear aberrations. We remove the minimum condition on $\alpha_0$ and express the beam emittance in terms of the minimum emittance

$$\frac{\epsilon_{dba}}{\epsilon_{dba,min}} = 8\frac{\beta_0}{\beta_0^*} - \sqrt{15}\,\alpha_0 + \frac{1}{2}\left(1 + \alpha_0^2\right)\frac{\beta_0^*}{\beta_0}. \tag{8.28}$$

In Fig. 8.4 this ratio of emittances is shown for different values of $\alpha_0$ as a function of $\beta_0/\beta_0^*$. It is apparent from Fig. 8.4 that the minimum emittance changes only little even for big variations of $\alpha_0$ about its optimum value allowing us to choose much more forgiving values for $\alpha_0$ without significant loss in beam emittance. This weak dependence can be used to lessen the problems caused by nonlinear aberrations.

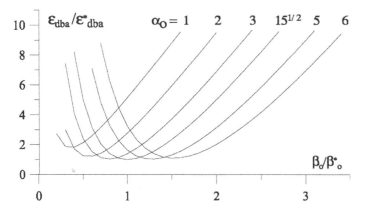

**Fig. 8.4.** Emittance and betatron functions in a dba-lattice

For arbitrary values of $\alpha_0$ still an optimum value for $\beta_0$ exists. We evaluate the derivative $\partial \langle \mathcal{H} \rangle / \partial \beta_0 = 0$ only and get for the optimum betatron function at the entrance to the bending magnet

$$\beta_0 = \ell_b \sqrt{\frac{3}{20}(1+\alpha_0^2)}. \tag{8.29}$$

The beam emittance in this case becomes

$$\frac{\epsilon_{dba}}{\epsilon_{dba,min}} = 4\sqrt{1+\alpha_0^2} - \sqrt{15}\,\alpha_0. \tag{8.30}$$

For the condition (8.29) the value of the betatron function $\beta(\ell_b)$ at the end of the bending magnet reaches a minimum for $\alpha_0 = \frac{4\sqrt{15}}{17} \approx 0.911$ at the expense of a loss in beam emittance by a factor of two. In this case

$$\frac{\alpha_0}{\alpha_0^*} = \frac{4}{17}, \tag{8.31}$$

$$\frac{\beta_0}{\beta_0^*} = \frac{23}{68}, \tag{8.32}$$

$$\frac{\beta(\ell_b)}{\beta^*(\ell_b)} = \frac{17}{32}, \tag{8.33}$$

$$\frac{\epsilon_{dba}}{\epsilon_{dba,min}} = \frac{32}{17}. \tag{8.34}$$

The betatron function at the end of the bending magnet has been reduced by almost a factor of two or by a factor of $17/32$ which is a great improvement with respect to instabilities. In a particular storage ring design one would therefore reduce the value of $\alpha_0$ although by not more than necessary for beam stability.

### 8.2.3 The Triple Bend Achromat (tba) Lattice

As a variation of the dba lattice a triple bend achromat lattice has become popular in recent synchrotron radiation source designs. In this case, three bending magnets are placed between each pair of insertion straight sections (Fig. 8.5). That results in a reduction of the circumference although at the expense of a similar reduction in available insertion straight sections. This lattice type serves well for smaller facilities and lower energies.

### 8.2.4 Limiting Effects

Given the usefulness of maximum photon beam brightness for experimenters one might wonder why don't we just design storage rings with a beam emittance below the diffraction limit. The answer has to do with limitations of beam stability due to nonlinear betatron oscillations. To reduce the beam

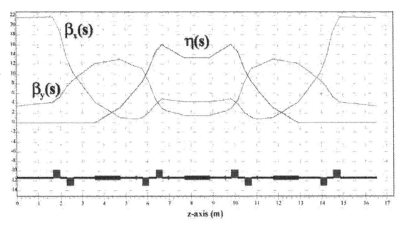

**Fig. 8.5.** Implementation of a tba-lattice at the National Synchrotron Laboratory, NSRL in Hefei, China

emittance, we require stronger and/or more quadrupole focusing. The energy spread in the beam causes a variation of focusing with lower energy particles being focused too much and higher energy particles focused too little as indicated in Fig. 8.6. The total amount of focusing in a storage ring is a measure for these chromatic aberrations, which can cause beam instability if not corrected. For this reason, we must compensate the chromatic aberrations which we call the storage ring chromaticity. Because the chromaticity derives from focusing and we have different focusing in both planes, there are two chromaticities, one for the horizontal and the other for the vertical plane.

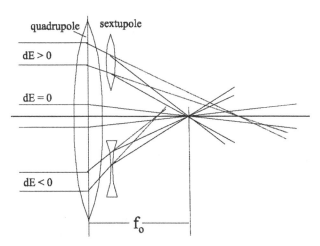

**Fig. 8.6.** Origin and correction of chromatic effects and chromaticity

Correction of the chromaticities can be accomplished by installing sextupole magnets into the storage ring at locations where the dispersion is not zero. The dispersion causes some degree of segregation between higher and lower energy particles with higher energy particles gathering more outside of the ideal orbit and lower energy particles more on the inside. Sextupoles can be considered as quadrupoles with varying focal strength across the horizontal aperture. A sextupole therefore can add some focusing for higher energy particles being outside of the ideal orbit ($x > 0$) and subtract some focusing for lower energy particles at $x < 0$ (Fig. 8.6). That compensates the under and over focusing these particles experience in the regular quadrupoles. Distributing sextupoles around the ring is therefore the preferred way to compensate the storage ring chromaticity.

Every coin has two sides, however. The sextupole field increases quadratically with $x$ and while we compensate the chromaticities, these same sextupoles generate nonlinear, quadratic perturbations especially for particles with large betatron oscillation amplitudes. These perturbations are known as geometric aberrations generating pillowcase perturbations in the images as is well known from light optics. The art of storage ring design is then to correct the chromatic aberrations while keeping the geometric aberrations at a minimum. This can be achieved up to a certain degree by distributing sextupoles along the orbit at properly selected locations. It would therefore be wrong to use just two sextupoles to correct the two chromaticities. The sextupole strengths would be too high generating serious geometric aberrations. However, even with carefully distributing the sextupoles around the ring lattice, we still deal with a nonlinear problem and we cannot expect to get perfect compensation. There will always be a limit on the maximum stable betatron oscillation amplitude in the storage ring. The design objective is to expand the limit for large amplitude betatron oscillations. This limit is called the dynamic aperture in contrast to the physical aperture defined by the vacuum chamber. There is no analytical solution for the dynamic aperture and it is determined by numerical particle tracking programs which follow individual particles for some thousands of turns through all nonlinear fields to probe stability limits.

For a stable beam with a long beam lifetime, we must have a minimum dynamic aperture to accommodate not only the beam proper but also a halo of particles around the beam. This halo is made-up of particles which have been deflected by a small angle during elastic collisions with a residual gas atom. Such collisions occur quite frequently, constantly populating the halo with new particles. By damping, these particles loose betatron oscillation amplitudes and leave slowly the halo again to join the beam proper. While there are only few particles in the halo at any one time, we cannot scrape off this halo by lack of sufficient dynamic aperture. The beam lifetime could be reduced considerable since there is a constant flow of particles into the halo and back to the beam. This flow cannot be interrupted.

# 9. Theory of Synchrotron Radiation

The phenomenon of synchrotron radiation has been introduced in a conceptual way in Chap. 3 and a number of basic relations have been derived. In this chapter we will approach the physics of synchrotron radiation in a more formal way to exhibit detailed characteristics. Specifically, we will derive expressions for the spatial and spectral distribution of photon emission in a way which is applicable later for special insertion devices.

The theory of synchrotron radiation is intimately related to the electromagnetic fields generated by moving charged particles. Wave equations can be derived from Maxwell's equations and we will find that any charged particle under the influence of external forces can emit radiation. We will formulate the characteristics of this radiation and apply the results to highly relativistic particles.

## 9.1 Radiation Field

The electromagnetic fields for a single moving point charge will be derived first and then applied to a large number of particles. The fields are determined by Maxwell's equations (C.1 to C.4) for moving charges in vacuum, $\epsilon_{\mathrm{r}} = 1$ and $\mu_{\mathrm{r}} = 1$. The magnetic field can be derived from a vector potential $\boldsymbol{A}$ defined by

$$\boldsymbol{B} = \boldsymbol{\nabla} \times \boldsymbol{A} . \tag{9.1}$$

Inserting the vector potential into Maxwell's curl equation (C.3) we have $\boldsymbol{\nabla} \times \left( \boldsymbol{E} + \frac{[c]}{c} \frac{\partial \boldsymbol{A}}{\partial t} \right) = 0$, or after integration

$$\boldsymbol{E} = -\frac{[c]}{c} \frac{\partial \boldsymbol{A}}{\partial t} - \boldsymbol{\nabla} \varphi , \tag{9.2}$$

where $\varphi$ is the scalar potential. We choose the scalar potential such that $[c] \boldsymbol{\nabla} \boldsymbol{A} + \frac{1}{c} \frac{\partial \varphi}{\partial t} = 0$, a condition known as the Lorentz gauge. With (B.24) applied to $\boldsymbol{A}$ the expression for the electric field together with (C.4) results in the wave equation

$$\nabla^2 \boldsymbol{A} - \frac{1}{c^2} \frac{\partial^2 \boldsymbol{A}}{\partial t^2} = \frac{4\pi}{[4\pi\epsilon_0]} \rho \boldsymbol{\beta} . \tag{9.3}$$

Similarly, we derive the wave equation for the scalar potential

$$\nabla^2\varphi - \frac{1}{c^2}\frac{\partial^2\varphi}{\partial t^2} = -\frac{4\pi}{[4\pi\epsilon_0]}\rho\,. \tag{9.4}$$

These are the well-known wave equations with the solutions

$$\boldsymbol{A}(t) = \frac{1}{[4\pi c\epsilon_0]}\frac{1}{c}\int \left.\frac{\boldsymbol{v}\rho(x,y,z)}{R}\right|_{t_{\mathrm{r}}} \mathrm{d}x\,\mathrm{d}y\,\mathrm{d}z \tag{9.5}$$

and

$$\varphi(t) = \frac{1}{[4\pi c\epsilon_0]}\frac{1}{c}\int \left.\frac{\rho(x,y,z)}{R}\right|_{t_{\mathrm{r}}} \mathrm{d}x\,\mathrm{d}y\,\mathrm{d}z\,. \tag{9.6}$$

Because of the finite velocity of light, all quantities under the integrals must be evaluated at the retarded time

$$t_{\mathrm{r}} = t - \frac{1}{c}R(t_{\mathrm{r}}) \tag{9.7}$$

when the radiation was emitted by the moving charge, in contrast to the time $t$ when the radiation is observed at a distant point. The quantity $R$ is the distance between the observation point $P(x,y,z)$ and the location of the charge element $\rho(x_{\mathrm{r}},y_{\mathrm{r}},z_{\mathrm{r}})\mathrm{d}x_{\mathrm{r}}\mathrm{d}y_{\mathrm{r}}\mathrm{d}z_{\mathrm{r}}$ at the retarded time $t_{\mathrm{r}}$. The vector

$$\boldsymbol{R} = (x_{\mathrm{r}} - x, y_{\mathrm{r}} - y, z_{\mathrm{r}} - z) \tag{9.8}$$

points away from the observation point to the charge element at the retarded time as shown in Fig. 9.1.

Special care must be exercised in performing the integrations. Although we consider only a point charge $q$, the integral in (9.6) cannot be replaced by $q/R$ but must be integrated over a finite volume followed by a transition to a point charge. As we will see this is a consequence of the fact that the velocity of light is finite and therefore the movement of charge elements must be taken into account.

To define the quantities involved in the integration we use Fig. 9.1. The combined field at the observation point $P$ at time $t$ comes from all charges located at a distance $R$ away from $P$. We consider the contribution from all charges contained within a spherical shell centered at $P$ with a radius $R$ and thickness $\mathrm{d}r$ to the radiation field at $P$ and time $t$. Radiation emitted at time $t_{\mathrm{r}}$ will reach $P$ at the time $t$. If $\mathrm{d}\sigma$ is a surface element of the spherical shell, the volume element of charge is $\mathrm{d}x\,\mathrm{d}y\,\mathrm{d}z = \mathrm{d}\sigma\mathrm{d}r$. The retarded time for the radiation from the outer surface of the shell is $t_{\mathrm{r}}$ and the retarded time for the radiation from the charge element on the inner surface of the shell is $t_{\mathrm{r}} - \frac{\mathrm{d}r}{c}$. From Fig. 9.1 we find the electromagnetic field observed at $P$ at time $t$ to originate from the fractional charges within the volume element $\mathrm{d}\sigma\mathrm{d}r$ or from the charge element $\mathrm{d}q = \rho\,\mathrm{d}\sigma\mathrm{d}r$.

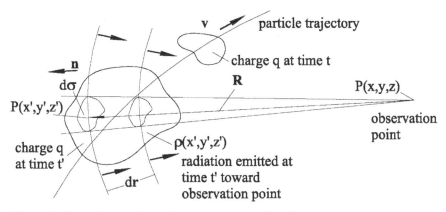

**Fig. 9.1.** Retarded position of a moving charge distribution

The radiation observed at point $P$ and time $t$ is the sum of all radiation arriving simultaneously at $P$. Elements of this radiation field may have been emitted by different charge elements and at different times and locations. In case of only one electrical charge moving with velocity $v$, we have to include in the integration those charge elements that move across the inner shell surface into the volume $d\sigma dr$ during the time $dr/c$. For a uniform charge distribution this additional charge is $dq = \rho v n\, dt\, d\sigma$ where $n$ is the vector normal to the surface of the shell and pointing away from the observer

$$n = \frac{R}{R}.\tag{9.9}$$

With $dt = dr/c$ and $\beta = v/c$, we get then for both contributions to the charge element

$$dq = \rho(1 + n\beta)\, dr\, d\sigma\,.\tag{9.10}$$

Depending on the direction of the velocity vector $\beta$, we find an increase or a reduction in the radiation field from moving charges. We solve (9.10) for $\rho\, dr\, d\sigma$ and insert into the integrals (9.5, 9.6). Now we may use the assumption that the electrical charge is a point charge and get for the retarded potentials of a moving point charge $q$ at time $t$ and observation point $P$

$$\boldsymbol{A}(P,t) = \frac{1}{[4\pi c\epsilon_0]}\frac{q}{R}\frac{\beta}{1+n\beta}\bigg|_{t_r}\tag{9.11}$$

and

$$\varphi(P,t) = \frac{1}{[4\pi c\epsilon_0]}\frac{1}{c}\frac{q}{R}\frac{1}{1+n\beta}\bigg|_{t_r}.\tag{9.12}$$

These equations are known as the Liénard Wiechert potentials and express the field potentials of a moving charge as functions of the charge parameters at the retarded time. To obtain the electric and magnetic fields we insert the retarded potentials into (9.1, 9.2) noting that the differentiation must be performed with respect to the time $t$ and location $P$ of the observer while the potentials are expressed at the retarded time $t_\mathrm{r}$.

In both equations for the vector and scalar potential we have the same denominator

$$r = R(1 + \boldsymbol{n}\boldsymbol{\beta}) \,. \tag{9.13}$$

It will become necessary to calculate the derivative of the retarded time with respect to the time $t$ and since $t_\mathrm{r} = t - R/c$ the time derivative of $t_\mathrm{r}$ is

$$\frac{\mathrm{d}t_\mathrm{r}}{\mathrm{d}t} = 1 - \frac{1}{c}\frac{\mathrm{d}R}{\mathrm{d}t_\mathrm{r}}\frac{\mathrm{d}t_\mathrm{r}}{\mathrm{d}t} \,. \tag{9.14}$$

The variation of the distance $R$ with the retarded time depends on the velocity $v$ of the moving charge and is the projection of the vector $\boldsymbol{v}\,\mathrm{d}\,t_\mathrm{r}$ onto the unity vector $\boldsymbol{n}$. Therefore,

$$\mathrm{d}R = \boldsymbol{v}\boldsymbol{n}\,\mathrm{d}t_\mathrm{r} \,. \tag{9.15}$$

and (9.14) becomes

$$\frac{\mathrm{d}t_\mathrm{r}}{\mathrm{d}t} = \frac{1}{1 + \boldsymbol{n}\boldsymbol{\beta}} = \frac{R}{r} \,. \tag{9.16}$$

The electric field (9.2) is with (9.11, 9.12) and (9.16) after a few manipulations expressed by

$$[4\pi\epsilon_0]\frac{\boldsymbol{E}}{q} = -\frac{1}{c}\frac{R}{r^2}\frac{\partial\boldsymbol{\beta}}{\partial t_\mathrm{r}} + \frac{\boldsymbol{\beta}\boldsymbol{R}}{cr^3}\frac{\partial r}{\partial t_\mathrm{r}} + \frac{1}{r^2}\boldsymbol{\nabla}_\mathrm{r}r \,. \tag{9.17}$$

In evaluating the nabla operator and other differentials we remember that all parameters on the r.h.s. must be taken at the retarded time (9.7) which itself depends on the location of the observation point $P$. To distinguish between the ordinary nabla operator and the case where the dependence of the retarded time on the position $P(x,y,z)$ must be considered, we add to the nabla symbol the index $_\mathrm{r}$ like $\boldsymbol{\nabla}_\mathrm{r}$. The components of this operator are then $\frac{\partial}{\partial x}\big|_\mathrm{r} = \frac{\partial}{\partial x} + \frac{\partial t_\mathrm{r}}{\partial x}\frac{\partial}{\partial t_\mathrm{r}}$, and similar for the other components. We evaluate first

$$\boldsymbol{\nabla}_\mathrm{r}r = \boldsymbol{\nabla}_\mathrm{r}R + \boldsymbol{\nabla}_\mathrm{r}(\beta\boldsymbol{R}) \tag{9.18}$$

and with $\boldsymbol{\nabla}R = -\boldsymbol{n}$ from (9.8)

$$\boldsymbol{\nabla}_\mathrm{r}R = -\boldsymbol{n} + \frac{\partial R}{\partial t_\mathrm{r}}\boldsymbol{\nabla}t_\mathrm{r} \,. \tag{9.19}$$

For the gradient of the retarded time, we get

$$\boldsymbol{\nabla} t_{\mathrm{r}} = \boldsymbol{\nabla}\left[ t - \frac{1}{c}R(t_{\mathrm{r}}) \right] = -\frac{1}{c}\boldsymbol{\nabla}_{\mathrm{r}}R = -\frac{1}{c}\left( -\boldsymbol{n} + \frac{\partial R}{\partial t_{\mathrm{r}}}\boldsymbol{\nabla} t_{\mathrm{r}} \right) \tag{9.20}$$

and performing the differentiation we get with $\frac{\partial x_{\mathrm{r}}}{\partial t_{\mathrm{r}}} = v_{x}$, ...

$$\frac{\partial R}{\partial t_{\mathrm{r}}} = \frac{\partial R}{\partial x_{\mathrm{r}}}\frac{\partial x_{\mathrm{r}}}{\partial t_{\mathrm{r}}} + \frac{\partial R}{\partial y_{\mathrm{r}}}\frac{\partial y_{\mathrm{r}}}{\partial t_{\mathrm{r}}} + \frac{\partial R}{\partial z_{\mathrm{r}}}\frac{\partial z_{\mathrm{r}}}{\partial t_{\mathrm{r}}} = \boldsymbol{n}\boldsymbol{v}\,. \tag{9.21}$$

Solving (9.20) for $\boldsymbol{\nabla} t_{\mathrm{r}}$ we get

$$\boldsymbol{\nabla} t_{\mathrm{r}} = \frac{\boldsymbol{R}}{cr} \tag{9.22}$$

and (9.19) becomes finally

$$\boldsymbol{\nabla}_{\mathrm{r}}R = -\boldsymbol{n} + \frac{\boldsymbol{R}}{r}(\boldsymbol{\beta}\boldsymbol{n})\,. \tag{9.23}$$

For the second term in (9.18) we note that the velocity $\boldsymbol{v}$ does not depend on the location of the observer and with $\boldsymbol{\nabla}_{\mathrm{r}}\boldsymbol{R}-1$, (9.22) and

$$\frac{\mathrm{d}\boldsymbol{R}}{\mathrm{d}t_{\mathrm{r}}} = \boldsymbol{v} \tag{9.24}$$

we get for the second term in (9.18)

$$\boldsymbol{\nabla}_{\mathrm{r}}(\boldsymbol{\beta}\boldsymbol{R}) = -\boldsymbol{\beta} + \frac{\partial(\boldsymbol{\beta}\boldsymbol{R})}{\partial t_{\mathrm{r}}}\boldsymbol{\nabla} t_{\mathrm{r}} = -\boldsymbol{\beta} + \left( \boldsymbol{R}\frac{\partial\boldsymbol{\beta}}{\partial t_{\mathrm{r}}} \right)\frac{\boldsymbol{R}}{cr} + \beta^{2}\frac{\boldsymbol{R}}{r}\,. \tag{9.25}$$

To complete the evaluation of the electric field in (9.17), we express the derivative $\frac{\partial r}{\partial t_{\mathrm{r}}}$ with

$$\frac{\partial r}{\partial t_{\mathrm{r}}} = \frac{\partial R}{\partial t_{\mathrm{r}}} + \frac{\partial(\boldsymbol{\beta}\boldsymbol{R})}{\partial t_{\mathrm{r}}} = c\boldsymbol{n}\boldsymbol{\beta} + c\beta^{2} + \boldsymbol{R}\frac{\partial\boldsymbol{\beta}}{\partial t_{\mathrm{r}}}\,, \tag{9.26}$$

where we made use of (9.21). Collecting all differential expressions required in (9.17) we get with (9.18, 9.23, 9.25, 9.26)

$$[4\pi\epsilon_{0}]\frac{\boldsymbol{E}}{q} = \frac{1}{r^{2}}\left[ -\boldsymbol{n} - \boldsymbol{\beta} + \frac{\boldsymbol{R}}{r}\left( \boldsymbol{n}\boldsymbol{\beta} + \beta^{2} + \tfrac{1}{c}\dot{\boldsymbol{\beta}}\boldsymbol{R} \right) \right]_{\mathrm{r}}$$
$$-\frac{\boldsymbol{R}}{cr^{2}}\dot{\boldsymbol{\beta}} + \boldsymbol{\beta}\frac{\boldsymbol{R}}{r^{3}}\left( \boldsymbol{n}\boldsymbol{\beta} + \beta^{2} + \tfrac{1}{c}\dot{\boldsymbol{\beta}}\boldsymbol{R} \right)_{\mathrm{r}}\,, \tag{9.27}$$

where $\dot{\boldsymbol{\beta}} = \mathrm{d}\boldsymbol{\beta}/\mathrm{d}t_{\mathrm{r}}$. After some manipulation and using (B.10), the equation for the electrical field of a charge $q$ moving with velocity $\boldsymbol{v}$ becomes

$$[4\pi\epsilon_{0}]\frac{\boldsymbol{E}}{q} = \frac{1-\beta^{2}}{r^{3}}(\boldsymbol{R} + R\boldsymbol{\beta})_{\mathrm{r}} + \frac{1}{cr^{3}}\left\{ \boldsymbol{R} \times \left[ (\boldsymbol{R} + R\boldsymbol{\beta})_{\mathrm{r}} \times \frac{\mathrm{d}\boldsymbol{\beta}}{\mathrm{d}t_{\mathrm{r}}} \right] \right\}\bigg|_{\mathrm{r}}\,, \tag{9.28}$$

where we have added the index $_r$ as a reminder that all quantities on the r.h.s. of (9.28) must be taken at the retarded time $t_r$.

This equation for the electric field of a moving charge has two distinct parts. The first part is inversely proportional to the square of the distance between radiation source and observer and depends only on the velocity of the charge. For a charge at rest $\beta=0$ this term reduces to the Coulomb field of a point charge $q$. The area close to the radiating charge where this term is dominant is called the Coulomb regime. The field is directed toward the observer for a positive charge at rest and tilts into the direction of propagation as the velocity of the charge increases. For highly relativistic particles we note the Coulomb field becomes very small.

We will not further consider this regime since we are interested only in the radiation field far away from the moving charge. The second term in (9.28) is inversely proportional to the distance from the charge and depends on the velocity as well as on the acceleration of the charge. This term scales linear with the distance $r$ falling off much slower than the Coulomb term and therefore reaches out to large distances from the radiation source. We call this regime the radiation regime and the remainder of this chapter will focus on the discussion of the radiation from moving charges. The electrical field in the radiation regime is

$$[4\pi\epsilon_0]\left.\frac{\boldsymbol{E}(t)}{q}\right|_{\text{rad}} = \frac{1}{cr^3}\left\{\boldsymbol{R}\times\left[(\boldsymbol{R}+R\boldsymbol{\beta})_{\text{r}}\times\frac{\mathrm{d}\boldsymbol{\beta}}{\mathrm{d}t_{\text{r}}}\right]\right\}\bigg|_{\text{r}}. \qquad (9.29)$$

The polarization of the electric field at the location of the observer is purely orthogonal to the direction of observation $\boldsymbol{R}$. Similar to the derivation of the electric field, we can derive the expression for the magnetic field and get from (9.1) with (9.11)

$$\boldsymbol{B} = \boldsymbol{\nabla}_{\text{r}}\times\boldsymbol{A} = q\left[\boldsymbol{\nabla}_{\text{r}}\times\frac{\boldsymbol{\beta}}{r}\right] = \frac{q}{r}\left[\boldsymbol{\nabla}_{\text{r}}\times\boldsymbol{\beta}\right] - \frac{q}{r^2}\left[\boldsymbol{\nabla}_{\text{r}}r\times\boldsymbol{\beta}\right], \qquad (9.30)$$

where again all parameters on the r.h.s. must be evaluated at the retarded time. The evaluation of the "retarded" curl operation $\boldsymbol{\nabla}_{\text{r}}\times\boldsymbol{\beta}$ becomes obvious if we evaluate one component only, for example, the $x$ component

$$\left(\frac{\partial}{\partial y}+\frac{\partial t_{\text{r}}}{\partial y}\frac{\partial}{\partial t_{\text{r}}}\right)\beta_z - \left(\frac{\partial}{\partial z}+\frac{\partial t_{\text{r}}}{\partial z}\frac{\partial}{\partial t_{\text{r}}}\right)\beta_y = [\boldsymbol{\nabla}\times\boldsymbol{\beta}]_x + \left[\boldsymbol{\nabla}t_{\text{r}}\times\frac{\mathrm{d}\boldsymbol{\beta}}{\mathrm{d}t_{\text{r}}}\right]_x. \quad (9.31)$$

In a similar way, we get the other components and find with (9.22) and the fact that the particle velocity $\boldsymbol{\beta}$ does not depend on the coordinates of the observation point ($\boldsymbol{\nabla}\times\boldsymbol{\beta}=0$),

$$[\boldsymbol{\nabla}_{\text{r}}r\times\boldsymbol{\beta}] = [\boldsymbol{\nabla}\times\boldsymbol{\beta}] + \left[\boldsymbol{\nabla}t_{\text{r}}\times\frac{\mathrm{d}\boldsymbol{\beta}}{\mathrm{d}t_{\text{r}}}\right] = \frac{1}{cr}\left[\boldsymbol{R}\times\frac{\mathrm{d}\boldsymbol{\beta}}{\mathrm{d}t_{\text{r}}}\right],$$

The gradient $\boldsymbol{\nabla}_{\text{r}}r$ has been derived earlier in (9.18) and inserting this into (9.30) we find the magnetic field of an electrical charge moving with velocity $\mathbf{v}$

$$[4\pi c\epsilon_0]\, \frac{\boldsymbol{B}}{q} = -\frac{1}{r^2}\,(\boldsymbol{\beta}\times\boldsymbol{n}) - \frac{R}{cr^2}\left[\frac{\mathrm{d}\boldsymbol{\beta}}{\mathrm{d}t}\times\boldsymbol{n}\right]\Bigg|_{\mathrm{r}}$$
$$+\frac{R}{r^3}\left(\boldsymbol{\beta}\boldsymbol{n}+\beta^2+\frac{1}{c}\frac{\mathrm{d}\boldsymbol{\beta}}{\mathrm{d}t}\,R\right)[\boldsymbol{\beta}\times\boldsymbol{n}]\Bigg|_{\mathrm{r}}. \tag{9.32}$$

Again, there are two distinct groups of field terms. In case of the electrical field the terms that fall off like the square of the distance are the Coulomb fields. For magnetic fields such terms appear only if the charge is moving $\beta \neq 0$ and are identical to the Biot Savart fields. Here we concentrate only on the far fields or radiation fields which decay inversely proportional to the distance from the source. The magnetic radiation field is then given by

$$[4\pi c\epsilon_0]\,\frac{\boldsymbol{B}(t)}{q}\Bigg|_{\mathrm{rad}} = -\frac{R}{cr^2}\left[\frac{\mathrm{d}\boldsymbol{\beta}}{\mathrm{d}t}\times\boldsymbol{n}\right]_{\mathrm{r}} + \frac{R}{cr^3}\left(\frac{\mathrm{d}\boldsymbol{\beta}}{\mathrm{d}t}\,R\right)[\boldsymbol{\beta}\times\boldsymbol{n}]_{\mathrm{r}}. \tag{9.33}$$

Comparing the magnetic field (9.33) with the electrical field (9.28) reveals a very simple correlation between both fields. The magnetic field can be obtained from the electric field, and vice versa, by mere vector multiplication with the unit vector $\boldsymbol{n}$

$$\boldsymbol{B} = \tfrac{1}{[c]}\,[\boldsymbol{E}\times\boldsymbol{n}]_{\mathrm{r}}. \tag{9.34}$$

From this equation we can deduce special properties for the field directions by noting that the electric and magnetic fields are orthogonal to each other and both are orthogonal to the direction of observation $\boldsymbol{n}$. The existence of electric and magnetic fields can give rise to radiation for which the Poynting vector is

$$\boldsymbol{S} = [4\pi c\epsilon_0]\,\frac{c}{4\pi}\,[\boldsymbol{E}\times\boldsymbol{B}]_{\mathrm{r}} = [4\pi c\epsilon_0]\,\frac{1}{4\pi}\,[\boldsymbol{E}\times(\boldsymbol{E}\times\boldsymbol{n})]_{\mathrm{r}}. \tag{9.35}$$

Using again the vector relation $(B.10)$ and noting that the electric field is normal to $\boldsymbol{n}$, we get for the Poynting vector or the radiation flux in the direction to the observer

$$\boldsymbol{S} = -[4\pi c\epsilon_0]\,\frac{c}{4\pi}\,E_{\mathrm{r}}^2\boldsymbol{n}\Big|_{\mathrm{r}}. \tag{9.36}$$

Equation (9.36) defines the energy flux density measured at the observation point $P$ and time $t$ in form of synchrotron radiation per unit cross section and parallel to the direction of observation $\boldsymbol{n}$. All quantities expressing this energy flux are still to be taken at the retarded time. For practical reasons it becomes desirable to express the Poynting vector at the retarded time as well. The energy flux at the observation point, in terms of the retarded time is then $\mathrm{d}W/\mathrm{d}t_{\mathrm{r}} = (\mathrm{d}W/\mathrm{d}t)\,(\mathrm{d}t/\mathrm{d}t_{\mathrm{r}})$ and instead of (9.36) we express the Poynting vector with (9.16) like

$$\boldsymbol{S}_{\mathrm{r}} = \boldsymbol{S}\frac{\mathrm{d}t}{\mathrm{d}t_{\mathrm{r}}} = -[4\pi c\epsilon_0]\,\frac{c}{4\pi}\boldsymbol{E}^2\,[\,(1+\boldsymbol{\beta}\boldsymbol{n})\,\boldsymbol{n}\,]_{\mathrm{r}}. \tag{9.37}$$

The Poynting vector in this form can be readily used for calculations like those determining the spatial distribution of the radiation power.

## 9.2 Total Radiation Power and Energy Loss

So far, no particular choice of the reference system has been assumed, but a particularly simple reference frame $\mathcal{L}^*$ is the one which moves uniformly with the charge before acceleration. From now on, we use a single particle with a charge $e$. To an observer in this reference system, the charge moves due to acceleration and the electric field in the radiation regime is from (9.29)

$$\boldsymbol{E}^*(t) = \frac{1}{[4\pi\epsilon_0]}\frac{e}{cR}\left[\boldsymbol{n}\times\left(\boldsymbol{n}\times\frac{\mathrm{d}\boldsymbol{\beta}^*}{\mathrm{d}t}\right)\right]\bigg|_{\mathrm{r}}. \tag{9.38}$$

The synchrotron radiation power per unit solid angle and at distance $R$ from the source is from (9.37) with $\boldsymbol{v} = 0$

$$\frac{\mathrm{d}P^*}{\mathrm{d}\Omega} = -\boldsymbol{n}\boldsymbol{S}^*\,R_{\mathrm{r}}^2 = [4\pi c\epsilon_0]\frac{c}{4\pi}\,\boldsymbol{E}^{*2}R^2\big|_{\mathrm{r}}. \tag{9.39}$$

Introducing the classical particle radius $r_c mc^2 = e^2/[4\pi\epsilon_0]$ to obtain expressions which are independent of electromagnetic units and with (9.38)

$$\frac{\mathrm{d}P^*}{\mathrm{d}\Omega} = \frac{r_c mc^2}{4\pi c}\left|\boldsymbol{n}\times\left(\boldsymbol{n}\times\frac{\mathrm{d}\boldsymbol{\beta}^*}{\mathrm{d}t}\right)\right|_{\mathrm{r}}^2 = \frac{r_c mc^2}{4\pi c}\left|\frac{\mathrm{d}\boldsymbol{\beta}^*}{\mathrm{d}t}\right|_{\mathrm{r}}^2\sin^2\vartheta_{\mathrm{r}}, \tag{9.40}$$

where $\vartheta_{\mathrm{r}}$ is the retarded angle between the direction of acceleration and the direction of observation $\boldsymbol{n}$. Integration over all solid angles gives the total radiated power. With $\mathrm{d}\Omega = \sin\vartheta_{\mathrm{r}}\mathrm{d}\vartheta_{\mathrm{r}}\mathrm{d}\phi$, where $\phi$ is the azimuthal angle with respect to the direction of acceleration, the total radiation power is in agreement with (3.1)

$$P^* = \frac{2}{3}r_c mc\left|\frac{\mathrm{d}\boldsymbol{\beta}^*}{\mathrm{d}t}\right|_{\mathrm{r}}^2. \tag{9.41}$$

This equation has been derived first by Larmor [49] within the realm of classical electrodynamics. The emission of a quantized photon, however, exerts a recoil on the electron varying its energy slightly. Schwinger [50] investigated this effect and derived a correction to the radiation power like

$$P^* = P^*_{\mathrm{classical}}\left(1 - \frac{55}{16\sqrt{3}}\frac{\epsilon_c}{E}\right), \tag{9.42}$$

where $\epsilon_c$ is the critical photon energy and $E$ the electron energy. The correction is generally very small and we ignore therefore this quantum mechanical effect in our discussions.

### 9.2.1 Transition Radiation

Digressing slightly from the discussion of synchrotron radiation we turn our attention to the solution of (9.39). Generally, we do not know the fields $\boldsymbol{E}^*$

and to solve (9.40) we need to know more about the particular trajectory of the particle motion. In the case of transition radiation, we have, however, all information to formulate a solution. Transition radiation is emitted when a charged particle passes through the boundary of two media with different dielectric constant. We will not go into the detailed general theory of transition radiation but concentrate on the case where a charged particle passes through a thin metallic foil in vacuum. As the particle passes through the foil backward transition radiation is emitted when the particle enters the foil and forward radiation is emitted when it appears on the other side. The emitted radiation energy can be derived directly from (9.39). First, we replace the electric radiation field by the magnetic field component and (9.39) becomes simply

$$\frac{\mathrm{d}\varepsilon(t)}{\mathrm{d}t} = [4\pi c\epsilon_0] \frac{c}{4\pi} \, \boldsymbol{B}^{*2}(t) R^2 \big|_{\mathrm{r}} \, \mathrm{d}\Omega \, . \tag{9.43}$$

From Parceval's theorem ($B.32$) we know that

$$\int_{-\infty}^{\infty} B^2(t) \, \mathrm{d}t = \frac{1}{2\pi} \int_{-\infty}^{\infty} B^2(\omega) \, \mathrm{d}\omega \, , \tag{9.44}$$

where $B(t) = \frac{1}{2\pi} \int B(\omega) \, \mathrm{e}^{-\mathrm{i}\omega t} \, \mathrm{d}\omega$ and $B(\omega) = \int B(t) \, \mathrm{e}^{\mathrm{i}\omega t} \, \mathrm{d}t$. The emission of transition radiation occurs in a very short time $\tau = \omega_{\mathrm{p}}^{-1}$, where $\omega_{\mathrm{p}}$ is the plasma frequency. For this reason, the transition radiation frequency reaches into the x-ray regime. We limit ourselves here to frequencies $\omega$, which are much lower such that $\tau \ll \omega^{-1}$. The magnetic field is nonzero only during the emission process and we can therefore set

$$B(\omega) = \int_{-\infty}^{\infty} B(t) \, e^{\mathrm{i}\omega t} \, \mathrm{d}t \approx \int_{-\tau/2}^{\tau/2} B(t) \, \mathrm{d}t \, . \tag{9.45}$$

To solve this integral we recall the definition of the vector potential $\boldsymbol{B}(t) = \nabla \times \boldsymbol{A}_{\mathrm{r}}$ and keep in mind that all quantities are to be taken at the retarded time. Expressing in component form
$\nabla \times \boldsymbol{A}_{\mathrm{r}} = \left\{ \frac{\partial A_z}{\partial y} - \frac{\partial A_y}{\partial z}, \frac{\partial A_x}{\partial z} - \frac{\partial A_z}{\partial x}, \frac{\partial A_x}{\partial z} - \frac{\partial A_z}{\partial x} \right\}_{t=t-\frac{1}{c}R(t)}$ the derivatives are
$\frac{\partial A_z}{\partial y} = \frac{\partial A_z}{\partial t_{\mathrm{r}}} \frac{\partial t_{\mathrm{r}}}{\partial y}$ etc. With $\frac{\partial t_{\mathrm{r}}}{\partial y} = \frac{1}{c} \frac{y_{\mathrm{r}} - y}{R} = \frac{n_y}{c}$ we get $\frac{\partial A_z}{\partial y} - \frac{\partial A_y}{\partial z} = \frac{1}{c} \frac{\partial A_z}{\partial t_{\mathrm{r}}} n_y - \frac{1}{c} \frac{\partial A_y}{\partial t_{\mathrm{r}}} n_z$ or finally

$$\boldsymbol{B}(t) = \nabla \times \boldsymbol{A}_{\mathrm{r}} = \frac{1}{c} \boldsymbol{n}_{\mathrm{r}} \times \frac{\partial}{\partial t_{\mathrm{r}}} \boldsymbol{A}_{\mathrm{r}} = \frac{1}{c} \frac{\partial}{\partial t_{\mathrm{r}}} \left[ \boldsymbol{n} \times \boldsymbol{A} \right]_{\mathrm{r}} \, . \tag{9.46}$$

The magnetic field spectrum (9.45) becomes then simply

$$B(\omega) = \int_{-\tau/2}^{\tau/2} B(t) \, \mathrm{d}t = \frac{1}{c} \left[ \boldsymbol{n} \times \boldsymbol{A} \right]_{\mathrm{r}} \big|_{\mathrm{initial}}^{\mathrm{final}} \, . \tag{9.47}$$

Initially, while the electron has not yet vanished into the metallic foil, the vector potential is made up of the Liénard Wiechert potentials of a free electron and its image charge (a positron) moving in the opposite direction. The vector potential is therefore

$$\frac{\boldsymbol{A}}{[4\pi c\epsilon_0]} = \underbrace{\frac{e\boldsymbol{\beta}}{R(1+\boldsymbol{\beta n})}}_{\text{electron}} + \underbrace{\frac{e\boldsymbol{\beta}}{R(1-\boldsymbol{\beta n})}}_{\text{positron}} . \tag{9.48}$$

Instead of (9.43) we use the spectral radiation energy $d\varepsilon(\omega) = [4\pi c\epsilon_0]\frac{c}{4\pi}R^2$ $d\Omega\frac{1}{2\pi}\boldsymbol{B}_r^{*2}(t)\,d\omega\,2$, where the extra factor of two comes from using only positive frequencies $\omega > 0$, and get with (9.48) and $e^2 = r_c mc^2\,[4\pi\epsilon_0]$

$$\frac{d^2\varepsilon}{d\omega d\Omega} = \frac{1}{4\pi^2}\frac{r_c mc^2}{c}\left\{\frac{\boldsymbol{n}\times\boldsymbol{\beta}}{1+\boldsymbol{\beta n}} + \frac{\boldsymbol{n}\times\boldsymbol{\beta}}{1-\boldsymbol{\beta n}}\right\}^2$$

$$= \frac{r_c mc^2}{\pi^2 c}\underbrace{|\boldsymbol{n}\times\boldsymbol{z}|^2}_{\sin^2\vartheta}\left(\frac{\beta}{1-\beta^2\underbrace{(\boldsymbol{nz})^2}_{\cos^2\vartheta}}\right)^2 ,$$

where we used $\boldsymbol{\beta}\approx\beta\boldsymbol{z}$ and where $\boldsymbol{z}$ is the unit vector along the $z$-axis. The emission angle $\vartheta$ is taken with respect to the $z$-axis. The spectral and spatial transition radiation distribution from a single electron is finally

$$\frac{d^2\varepsilon}{d\omega d\Omega} = \frac{r_c mc^2}{\pi^2 c}\frac{\beta^2\sin^2\vartheta}{\left(1-\beta^2\cos^2\vartheta\right)^2} . \tag{9.49}$$

The spatial radiation distribution of transition radiation is shown in Fig. 9.2. No radiation is emitted along the axis $\vartheta = 0$ while the radiation intensity reaches a maximum at an emission angle of $1/\gamma$. Equation (9.49) does not exhibit any frequency dependence, which is due to the fact that the emission process occurs in a very short time generating a uniform spectrum. Very high frequencies in the x-ray regime, where the spectral intensity is expected to drop, have been excluded in this derivation.

Integrating (9.49) over a half space, we get

$$\frac{d\varepsilon}{d\omega} = \frac{2r_c mc^2}{\pi c}\int_0^{\pi/2}\frac{\beta^2\sin^2\vartheta}{\left(1-\beta^2\cos^2\vartheta\right)^2}\sin\vartheta\,d\vartheta$$

$$= \frac{2r_c mc^2}{\pi c}\frac{1}{4\beta}\left[(1+\beta^2)\ln\frac{1+\beta}{1-\beta} - 2\beta\right] , \tag{9.50}$$

which is for relativistic particles $\gamma \gg 1$

$$\frac{d\varepsilon(\omega)}{d\omega} \approx \frac{2r_c mc^2}{\pi c}\ln\gamma . \tag{9.51}$$

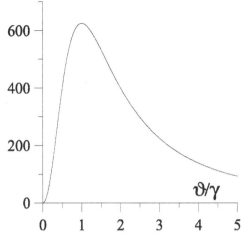

**Fig. 9.2.** Intensity distribution $\frac{\mathrm{d}^2\varepsilon}{\mathrm{d}\omega\mathrm{d}\Omega}\frac{\pi^2 c}{r_c m c^2}$ of transition radiation

The spectral energy emitted into one half space by a single electron in form of transition radiation is uniform for all frequencies reaching up into the soft x-ray regime and depends only logarithmically on the particle energy.

### 9.2.2 Synchrotron Radiation Power

Coming back to synchrotron radiation we must define the electron motion in great detail. It is this motion which determines many of the photon beam characteristics. The radiation power and spatial distribution in the electron system is identical to that from a linear microwave antenna being emitted normal to the direction of acceleration with a $\sin^2$-distribution.

In Sect. C.8 we have shown that the radiation power is invariant to Lorentz transformations, we may set $P = P^*$ and the total radiation power in the laboratory system is

$$P = \tfrac{2}{3} r_c m c \gamma^6 \left[\dot{\boldsymbol{\beta}}^2 - (\boldsymbol{\beta} \times \dot{\boldsymbol{\beta}})^2\right] , \tag{9.52}$$

which has been discussed before leading to (3.2). Equation (9.52) expresses the radiation power in a simple way and allows us to calculate other radiation characteristics based on beam parameters in the laboratory system. Specifically, we will distinguish between acceleration parallel $\left.\frac{\mathrm{d}\boldsymbol{\beta}}{\mathrm{d}t}\right|_{\parallel}$ and perpendicular $\left.\frac{\mathrm{d}\boldsymbol{\beta}}{\mathrm{d}t}\right|_{\perp}$ to the propagation $\boldsymbol{\beta}$ of the charge and set therefore

$$\frac{\mathrm{d}\boldsymbol{\beta}}{\mathrm{d}t} = \left.\frac{\mathrm{d}\boldsymbol{\beta}}{\mathrm{d}t}\right|_{\parallel} + \left.\frac{\mathrm{d}\boldsymbol{\beta}}{\mathrm{d}t}\right|_{\perp} . \tag{9.53}$$

Insertion into (9.52) shows the total radiation power to consist of separate contributions from parallel and orthogonal acceleration. Separating both contributions, we get the synchrotron radiation power for both parallel and transverse acceleration, respectively

$$P_\parallel = \tfrac{2}{3} r_c m c \gamma^6 \left. \frac{\mathrm{d}\boldsymbol{\beta}}{\mathrm{d}t} \right|_\parallel^2 , \tag{9.54}$$

$$P_\perp = \tfrac{2}{3} r_c m c \gamma^4 \left. \frac{\mathrm{d}\boldsymbol{\beta}}{\mathrm{d}t} \right|_\perp^2 . \tag{9.55}$$

Expressions have been derived that define the radiation power for parallel acceleration like in a linear accelerator or orthogonal acceleration found in circular accelerators or deflecting systems. We note a similarity for both contributions except for the energy dependence. At relativistic energies, the same acceleration force leads to much less radiation if the acceleration is parallel to the motion of the particle compared to orthogonal acceleration. Parallel acceleration is related to the accelerating force $F_\parallel$ by $\boldsymbol{v}_\parallel = \frac{1}{\gamma^3} \frac{\mathrm{d}\boldsymbol{p}_\parallel}{\mathrm{d}t}$ and after insertion into (9.54) the radiation power due to parallel acceleration becomes

$$P_\parallel = \frac{2}{3} \frac{r_c \, c}{m c^2} \left( \frac{\mathrm{d}\boldsymbol{p}_\parallel}{\mathrm{d}t} \right)^2 . \tag{9.56}$$

The radiation power for acceleration along the propagation of the charged particle is therefore independent of the energy of the particle and depends only on the accelerating force or with $\mathrm{d}\boldsymbol{p}_\parallel/\mathrm{d}t = \beta c \, \mathrm{d}E/\mathrm{d}x$ on the energy increase per unit length, $\mathrm{d}E/\mathrm{d}x$, of the accelerator.

In contrast, we find very different radiation characteristics for transverse acceleration as it happens, for example, during the transverse deflection of a charged particle in a magnetic field. The transverse acceleration $\boldsymbol{v}_\perp$ is expressed by the Lorentz force

$$\frac{\mathrm{d}\boldsymbol{p}_\perp}{\mathrm{d}t} = \gamma m \dot{\boldsymbol{v}}_\perp = [c] \, e \, [\boldsymbol{\beta} \times \boldsymbol{B}] \tag{9.57}$$

and after insertion into (9.55) the radiation power from transversely accelerated particles becomes

$$P_\parallel = \tfrac{2}{3} r_c m c \gamma^2 \left( \frac{\mathrm{d}\boldsymbol{p}_\perp}{\mathrm{d}t} \right)^2 . \tag{9.58}$$

Comparing (9.56) with (9.58) we find that the same accelerating force leads to a much higher radiation power by a factor $\gamma^2$ for transverse acceleration with respect to longitudinal acceleration. For all practical purposes technical limitations prevent the occurrence of sufficient longitudinal acceleration to generate noticeable radiation. We express the deflecting magnetic field $\boldsymbol{B}$ by the bending radius $\rho$ and get the instantaneous synchrotron radiation power

$$P_\gamma = \tfrac{2}{3} r_c mc^2 \frac{c\,\beta^4\gamma^4}{\rho^2}, \tag{9.59}$$

or in more practical units

$$P_\gamma(\text{GeV/s}) = \frac{c\,C_\gamma}{2\pi} \frac{E^4}{\rho^2}, \tag{9.60}$$

where we use *Sands'* definition of the radiation constant [14]

$$C_\gamma = \frac{4\pi}{3} \frac{r_c}{(mc^2)^3} = 8.8575 \times 10^{-5} \quad \text{m/GeV}^3. \tag{9.61}$$

This numerical value is correct for relativistic electrons and positrons and must be modified for other particles.

From here on we will stop considering longitudinal acceleration unless specifically mentioned and replace therefore the index $\perp$ by setting $P_\perp = P_\gamma$. We also restrict from now on the discussion to singly charged particles and set $q = e$ ignoring extremely high energies where multiple charged ions start to radiate.

The electromagnetic radiation of charged particles in transverse magnetic fields is proportional to the fourth power of the particle momentum $\beta\gamma$ and inversely proportional to the square of the bending radius $\rho$. The radiation emitted by charged particles being deflected in magnetic fields is called synchrotron radiation. The synchrotron radiation power increases very fast for high energy particles and provides the most severe limitation to the maximum energy achievable in circular accelerators. We note also a strong dependence on the kind of particles involved in the process of radiation. Because of the much heavier mass of protons compared to the lighter electrons, we find appreciable synchrotron radiation only in circular electron accelerators. The radiation power of protons actually is smaller compared to that for electrons by the fourth power of the mass ratio or by the factor

$$\frac{P_e}{P_p} = 1836^4 = 1.36 \; 10^{13}. \tag{9.62}$$

In spite of this enormous difference measurable synchrotron radiation has been predicted by Coisson [15] and was indeed detected at the 400 GeV proton synchrotron SPS at CERN [16, 17]. Substantial synchrotron radiation is expected in circular proton accelerators at a beam energy of 10 TeV and more.

The knowledge of the synchrotron radiation power allows us now to calculate the energy loss of a particle per turn in a circular accelerator by integrating the radiation power along the circumference $L_0$ of the circular accelerator

$$\Delta E = \oint P_\gamma \, \mathrm{d}t = \tfrac{2}{3} r_c \, mc^2 \beta^3 \gamma^4 \int_{L_0} \frac{\mathrm{d}s}{\rho^2}. \tag{9.63}$$

If we assume an isomagnetic lattice where the bending radius is the same for all bending magnets $\rho = $ const, and integrate around a circular accelerator, the energy loss per turn due to synchrotron radiation is given by

$$\Delta E = \tfrac{4\pi}{3} r_c \, mc^2 \beta^3 \frac{\gamma^4}{\rho} \, . \tag{9.64}$$

The integration obviously is to be performed only along those parts of the circular accelerator where synchrotron radiation occurs or along bending magnets only. In more practical units, the energy loss of relativistic electrons per revolution in a circular accelerator with an isomagnetic lattice and a bending radius $\rho$ is given by

$$\Delta E = C_\gamma \frac{E^4}{\rho} \, . \tag{9.65}$$

From this energy loss per particle in each turn we calculate the total synchrotron radiation power for a beam of $N_e$ particles. The total synchrotron radiation power for a single particle is its energy loss multiplied by the revolution frequency of the particle around the circular orbit. If $L_0$ is the circumference of the orbit we have for the revolution frequency $f_{\mathrm{rev}} = \beta c / L_0$ and for the circulating particle current $I = e f_{\mathrm{rev}} N_e$. The total synchrotron radiation power is then

$$P_\gamma(\mathrm{MW}) = C_\gamma \frac{E^4(\mathrm{GeV})}{\rho(\mathrm{m})} \, I(\mathrm{A}) \, . \tag{9.66}$$

The total synchrotron radiation power scales like the fourth power of energy and is inversely proportional to the bending radius. The strong dependence of the radiation on the particle energy causes severe practical limitations on the maximum achievable energy in a circular accelerator.

## 9.3 Radiation Lobes

Expressions for the radiation fields and Poynting vector exhibit strong vectorial dependencies on the directions of motion and acceleration of the charged particles and on the direction of observation. These vectorial dependencies indicate that the radiation may not be emitted isotropic but rather into specific directions forming characteristic radiation patterns. In this section we will derive these spatial radiation characteristics and determine the direction of preferred radiation emission.

In (9.40) the radiation power per unit solid angle is expressed in the reference frame of the particle

$$\frac{\mathrm{d}P}{\mathrm{d}\Omega} = \frac{r_c mc}{4\pi} \dot{\beta}_{\mathrm{r}}^{*2} \sin^2 \Theta \tag{9.67}$$

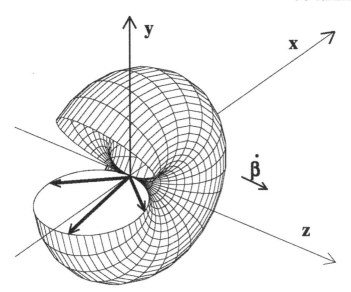

**Fig. 9.3.** Radiation pattern in the particle frame of reference or for nonrelativistic particles in the laboratory system

showing a particular directionality of the radiation as shown in Fig. 9.3. The radiation power is mainly concentrated in the $x, y$-plane and is proportional to $\sin^2 \Theta$ where $\Theta$ is the angle between the direction of acceleration, in this case the $z$-axis, and the direction of observation $n$. The radiation pattern in Fig. 9.3 is formed by the end points of vectors with the length $dP/d\Omega$ and angles $\Theta$ with respect to the $z$-axis. Because of symmetry, the radiation is isotropic with respect to the polar angle $\varphi$ and therefore, the radiation pattern is rotation symmetric about the direction of acceleration or in this case about the $z$-axis.

This pattern is the correct representation of the radiation for the reference frame of the radiating particle. We may, however, also consider this pattern as the radiation pattern from non relativistic particles like that from a linear radio antenna. For relativistic particles the radiation pattern differs significantly from the non relativistic case. The Poynting vector in the form of (9.37) can be used to calculate the radiation power per unit solid angle in the direction to the observer $-n$

$$\frac{dP}{d\Omega} = -\, nS\, R^2 \big|_{\mathrm{r}} = [4\pi c\epsilon_0]\, \frac{c}{4\pi}\, \boldsymbol{E}^2\, (1 + \boldsymbol{\beta n})\, R^2 \big|_{\mathrm{r}} \, . \tag{9.68}$$

We calculate the spatial distribution of the synchrotron radiation for the case of acceleration orthogonal to the propagation of the particle as it happens in beam transport systems where the particles are deflected by a transverse magnetic fields. The particle is assumed to be located at the origin of a right-handed coordinate system as shown in Fig. 9.4 propagating in the $z$-direction

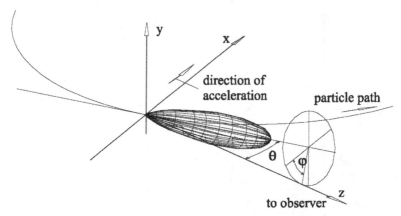

**Fig. 9.4.** Radiation geometry in the laboratory frame of reference for highly relativistic particles

and the orthogonal acceleration in this coordinate system occurs along the $x$-axis.

With the expression (9.29) for the electric fields in the radiation regime the spatial radiation power distribution (9.68) becomes

$$\frac{\mathrm{d}P}{\mathrm{d}\Omega} = \frac{c}{4\pi}r_\mathrm{c}mc^2\frac{R^5}{c^3r^5}\left\{\boldsymbol{n}\times\left[(\boldsymbol{n}+\boldsymbol{\beta})\times\dot{\boldsymbol{\beta}}\right]\right\}^2 . \tag{9.69}$$

We will now replace all vectors by their components to obtain the directional dependency of the synchrotron radiation. The vector $\boldsymbol{n}$ pointing from the observation point to the source point of the radiation has from Fig. 9.4 the components

$$\boldsymbol{n} = (-\sin\theta\cos\varphi, -\sin\theta\sin\varphi, \cos\theta) , \tag{9.70}$$

where the angle $\theta$ is the angle between the direction of particle propagation and the direction of emission of the synchrotron light $-\boldsymbol{n}$. The $x$-component of the acceleration can be derived from the Lorentz equation

$$\gamma m\dot{\boldsymbol{v}}_x = \frac{\mathrm{d}p_x}{\mathrm{d}t} = [c]\, e\beta_z B_y . \tag{9.71}$$

With $v_z \approx v$ we have $1/\rho = [c]\, eB_y/cp = [c]\, eB_y/(\gamma mcv)$ and the acceleration vector is

$$\dot{\boldsymbol{v}}_\perp = (\dot{v}, 0, 0) = \left(\frac{v^2}{\rho}, 0, 0\right) . \tag{9.72}$$

The velocity vector is

$$\boldsymbol{v} = (0, 0, v) \tag{9.73}$$

and after replacing the double vector product in (9.69) by a single vector sum

$$\boldsymbol{n} \times [(\boldsymbol{n} + \boldsymbol{\beta}) \times \boldsymbol{\beta}] = (\boldsymbol{n} + \boldsymbol{\beta})\,(\boldsymbol{n}\,\boldsymbol{\beta}) - \boldsymbol{\beta}(1 + \boldsymbol{n}\,\boldsymbol{\beta})\,, \tag{9.74}$$

we may now square the r.h.s. of (9.69) and replace all vectors by their components. The denominator in (9.69) then becomes

$$r^5 = R^5(1 + \boldsymbol{n}\boldsymbol{\beta})^5 = R^5(1 - \beta\cos\theta)^5\,, \tag{9.75}$$

and the full expression for the radiation power exhibiting the spatial distribution is finally

$$\frac{\mathrm{d}P}{\mathrm{d}\Omega} = \frac{r_c m c^2 c}{4\pi}\frac{\beta^4}{\rho^2}\frac{(1 - \beta\cos\theta)^2 - (1 - \beta^2)\sin^2\theta\cos^2\varphi}{(1 - \beta\cos\theta)^5}\,. \tag{9.76}$$

This equation describes the instantaneous synchrotron radiation power per unit solid angle from charged particles moving with velocity $v$ and being accelerated normal to the propagation by a magnetic field. The angle $\theta$ is the angle between the direction of observation $-\boldsymbol{n}$ and propagation $\boldsymbol{v}/v$. Integration over all angles results again in the total synchrotron radiation power (9.59).

In Fig. 9.5 the radiation power distribution is shown in real space as derived from (9.76). We note that the radiation is highly collimated in the forward direction along the $z$-axis which is also the direction of particle propagation. Synchrotron radiation in particle accelerators or beam lines is emitted whenever there is a deflecting electromagnetic field and emerges mostly tangentially from the particle trajectory. An estimate of the typical opening angle can be derived from (9.76). We set $\varphi = 0$ and expand the cosine function for small angles $\cos\theta \approx 1 - \frac{1}{2}\theta^2$. With $\beta \approx 1 - \frac{1}{2}\gamma^{-2}$ we find the radiation power to scale like $(\gamma^{-2} + \theta^2)^{-3}$. The radiation power therefore is reduced to about one eighth the peak intensity at an emission angle of $\theta_\gamma = 1/\gamma$ or virtually all synchrotron radiation is emitted within an angle of

$$\theta_\gamma = \pm\frac{1}{\gamma} \tag{9.77}$$

with respect to the direction of the particle propagation.

From Fig. 9.5 we observe a slightly faster fall off for an azimuthal angle of $\varphi = 0$ which is in the plane of particle acceleration and propagation. Although the synchrotron radiation is emitted symmetrically within a small angle of the order of $\pm\frac{1}{\gamma}$ with respect to the direction of particle propagation, the radiation pattern from a relativistic particle as observed in the laboratory is very different in the deflecting plane from that in the nondeflecting plane. While the particle radiates from every point along its path, the direction of this path changes in the deflecting plane but does not in the nondeflecting plane. The synchrotron radiation pattern from a bending magnet therefore

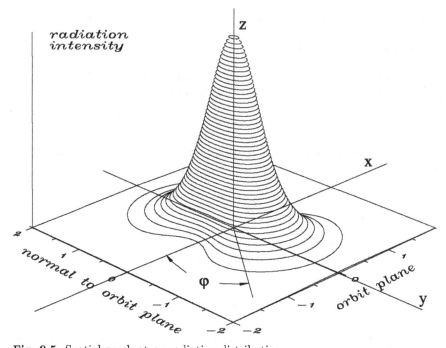

**Fig. 9.5.** Spatial synchrotron radiation distribution

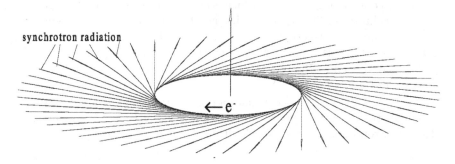

**Fig. 9.6.** Synchrotron radiation from a circular particle accelerator

resembles the form of a swath where the radiation is emitted evenly and tangentially from every point of the particle trajectory as shown in Fig. 9.6.

The extreme collimation of the synchrotron radiation and its high intensity in high energy electron accelerators can cause significant heating problems as well as desorption of gas molecules from the surface of the vacuum chamber. In addition, the high density of thermal energy deposition on the vacuum chamber walls can cause significant mechanical stresses causing cracks in the material. A careful design of the radiation absorbing surfaces to avoid damage to the integrity of the material is required. On the other

hand, this same radiation is a valuable source of photons for a wide variety
of research applications where, specifically, the collimation of the radiation
together with the small source dimensions are highly desired features of the
radiation.

## 9.4 Synchrotron Radiation Spectrum

Synchrotron radiation from relativistic charged particles is emitted over a
wide spectrum of photon energies. The basic characteristics of this spectrum
can be derived from simple principles as suggested in [19] and discussed in
Chap. 3. The spectrum extends from very low photon energies up to about
the critical photon energy

$$\varepsilon_c = \tfrac{3}{2}\hbar c \frac{\gamma^3}{\rho}. \tag{9.78}$$

The significance of the critical photon energy is its definition for the upper
bound for the synchrotron radiation spectrum. The spectral intensity falls
off rapidly for photon energies above the critical photon energy. In practical
units, the critical photon energy is

$$\epsilon_c(\mathrm{keV}) = 2.218 \frac{E^3(\mathrm{GeV})}{\rho(\mathrm{m})} = 0.665 \, E^2(\mathrm{GeV}) \, B(\mathrm{T}). \tag{9.79}$$

The synchrotron radiation spectrum from relativistic particles in a circular accelerator is made up of harmonics of the particle revolution frequency
$\omega_0$ and extends to values up to and beyond the critical frequency (9.78).
Generally, a real synchrotron radiation beam from say a storage ring will
not display this harmonic structure. The distance between the harmonics is
extremely small compared to the extracted photon frequencies in the VUV
and x-ray regime while the line width is finite due to the energy spread in a
beam of many particles and the spectrum becomes therefore continuous. For
a single pass of particles through a bending magnet in a beam transport line,
we observe the same spectrum, although now genuinely continuous as can be
derived with the use of Fourier transforms of a single light pulse. Specifically,
the maximum frequency is the same assuming similar parameters.

## 9.5 Radiation Field in the Frequency Domain

Synchrotron radiation is emitted within a wide range of frequencies. As we
have seen in the previous paragraph, a particle orbiting in a circular accelerator emits light flashes at the revolution frequency. We expect therefore in
the radiation frequency spectrum all harmonics of the revolution frequency

up to very high frequencies limited only by the very short duration of the radiation pulse being sent into a particular direction toward the observer. The number of harmonics increases with beam energy and reaches at the critical frequency the order of $\gamma^3$.

The frequency spectrum of synchrotron radiation has been derived by many authors. In this text, we will stay closer to the derivation by Jackson [19] than others. The general method to derive the frequency spectrum is to transform the electric field from the time domain to the frequency domain by the use of Fourier transforms. Applying this method, we will determine the radiation characteristics of the light emitted by a single pass of a particle in a circular accelerator at the location of the observer. The electric field at the observation point has a strong time dependence and is given by (9.29) while the total radiation energy for one pass is from (9.38)

$$\frac{dW}{d\Omega} = -\int_{-\infty}^{\infty} \frac{dP}{d\Omega}\, dt = \int_{-\infty}^{\infty} \boldsymbol{S}_r \boldsymbol{n}\, R^2 dt = [4\pi c\epsilon_0]\frac{cR^2}{4\pi}\int_{-\infty}^{\infty} \boldsymbol{E}_r^2(t)\, dt\,. \quad (9.80)$$

The transformation from the time domain to the frequency domain is performed by a Fourier transform or an expansion into Fourier harmonics. This is the point where the particular characteristics of the transverse acceleration depend on the magnetic field distribution and are, for example, different in a single bending magnet as compared to an oscillatory wiggler magnet. We use here the method of Fourier transforms to describe the electric field of a single particle passing only once through a homogeneous bending magnet. In case of a circular accelerator the particle will appear periodically with the period of the revolution time and we expect a correlation of the frequency spectrum with the revolution frequency. This is indeed the case and we will later discuss the nature of this correlation. Expressing the electrical field $\boldsymbol{E}_r(t)$ by its Fourier transform, we set

$$\boldsymbol{E}_r(\omega) = \int_{-\infty}^{\infty} \boldsymbol{E}_r(t)\, e^{-i\omega t} dt\,, \quad (9.81)$$

where $-\infty < \omega < \infty$. Applying Parseval's theorem we have

$$\int_{-\infty}^{\infty} |\boldsymbol{E}_r(\omega)|^2\, d\omega = 2\pi \int_{-\infty}^{\infty} |\boldsymbol{E}_r(t)|^2\, dt \quad (9.82)$$

and the total absorbed radiation energy from a single pass of a particle is therefore

$$\frac{dW}{d\Omega} = [4\pi c\epsilon_0]\frac{cR^2}{8\pi^2}\int_{-\infty}^{\infty} |\boldsymbol{E}_r(\omega)|^2\, d\omega\,. \quad (9.83)$$

Evaluating the electrical field by its Fourier components, we derive an expression for the spectral distribution of the radiation energy

$$\frac{d^2W}{d\Omega d\omega} = [4\pi c \epsilon_0] \frac{c}{4\pi^2} |\boldsymbol{E}_r(\omega)|^2 R_r^2, \tag{9.84}$$

where we have implicitly used the fact that $\boldsymbol{E}_r(\omega) = \boldsymbol{E}_r(-\omega)$ since $\boldsymbol{E}_r(t)$ is real. To calculate the Fourier transform, we use (9.29) and note that the electrical field is expressed in terms of quantities at the retarded time. The calculation is simplified if we express the whole integrand in (9.81) at the retarded time and get with $t_r = 1 - \frac{1}{c}R(t_r)$ and $dt_r = \frac{R(t_r)}{r}dt$ instead of (9.81)

$$\boldsymbol{E}_r(\omega) = \frac{1}{[4\pi\epsilon_0]}\frac{e}{c}\int_{-\infty}^{\infty} \left.\frac{\boldsymbol{R}\times\left[(\boldsymbol{R}+\beta R)\times\dot{\boldsymbol{\beta}}\right]}{r^2 R}\right|_r e^{-i\omega\left(t_r+\frac{R_r}{c}\right)}\, dt_r. \tag{9.85}$$

We require now that the radiation be observed at a point sufficiently far away from the source that during the time of emission the vector $\boldsymbol{R}(t_r)$ does not change appreciably in direction. This assumption is generally justified since the duration of the photon emission is of the order of $1/(\omega_L\gamma)$, where $\omega_L = c/\rho$ is the Larmor frequency. The observer therefore should be at a distance from the source large compared to $\rho/\gamma$. Equation (9.85) together with (9.14) may then be written like

$$\boldsymbol{E}_r(\omega) = \frac{1}{[4\pi\epsilon_0]}\frac{e}{cR}\int_{-\infty}^{\infty} \left.\frac{\boldsymbol{n}\times\left[(\boldsymbol{n}+\beta)\times\dot{\boldsymbol{\beta}}\right]}{(1+\boldsymbol{n}\beta)^2}\right|_r e^{-i\omega\left(t_r+\frac{R_r}{c}\right)}\, dt_r. \tag{9.86}$$

With

$$\frac{\boldsymbol{n}\times\left[(\boldsymbol{n}+\beta)\times\dot{\boldsymbol{\beta}}\right]}{(1+\boldsymbol{n}\beta)^2} = \frac{d}{dt_r}\frac{\boldsymbol{n}\times(\boldsymbol{n}\times\beta)}{1+\boldsymbol{n}\beta}, \tag{9.87}$$

we integrate (9.86) by parts while noting that the integrals vanish at the boundaries and get

$$\boldsymbol{E}_r(\omega) = \frac{1}{[4\pi\epsilon_0]}\frac{-i e\omega}{cR}\int_{-\infty}^{\infty} [\boldsymbol{n}\times(\boldsymbol{n}\times\beta)]_r\, e^{-i\omega\left(t_r+\frac{R_r}{c}\right)}\, dt_r. \tag{9.88}$$

After insertion into (9.84) the spectral and spatial intensity distribution is

$$\frac{d^2W}{d\Omega\, d\omega} = \frac{r_c mc^2}{4\pi c}\omega^2 \left|\int_{-\infty}^{\infty} [\boldsymbol{n}\times(\boldsymbol{n}\times\beta)]\, e^{-i\omega\left(t_r+\frac{R_r}{c}\right)}\, dt_r\right|_r^2. \tag{9.89}$$

The spectral and spatial radiation distribution depends on the Fourier transform of the particle trajectory which itself is a function of the magnetic field distribution. The trajectory in a uniform dipole field is different from say the step function of real lumped bending magnets or oscillating deflecting fields from wiggler magnets and the radiation characteristics may therefore be different. In this chapter, we will concentrate only on a uniform dipole field and postpone the discussion of specific radiation characteristics for insertion devices to Chap. 10.

The integrand in (9.89) can be expressed in component form to simplify integration. For that we consider a fixed coordinate system $(x, y, z)$ as shown in Fig. 9.7. The observation point is far away from the source point and we focus on the radiation that is centered about the tangent to the orbit at the source point. The observation point $P$ and the vectors $\boldsymbol{R}$ and $\boldsymbol{n}$ are therefore within the $(y, z)$-plane and radiation is emitted at angles $\theta$ with respect to the $z$-axis.

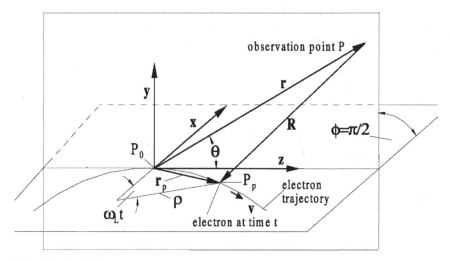

**Fig. 9.7.** Radiation geometry

The vector from the origin of the coordinate system $P_0$ to the observation point $P$ is $\boldsymbol{r}$, the vector $\boldsymbol{R}$ is the vector from $P$ to the particle at $P_\mathrm{p}$ and $\boldsymbol{r}_\mathrm{p}$ is the vector from the origin to $P_\mathrm{p}$. With this we have

$$\boldsymbol{r} = \boldsymbol{r}_\mathrm{p} - \boldsymbol{R}(t_\mathrm{r}) \,, \tag{9.90}$$

where $\boldsymbol{r}_\mathrm{p}$ and $\boldsymbol{R}_\mathrm{r}$ are taken at the retarded time. The exponent in (9.89) is then

$$\omega(t_\mathrm{r} + R_\mathrm{r}/c) = \omega(t_\mathrm{r} + \boldsymbol{n}\boldsymbol{R}_\mathrm{r}/c) = \omega\left(t_\mathrm{r} + \frac{\boldsymbol{n}\boldsymbol{r}_\mathrm{p}}{c} - \frac{\boldsymbol{n}\boldsymbol{r}}{c}\right) \tag{9.91}$$

and the term $-\omega\frac{\boldsymbol{n}\boldsymbol{r}}{c}$ is independent of the time generating only a constant phase factor which is completely irrelevant for the spectral distribution and may therefore be ignored.

In determining the vector components, we note from Fig. 9.7 that now the coordinate system is fixed in space. Following the above discussion the azimuthal angle is constant and set to $\varphi = \frac{1}{2}\pi$ because we are interested only in the vertical radiation distribution. the horizontal distribution is uniform by virtue of the tangential emission along the orbit. With these assumptions, we get the vector components for the vector $\boldsymbol{n}$ from (9.70)

$$\boldsymbol{n} = (0, -\sin\theta, -\cos\theta). \tag{9.92}$$

The vector $\boldsymbol{r}_\mathrm{p}$ is defined by Fig. 9.7 and depends on the exact variation of the deflecting magnetic field along the path of the particles. Here we assume a constant bending radius $\rho$ and have

$$\boldsymbol{r}_\mathrm{p} = [-\rho\cos(\omega_\mathrm{L}t_\mathrm{r}), 0, \rho\sin(\omega_\mathrm{L}t_\mathrm{r})], \tag{9.93}$$

where $\omega_\mathrm{L} = \beta c/\rho$ is the Larmor frequency. From these component representations the vector product

$$\frac{\boldsymbol{n}\boldsymbol{r}_\mathrm{p}}{c} = -\frac{\rho}{c}\sin(\omega_\mathrm{L}t_\mathrm{r})\cos\theta. \tag{9.94}$$

Noting that both arguments of the trigonometric functions in (9.94) are very small, we may expand the r.h.s. of (9.94) up to third order in $t_\mathrm{r}$ and the factor $t_\mathrm{r} + \boldsymbol{n}\boldsymbol{r}_\mathrm{p}/c$ in (9.91) becomes

$$t_\mathrm{r} + \frac{\boldsymbol{n}\boldsymbol{r}_\mathrm{p}}{c} = t_\mathrm{r} - \frac{\rho}{c}\left[\omega_\mathrm{L}t_\mathrm{r} - \frac{1}{6}(\omega_\mathrm{L}t_\mathrm{r})^3\left(1 - \frac{1}{2}\theta^2\right)\right]. \tag{9.95}$$

With $\omega_\mathrm{L} = \beta c/\rho$ we get $t_\mathrm{r}(1 - \rho\omega_\mathrm{L}/c) = (1-\beta)t_\mathrm{r} \approx t_\mathrm{r}/(2\gamma^2)$. Keeping only up to third order terms in $\omega_\mathrm{L}t_\mathrm{r}$ and $\theta$ we have finally for high energetic particles $\beta \approx 1$

$$t_\mathrm{r} + \frac{\boldsymbol{n}\boldsymbol{r}_\mathrm{p}}{c} = \frac{1}{2}\left(\gamma^{-2} + \theta^2\right)t_\mathrm{r} + \frac{1}{6}\omega_\mathrm{L}^2 t_\mathrm{r}^3. \tag{9.96}$$

The triple vector product in (9.89) can be evaluated in a similar way. For the velocity vector we derive from Fig. 9.7

$$\boldsymbol{\beta} = \beta\left[-\mathrm{sign}(1/\rho)\sin(\omega_\mathrm{L}t_\mathrm{r}), 0, \cos(\omega_\mathrm{L}t_\mathrm{r})\right]. \tag{9.97}$$

Consistent with the definition of the curvature in (6.7), the sign of the curvature $\mathrm{sign}(1/\rho)$ is positive for a positive charge and a positive magnetic field vector $B_y$. The vector relation (B.10) and (9.92, 9.97) can be used to express the triple vector product in terms of its components

$$\boldsymbol{n}\times(\boldsymbol{n}\times\boldsymbol{\beta}) = \beta\left[\mathrm{sign}(1/\rho)\sin(\omega_\mathrm{L}t_\mathrm{r}), \tfrac{1}{2}\sin 2\theta\,\cos(\omega_\mathrm{L}t_\mathrm{r}), -\sin^2\theta\,\cos(\omega_\mathrm{L}t_\mathrm{r})\right]. \tag{9.98}$$

Splitting this three-dimensional vector into two parts will allow us to characterize the polarization states of the radiation. To do this, we take the unit vector $\boldsymbol{u}_\perp$ in the $x$-direction and $\boldsymbol{u}_\parallel$ a unit vector normal to $\boldsymbol{u}_\perp$ and normal to $\boldsymbol{r}$. The $y$ and $z$ − components of (9.98) are then also the components of $\boldsymbol{u}_\parallel$ and we may express the vector (9.98) by

$$\boldsymbol{n} \times (\boldsymbol{n} \times \boldsymbol{\beta}) = \beta \operatorname{sign}(1/\rho) \sin(\omega_L t_r) \, \boldsymbol{u}_\perp + \beta \sin\theta \, \cos(\omega_L t_r) \boldsymbol{u}_\parallel . \qquad (9.99)$$

Inserting (9.96) and (9.99) into the integrand (9.88) we get with $\beta \approx 1$

$$\boldsymbol{E}_r(\omega) = \frac{-1}{[4\pi\epsilon_0]} \frac{e\omega}{cR} \int_{-\infty}^{\infty} \left[ \operatorname{sign}(1/\rho) \sin(\omega_L t_r) \, \boldsymbol{u}_\perp + \sin\theta \, \cos(\omega_L t_r) \boldsymbol{u}_\parallel \right] \mathrm{e}^X \, \mathrm{d}t_r ,$$

$$(9.100)$$

where

$$X = -\mathrm{i} \frac{\omega}{2\gamma^2} \left[ (1 + \gamma^2\theta^2) \, t_r + \tfrac{1}{3}\gamma^2\omega_L^2 t_r^3 \right] .$$

Two polarization directions have been defined for the electric radiation field. One of which, $\boldsymbol{u}_\perp$, is in the plane of the particle path being perpendicular to the particle velocity and to the deflecting magnetic field. Following Sokolov and Ternov [51] we call this the $\sigma$-mode ($\boldsymbol{u}_\perp = \boldsymbol{u}_\sigma$). The other polarization direction in the plane containing the deflecting magnetic field and the observation point is perpendicular to $\boldsymbol{n}$ and is called the $\pi$-mode ($\boldsymbol{u}_\parallel = \boldsymbol{u}_\pi$). Since the emission angle $\theta$ is very small, we find this polarization direction to be mostly parallel to the magnetic field. Noting that most accelerators or beam lines are constructed in the horizontal plane, the polarizations are also often referred to as the horizontal polarization for the $\sigma$-mode and as the vertical polarization for the $\pi$-mode.

### 9.5.1 Spectral Distribution in Space and Polarization

As was pointed out by Jackson [19], the mathematical need to extend the integration over infinite times does not invalidate our expansion of the trigonometric functions where we assumed the argument $\omega_L t_r$ to be small. Although the integral (9.100) extends over all past and future times, the integrand oscillates rapidly for all but the lowest frequencies and therefore only times of the order $ct_r = \pm \rho/\gamma$ centered about $t_r$ contribute to the integral. This is a direct consequence of the fact that the radiation is emitted in the forward direction and therefore only photons from a very small segment of the particle trajectory reach the observation point. For very small frequencies of the order of the Larmor frequency, however, we must expect considerable deviations from our results. In practical circumstances such low harmonics will, however, not propagate in the vacuum chamber [27] and the observed photon spectrum therefore is described accurately for all practical purposes.

The integral in (9.100) can be expressed by modified Bessel's functions in the form of Airy's integrals as has been pointed out by Schwinger [21].

Since the deflection angle $\omega_L t_r$ is very small, we may use linear expansions $\sin(\omega_L t_r) \approx \omega_L t_r$ and $\cos(\omega_L t_r) \approx 1$. Inserting the expression for the electric field (9.100) into (9.83) we note that cross terms of both polarizations vanish $\boldsymbol{u}_\perp \boldsymbol{u}_\parallel = 0$ and the radiation intensity can therefore be expressed by two separate orthogonal polarization components. Introducing in (9.100) the substitutions [21]

$$\omega_L t_r = \sqrt{\frac{1}{\gamma^2} + \theta^2}\, x \,, \tag{9.101}$$

$$\xi = \tfrac{1}{3}\frac{\omega}{\omega_L}\frac{1}{\gamma^3}(1 + \gamma^2\theta^2)^{3/2} = \tfrac{1}{2}\frac{\omega}{\omega_c}(1 + \gamma^2\theta^2)^{3/2} \,, \tag{9.102}$$

where $\hbar\omega_c$ is the critical photon energy, the argument in the exponential factor of (9.100) becomes

$$\frac{\omega}{2\gamma^2}\left[(1 + \gamma^2\theta^2)\, t_r + \tfrac{1}{3}\gamma^2\omega_L^2 t_r^3\right] = \tfrac{1}{2}\,\xi(3x + x^3) \,. \tag{9.103}$$

With these substitutions, (9.100) can be evaluated noting that only even terms contribute to the integral. With $\omega_L t_r$ and $\theta$ being small quantities we get integrals of the form [52]

$$\int_0^\infty \cos\left[\tfrac{1}{2}\,\xi(3x + x^3)\right]\, dx = \tfrac{1}{\sqrt{3}}K_{1/3}(\xi) \,, \tag{9.104a}$$

$$\int_0^\infty \sin\left[\tfrac{1}{2}\,\xi(3x + x^3)\right]\, dx = \tfrac{1}{\sqrt{3}}K_{2/3}(\xi) \,, \tag{9.104b}$$

where the functions $K_\nu$ are modified Bessel's functions of the second kind. These functions assume finite values for small arguments but vanish exponentially for large arguments as shown in Fig. 9.8. Fast converging series for these modified Bessel's functions with fractional index have been derived by Kostroun [53]. The Fourier transform of the electrical field (9.100) finally becomes

$$\boldsymbol{E}_r(\omega) = \frac{-1}{[4\pi\epsilon_0]}\frac{\sqrt{3}e}{cR}\frac{\omega}{\omega_c}\gamma(1 + \gamma^2\theta^2)\left[\text{sign}(1/\rho)\, K_{2/3}(\xi)\, \boldsymbol{u}_\sigma - \mathrm{i}\,\frac{\gamma\theta K_{1/3}(\xi)}{\sqrt{1 + \gamma^2\theta^2}}\, \boldsymbol{u}_\pi\right], \tag{9.105}$$

describing the spectral radiation field far from the source for particles traveling through a uniform magnetic dipole field. Later, we will modify this expression to make it suitable for particle motion in undulators or other nonuniform fields.

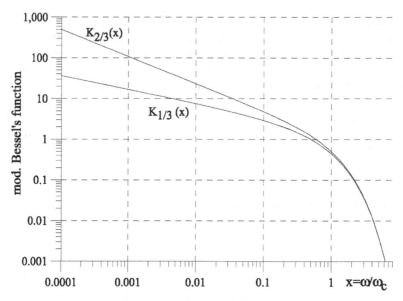

**Fig. 9.8.** Modified Bessel's functions $K_{1/3}(x)$ and $K_{2/3}(x)$

The spectral synchrotron radiation energy emitted by one electron per pass is proportional to the square of the electrical field (9.105) and is from (9.84)

$$\frac{\mathrm{d}^2 W}{\mathrm{d}\Omega\mathrm{d}\omega} = \frac{3\,r_c mc}{4\pi^2}\gamma^2\left(\frac{\omega}{\omega_c}\right)^2 (1+\gamma^2\theta^2)^2\left[K_{2/3}^2(\xi)\,\boldsymbol{u}_\sigma^2 + \frac{\gamma^2\theta^2 K_{1/3}^2(\xi)}{1+\gamma^2\theta^2}\boldsymbol{u}_\pi^2\right].$$

$$(9.106)$$

The radiation spectrum has two components of orthogonal polarization, one in the plane of the particle trajectory and the other almost parallel to the deflecting magnetic field. In (9.105) both polarizations appear explicitly through the orthogonal unit vectors. Forming the square of the electrical field to get the radiation intensity, cross terms disappear because of the orthogonality of the unit vectors $\boldsymbol{u}_\sigma$ and $\boldsymbol{u}_\pi$. The expression for the radiation intensity therefore preserves separately the two polarization modes in the square brackets of (9.106) representing the $\sigma$-mode and $\pi$-mode of polarization, respectively.

It is interesting to study the spatial distribution for the two polarization modes in more detail. Not only are the intensities very different but the spatial distribution is different too. The spatial distribution of the $\sigma$-mode is directed mainly in the forward direction while the $\pi$-mode radiation is emitted into two lobes at finite angles and zero intensity in the forward direction $\theta = 0$. In Fig. 9.9 the instantaneous radiation lobes are shown for both the $\sigma$- and the

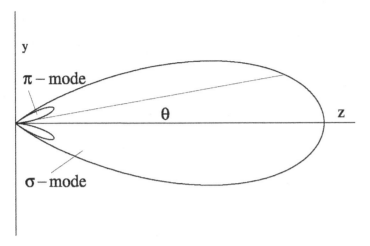

**Fig. 9.9.** Radiation lobes for $\sigma$- and $\pi$-mode polarization

$\pi$-mode at the critical photon energy and being emitted tangentially from the orbit at the origin of the coordinate system. A more detailed discussion of elliptical polarization properties can be found in Sect. 10.3.

### 9.5.2 Spectral and Spatial Photon Flux

The radiation intensity $W$ from a single electron and for a single pass may not always be the most useful parameter. A more useful parameter is the spectral photon flux per unit solid angle into a frequency bin $\Delta\omega/\omega$ and for a circulating beam current $I$

$$\frac{\mathrm{d}^2 \dot{N}_{\mathrm{ph}}(\omega)}{\mathrm{d}\theta\,\mathrm{d}\psi} = \frac{\mathrm{d}W(\omega)}{\mathrm{d}\omega\,\mathrm{d}\Omega}\frac{1}{\hbar}\frac{I}{e}\frac{\Delta\omega}{\omega}. \tag{9.107}$$

Here we have replaced the solid angle by its components, the vertical angle $\theta$ and the bending angle $\psi$. In more practical units the differential photon flux is

$$\frac{\mathrm{d}^2 \dot{N}_{\mathrm{ph}}(\omega)}{\mathrm{d}\theta\,\mathrm{d}\psi} = C_\Omega E^2 I \frac{\Delta\omega}{\omega}\left(\frac{\omega}{\omega_{\mathrm{c}}}\right)^2 K_{2/3}^{\,2}(\xi)\,F(\xi,\theta)\,, \tag{9.108}$$

where

$$C_\Omega = \frac{3\alpha}{4\pi^2 e(mc^2)^2} = 1.3273\;10^{16}\;\frac{\text{photons}}{\text{s mrad}^2\;\text{GeV}^2\text{A}}\,, \tag{9.109}$$

$I$ the circulating particle beam current, $\alpha$ the fine structure constant, and

$$F(\xi,\theta) = (1+\gamma^2\theta^2)^2\left[1 + \frac{\gamma^2\theta^2}{1+\gamma^2\theta^2}\frac{K_{1/3}^{\,2}(\xi)}{K_{2/3}^{\,2}(\xi)}\right]\,. \tag{9.110}$$

For approximate numerical calculations of photon fluxes, we may use the graphic representation in Fig. 9.8 of the modified Bessel's function.

The spatial radiation pattern varies with the frequency of the radiation. Specifically, the angular distribution concentrates more and more in the forward direction as the radiation frequency increases. The radiation distribution in frequency and angular space is shown for both the $\sigma$- (Fig. 9.10) and the $\pi$-mode (Fig. 9.11) at the fundamental frequency. The high collimation of synchrotron radiation in the forward direction makes it a prime research tool to probe materials and its atomic and molecular properties.

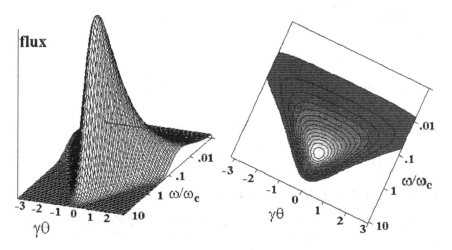

**Fig. 9.10.** Distribution in frequency and angular space for $\sigma$-mode radiation

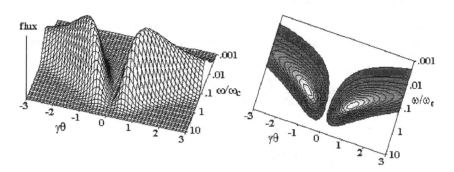

**Fig. 9.11.** Distribution in frequency and angular space for $\pi$-mode radiation

### 9.5.3 Harmonic Representation

Expression (9.106) can be transformed into a different formulation empha-sizing the harmonic structure of the radiation spectrum. The equivalence between both formulations has been shown by Sokolov and Ternov [51] ex-pressing the modified Bessel's functions $K_{1/3}$ and $K_{2/3}$ by regular Bessel's functions of high order. With $\nu = \frac{\omega}{\omega_L}$ the asymptotic formulas for $\nu \gg 1$ are

$$K_{1/3}(\xi) = \frac{\sqrt{3}\pi}{\sqrt{1 - \beta^2 \cos^2 \theta}} J_\nu(\nu\beta\cos\theta) , \qquad (9.111)$$

$$K_{2/3}(\xi) = \frac{\sqrt{3}\pi}{1 - \beta^2 \cos^2 \theta} J'_\nu(\nu\beta\cos\theta) , \qquad (9.112)$$

where $\xi = \frac{\nu}{3}\left(1 - \beta^2 \cos^2 \theta\right)^{3/2} \approx \frac{\nu}{3}\left(\gamma^{-2} + \beta^2\theta^2\right)^{3/2}$ for small angles. These approximations are justified since we are only interested in very large har-monics of the revolution frequency. The harmonic number $\nu$ for the critical photon frequency, for example, is given by $\nu_c = \omega_c/\omega_L = \frac{3}{2}\gamma^3$ which for practical cases is generally a very large number. Inserting these approxima-tions into (9.106) gives the the formulation that has been derived first by Schott [3, 4, 5] in 1907 long before synchrotron radiation was discovered in an attempt to calculate the radiation intensity of atomic spectral lines

$$\frac{d^2 P}{d\Omega d\nu} = \frac{r_c m c^3}{2\pi\rho^2}\nu^2 \left[ J'^2_\nu(\nu\cos\theta) + \theta^2 J^2_\nu(\nu\cos\theta) \right] , \qquad (9.113)$$

where we have introduced the radiation power $P = W\frac{c}{2\pi\rho}$. This form still exhibits the separation of the radiation into the two polarization modes.

## 9.6 Spatial Radiation Power Distribution

Integrating over all frequencies we obtain the angular distribution of the synchrotron radiation. From (9.106) we note the need to perform integrals of the form $\int_{-\infty}^{\infty} \omega^2 K_\mu^2(a\omega)\, d\omega$, where $a\omega = \xi$. The solution can be found in the integral tables of Gradshteyn and Ryzhik [54] as solution number GR(6.576.4)[1]

$$\int_0^\infty \omega^2 K_\mu^2(a\omega)\, d\omega = \frac{\pi^2}{32a^3}\frac{1 - 4\mu}{\cos\pi\mu} . \qquad (9.114)$$

Applying this solution to (9.106) and integrating over all frequencies, we get for the angular distribution of the synchrotron radiation

---

[1] In this chapter we will need repeatedly results from mathematical tables. We abbreviate such solutions with the first letters of the authors names and the formula number.

$$\frac{dW}{d\Omega} = \frac{21}{32} \frac{r_c mc^2}{\rho} \frac{\gamma^5}{(1+\gamma^2\theta^2)^{5/2}} \left(1 + \frac{5}{7}\frac{\gamma^2\theta^2}{1+\gamma^2\theta^2}\right). \tag{9.115}$$

This result is consistent with the angular radiation power distribution (9.76) where we found that the radiation is collimated very much in the forward direction with most of the radiation energy being emitted within an angle of $\pm 1/\gamma$. There are two contributions to the total radiation intensity, the $\sigma$-mode and the $\pi$-mode. The $\sigma$-mode has a maximum intensity in the forward direction, while the maximum intensity for the $\pi$-mode occurs at an angle of $\theta_\pi = 1/(\sqrt{5/2}\,\gamma)$. The quantity $W$ is the radiation energy per unit solid angle from a single electron and a single pass and the average radiation power is therefore $P_\gamma = W/T_{\rm rev}$ or (9.115) becomes

$$\frac{dP_\gamma}{d\Omega} = \frac{2r_c mc^3}{6\pi\rho^2} \frac{\gamma^5}{(1+\gamma^2\theta^2)^{5/2}} \left(1 + \frac{5}{7}\frac{\gamma^2\theta^2}{1+\gamma^2\theta^2}\right). \tag{9.116}$$

Integrating (9.116) over all angles $\theta$ we find the synchrotron radiation power into both polarization modes, the $\sigma$-mode perpendicular to the magnetic field and the $\pi$-mode parallel to the magnetic field. In doing so, we note first that (9.116) can be simplified with (9.59) and $\beta = 1$

$$\frac{dP_\gamma}{d\Omega} = \frac{21}{32} \frac{P_\gamma}{2\pi} \frac{\gamma}{(1+\gamma^2\theta^2)^{5/2}} \left(1 + \frac{5}{7}\frac{\gamma^2\theta^2}{1+\gamma^2\theta^2}\right). \tag{9.117}$$

This result is consistent with (9.76) although it should be noted that (9.117) gives the average radiation power from a circular accelerator with uniform intensity in $\psi$, while (9.76) is the instantaneous power into the forward lobe. Equation (9.117) exhibits the power into each polarization mode for which the total power can be obtained by integration over all angles. First, we integrate over all points along the circular orbit and get a factor $2\pi$ since the observed radiation power does not depend on the location along the orbit. Continuing the integration over all angles of $\theta$, we find the contributions to the integral to become quickly negligible for angles larger than $1/\gamma$. If it were not so, we could not have used (9.117) where the trigonometric functions have been replaced by their small arguments. Both terms in (9.117) can be integrated readily and the first term becomes with GR(2.271.6) [54]

$$\int_{\theta_{\max}\gamma \ll 1}^{\theta_{\max}\gamma \gg 1} \frac{\gamma\, d\theta}{(1+\gamma^2\theta^2)^{5/2}} = \frac{4}{3}. \tag{9.118}$$

The second term is with GR[2.272.7] [54]

$$\int_{\theta_{\max}\gamma \ll 1}^{\theta_{\max}\gamma \gg 1} \frac{\gamma^3\theta^2\, d\theta}{(1+\gamma^2\theta^2)^{7/2}} = \frac{4}{15}. \tag{9.119}$$

With these integrals and (9.117) we express the radiation power into the $\sigma$-and $\pi$-mode with $P_\gamma$ from (9.59) by

$$P_\sigma = \tfrac{7}{8}P_\gamma, \tag{9.120a}$$
$$P_\pi = \tfrac{1}{8}P_\gamma. \tag{9.120b}$$

The horizontally polarized component of synchrotron radiation greatly dominates the photon beam characteristics and only 12.5% of the total intensity is polarized in the vertical plane. In the forward direction the $\sigma$-polarization even approaches 100%. Obviously, the sum of both components is equal to the total radiation power. This high polarization of the radiation provides a valuable characteristic for experimentation with synchrotron radiation. In addition, the emission of polarized light generates a slow polarizing reaction on the particle beam orbiting in a circular accelerator like in a storage ring [55].

### 9.6.1 Asymptotic Solutions

Expressions for the radiation distribution can be greatly simplified if we restrict the discussion to very small or very large arguments of the modified Bessel's functions for which approximate expressions exist [24]. Knowledge of the radiation distribution at very low photon frequencies becomes important for experiments using such radiation or for beam diagnostics where the beam cross section is being imaged to a TV camera using the visible part of the radiation spectrum. To describe this visible part of the spectrum, we may in most cases assume that the photon frequency is much lower than the critical photon frequency.

**Low frequencies and small observation angles.** For very small arguments or low frequencies and small angles, we find the following approximations AS(9.6.9) [24]

$$K_{1/3}(\xi \longrightarrow 0) \approx \frac{\Gamma^2(1/3)}{2^{2/3}} \left(\frac{\omega}{\omega_c}\right)^{-2/3} \frac{1}{1 + \gamma^2\theta^2}, \tag{9.121a}$$

$$K_{2/3}(\xi \longrightarrow 0) \approx 2^{2/3}\Gamma^2(2/3) \left(\frac{\omega}{\omega_c}\right)^{-4/3} \frac{1}{(1 + \gamma^2\theta^2)^2}, \tag{9.121b}$$

where the Gamma functions $\Gamma(1/3) = 2.6789385$ and $\Gamma(2/3) = 1.351179$ and from (9.103)

$$\xi = \frac{1}{2}\frac{\omega}{\omega_c}(1 + \gamma^2\theta^2)^{3/2}. \tag{9.122}$$

Inserting this into (9.108) the photon flux spectrum in the forward direction becomes for $\theta = 0$ and $\frac{\omega}{\omega_c} \ll 1$

$$\frac{d^2\dot{N}_{ph}}{d\theta d\psi} \approx C_\Omega E^2 I \, \Gamma^2(2/3) \left(\frac{2\omega}{\omega_c}\right)^{2/3} \frac{\Delta\omega}{\omega}. \tag{9.123}$$

The photon spectrum at very low frequencies is independent of the particle energy since $\omega_c \propto E^3$. Clearly, in this approximation there is no angular dependence for the $\sigma$-mode radiation and the intensity increases with frequency. The $\pi$-mode radiation on the other hand is zero for $\theta = 0$ and increases in intensity with the square of $\theta$ as long as the approximation is valid.

**High frequencies or large observation angles.** For large arguments of the modified Bessel's functions or for high frequencies and large emission angles different approximations hold. In this case, the approximate expressions are actually the same for both Bessel's functions indicating the same exponential drop off for high energetic photons AS(9.7.2)

$$K_{1/3}^2(\xi \longrightarrow \infty) \approx \frac{\pi}{2}\frac{e^{-2\xi}}{\xi}\,, \tag{9.124a}$$

$$K_{2/3}^2(\xi \longrightarrow \infty) \approx \frac{\pi}{2}\frac{e^{-2\xi}}{\xi}\,. \tag{9.124b}$$

The photon flux distribution in this approximation becomes from (9.106)

$$\frac{d^2 N_{ph}}{d\theta d\psi} \approx \frac{3r_c mc^2}{4\pi\hbar c}\gamma^2\frac{\omega}{\omega_c}e^{-2\xi}\sqrt{1+\gamma^2\theta^2}\frac{\Delta\omega}{\omega}\,, \tag{9.125}$$

where $N_{ph}'$ is the number of photons emitted per pass. The spatial radiation distribution is greatly determined by the exponential factor and the relative amplitude with respect to the forward direction scales therefore like

$$\exp\left\{-\frac{\omega}{\omega_c}\left[(1+\gamma^2\theta^2)^{3/2}-1\right]\right\}\,. \tag{9.126}$$

We look now for the specific angle for which the intensity has fallen to $1/e$. Since $\omega \gg \omega_c$, this angle must be very small $\gamma\theta \ll 1$ and we can ignore other $\theta$-dependent factors. The exponential factor becomes equal to $1/e$ for

$$\tfrac{3}{2}\frac{\omega}{\omega_c}\gamma^2\theta_{1/e}^2 \approx 1 \tag{9.127}$$

and solving for $\theta_{1/e}$ we get finally

$$\theta_{1/e} = \sqrt{\tfrac{2}{3}}\frac{1}{\gamma}\frac{\omega_c}{\omega} \qquad \text{for} \ \ \omega \gg \omega_c\,. \tag{9.128}$$

The high energy end of the synchrotron radiation spectrum is more and more collimated into the forward direction. The angular distribution is graphically illustrated for both polarization modes in Figs. 9.10 and 9.11.

## 9.7 Angle-Integrated Spectrum

Synchrotron radiation is emitted over a wide range of frequencies and it is of great interest to know the exact frequency distribution of the radiation.

Since the radiation is very much collimated in the forward direction, it is useful to integrate over all angles of emission to obtain the total spectral photon flux that might be accepted by a beam line with proper aperture. To that goal, (9.106) will be integrated with respect to the emission angles to obtain the frequency spectrum of the radiation. The emission angle $\theta$ appears in (9.106) in a rather complicated way which makes it difficult to perform the integration directly. We replace therefore the modified Bessel's functions by Airy's functions defined by AS(10.4.14) and AS(10.4.31) [24]

$$Ai\left(z\right) = \frac{\sqrt{z}}{\sqrt{3}\pi}K_{1/3}(\xi)\,, \tag{9.129a}$$

$$A'i\left(z\right) = -\frac{z}{\sqrt{3}\pi}K_{2/3}(\xi)\,. \tag{9.129b}$$

With the definition

$$\eta = \frac{3}{4}\frac{\omega}{\omega_{\mathrm{c}}} \tag{9.130}$$

we get from (9.103)

$$z = \left(\tfrac{3}{2}\xi\right)^{2/3} = \eta^{2/3}\left(1 + \gamma^2\theta^2\right)\,. \tag{9.131}$$

We apply this to the periodic motion of particles orbiting in a circular accelerator. In this case the spectral distribution of the radiation power can be obtained by noting that the differential radiation energy (9.106) is emitted every time the particle passes by the source point. A short pulse of radiation is sent towards the observation point at periodic time intervals equal to the revolution time $T_{\mathrm{rev}} = \frac{c}{2\pi\rho}$. The spectral power distribution (9.106) expressed by Airy functions is

$$\frac{\mathrm{d}^2 P_\gamma}{\mathrm{d}\omega\,\mathrm{d}\Omega} = \frac{9 P_\gamma}{2\pi}\frac{\gamma}{\omega_{\mathrm{c}}}\left[\eta^{2/3}A'i^2\left(z\right) + \eta^{4/3}\gamma^2\theta^2 Ai^2\left(z\right)\right]\,. \tag{9.132}$$

To obtain the photon frequency spectrum, we integrate over all angles of emission which is accomplished by integrating along the orbit contributing a mere factor of $2\pi$ and over the angle $\theta$. Although this latter integration is to be performed between $-\pi$ and $+\pi$, we choose the mathematically easier integration from $-\infty$ to $+\infty$ because the Airy functions fall off very fast for large arguments. In fact, we have seen already that most of the radiation is emitted within a very small angle of $\pm 1/\gamma$. The integrals to be solved are of the form $\int_0^\infty \theta^n Ai^2\left[\eta^{2/3}(1 + \gamma^2\theta^2)\right]\mathrm{d}\theta$ where $n = 0$ or $2$. We concentrate first on the second term in (9.132), and form with (9.104) and (9.129a) the square of the Airy function

$$\theta^2 Ai^2(z) = \frac{1}{\pi^2}\int_0^\infty \theta^2 \cos\left[\tfrac{1}{3}x^3 + zx\right]\mathrm{d}x\int_0^\infty \theta^2 \cos\left[\tfrac{1}{3}y^3 + z\,y\right]\mathrm{d}y\,. \tag{9.133}$$

We solve these integrals by making use of the trigonometric relation

$$\cos(\alpha + \tfrac{1}{2}\beta) \cos(\alpha - \tfrac{1}{2}\beta) = \cos\alpha \, \cos\beta \,. \tag{9.134}$$

After introducing the substitutions $x + y = s$ and $x - y = t$, we obtain integrals over two terms which are symmetric in $s$ and $t$ and therefore can be set equal to get

$$\theta^2 Ai^2(z) = \frac{1}{2\pi^2} \int_0^\infty \int_0^\infty \theta^2 \cos\left[\tfrac{1}{12}s^3 + 3\,s\,t^2 + z\,s\right] ds\, dy \,, \tag{9.135}$$

where the factor $\frac{1}{2}$ comes from the transformation of the area element $ds\, dy = \frac{ds}{\sqrt{2}}\frac{dt}{\sqrt{2}}$. In our problem we replace the argument $z$ by the expression $z = \eta^{2/3}\left(1 + \gamma^2\theta^2\right)$ and integrate over the angle $\theta$

$$\pi^2 \int_{-\infty}^{\infty} \theta^2 Ai^2(z)\, d\theta$$

$$= \int_{-\infty}^{\infty} \int_{-\infty}^{\infty} \int_{-\infty}^{\infty} \theta^2 \cos\left[\tfrac{1}{12}s^3 + 3\,s\,t^2 + s\,\eta^{2/3}\left(1 + \gamma^2\theta^2\right)\right] ds\, dy\, d\theta \,. \tag{9.136}$$

The integrand is symmetric with respect to $\theta$ and the integration therefore needs to be performed only from $0$ to $\infty$ with the result being doubled. We also note that the integration is taken over only one quadrant of the $(s,t)$-space. Further simplifying the integration, the number of variables in the argument of the cosine function can be reduced in the following way. We note the coefficient $\frac{1}{4}t^2 + \eta^{2/3}\gamma^2\theta^2$ which is the sum of squares. Setting $\frac{1}{2}t = r\cos\varphi$ and $\eta^{1/3}\gamma\theta = r\sin\varphi$ this term becomes simply $r^2$. The area element transforms like $dt\, d\theta = 2/(\eta^{1/3}\gamma)\, r\, dr\, d\varphi$ and integrating over $\varphi$ from $0$ to $\pi/2$, since we need to integrate only over one quarter plane, (9.136) becomes finally

$$\int_{-\infty}^{\infty} \theta^2 Ai^2(z)\, d\theta = \frac{1}{2\pi\eta\gamma^3} \int_0^\infty \int_0^\infty r^2 \cos\left[\tfrac{1}{12}s^3 + s\,\eta^{2/3} + r^2\right] r\, dr\, ds \,. \tag{9.137}$$

The integrand of (9.137) has now a form close to that of an Airy integral and we will try to complete that similarity. With $q = (3\xi/2)^{1/3}x$ the definition of the Airy functions AS(10.4.31)[24] are consistent with (9.129)

$$Ai(z) = \frac{1}{\pi} \int_0^\infty \cos\left[\tfrac{1}{3}q^3 + z\, q\right] dq \,. \tag{9.138}$$

Equation (9.137) can be modified into a similar form by setting

$$w^3 = \tfrac{1}{4}s^3 \quad\text{and}\quad s\left(\eta^{2/3} + r^2\right) = y\, w \,. \tag{9.139}$$

Solving for $w$ we get $w = s\,/\,2^{2/3}$ and with $y = 2^{2/3}(\eta^{2/3} + r^2)$, $ds = 2^{2/3}dw$ and $dy = 2^{5/3}r\, dr$ equation (9.137) becomes

$$\int_{-\infty}^{\infty} \theta^2 \mathcal{A}i^2 \left(z\right) d\theta = \frac{1}{4\eta\gamma^3} \int_{y_0}^{\infty} \left(\frac{y}{2^{2/3}} - \eta^{2/3}\right) \mathcal{A}i \left(y\right) dy , \qquad (9.140)$$

where we have used the definition of Airy's function and where the integration starts at

$$y_0 = (2\eta)^{2/3} = \left(\frac{3}{2}\frac{\omega}{\omega_c}\right)^{2/3} \qquad (9.141)$$

corresponding to $r = 0$.

We may separate this integral into two parts and get a term $y\mathcal{A}i \left(y\right)$ under one of the integrals. This term is by definition of the Airy's functions AS(10.4.1) [24] equal to $\mathcal{A}i''$. Integration of this second derivative gives

$$\int_{y_0}^{\infty} \mathcal{A}i''(y) \, dy = -\mathcal{A}i'(y_0) \qquad (9.142)$$

and collecting all terms in (9.140) we have finally

$$\int_{-\infty}^{\infty} \theta^2 \mathcal{A}i^2 \left(z\right) d\theta = -\frac{1}{4\eta^{1/3}\gamma^3} \left[\frac{\mathcal{A}i'(y_0)}{y_0} + \int_{y_0}^{\infty} \mathcal{A}i \left(y\right) dy\right] . \qquad (9.143)$$

The derivation of the complete spectral radiation power distribution (9.132) requires also the evaluation of the integral $\int \mathcal{A}i'(z) \, d\theta$. This can be done with the help of the integral $\int \mathcal{A}i(z) \, d\theta$ and the integral we have just derived. We follow a similar derivation that led us just from (9.136) to (9.137) and get instead of (9.143)

$$\int_{-\infty}^{\infty} \mathcal{A}i^2 \left(z\right) d\theta = -\frac{1}{2\eta^{1/3}\gamma} \int_{y_0}^{\infty} \mathcal{A}i \left(y\right) dy . \qquad (9.144)$$

Recalling the definition of the argument $z = \eta^{2/3} \left(1 + \gamma^2\theta^2\right)$, we differentiate (9.144) twice with respect to $\eta^{2/3}$ to get

$$2\int_{-\infty}^{\infty} \left[\mathcal{A}i''(z) + \mathcal{A}i'^2(z)\right] d\theta = -\frac{2^{1/3}}{\eta^{1/3}\gamma}\mathcal{A}i'(y_0) . \qquad (9.145)$$

Using the relation $\mathcal{A}i''(z) = z\mathcal{A}i \left(z\right)$ and the results (9.142, 9.143) in (9.145) we get

$$\int_{-\infty}^{\infty} \mathcal{A}i'^2(z) \, d\theta = -\frac{\eta^{1/3}}{4\gamma} \left[\frac{3\mathcal{A}i'(y_0)}{y_0} + \int_{y_0}^{\infty} \mathcal{A}i \left(y\right) dy\right] . \qquad (9.146)$$

At this point, all integrals have been derived that are needed to describe the spectral radiation power separately in both polarization modes and the spectral radiation power from (9.132) becomes

$$\frac{\mathrm{d}P_\gamma}{\mathrm{d}\omega} = \frac{27 P_\gamma \omega}{16\,\omega_\mathrm{c}^2} \left[ \left( -\frac{3 A i'\,(y_0)}{y_0} - \int\limits_{y_0}^\infty A i\,(y)\,\mathrm{d}y \right) \right.$$

$$\left. \times \left( \frac{A i'\,(y_0)}{y_0} + \int\limits_{y_0}^\infty A i\,(y)\,\mathrm{d}y \right) \right]. \tag{9.147}$$

The first term describes the $\sigma$-mode of polarization and the second term the $\pi$-mode. Combining both polarization modes, we may derive a comparatively simple expression for the spectral radiation power. To this goal, we replace the Airy's functions by modified Bessel's functions

$$\frac{A i'\,(y_0)}{y_0} = -\tfrac{1}{\sqrt{3}\pi} K_{2/3}(x_0)\,, \tag{9.148}$$

where from (9.129, 9.130), and (9.140) $x_0 = \omega/\omega_\mathrm{c}$. With $\sqrt{y}\,\mathrm{d}y = \mathrm{d}x$, the recurrence formula $2 K'_{2/3} = -K_{1/3} - K_{5/3}$ and (9.129) the Airy integral is

$$\int_{y_0}^\infty A i\,(y)\,\mathrm{d}y = -\frac{2}{\sqrt{3}\pi}\int_{x_0}^\infty K'_{2/3} x\,\mathrm{d}x - \frac{1}{\sqrt{3}\pi}\int_{x_0}^\infty K_{5/3}(\xi)\,\mathrm{d}\xi$$

$$= \frac{2}{\sqrt{3}\pi} K_{2/3}(\xi) - \frac{1}{\sqrt{3}\pi}\int_{x_0}^\infty K_{5/3}(\xi)\,\mathrm{d}\xi\,. \tag{9.149}$$

We use (9.148) and (9.149) in (9.147) and get the simple expression for the synchrotron radiation spectrum

$$\frac{\mathrm{d}P_\gamma}{\mathrm{d}\omega} = \frac{P_\gamma}{\omega_\mathrm{c}} \tfrac{9\sqrt{3}}{8\pi} \frac{\omega}{\omega_\mathrm{c}} \int_{x_0}^\infty K_{5/3}(x)\,\mathrm{d}x = \frac{P_\gamma}{\omega_\mathrm{c}} S\left(\frac{\omega}{\omega_\mathrm{c}}\right)\,, \tag{9.150}$$

where we defined the universal function

$$S\left(\frac{\omega}{\omega_\mathrm{c}}\right) = \tfrac{9\sqrt{3}}{8\pi} \frac{\omega}{\omega_\mathrm{c}} \int_{\omega/\omega_\mathrm{c}}^\infty K_{5/3}(x)\,\mathrm{d}x\,. \tag{9.151}$$

The spectral distribution depends only on the critical frequency $\omega_\mathrm{c}$, the total radiation power and a purely mathematical function. This result has been derived originally by Ivanenko and Sokolov [20] and independently by Schwinger [21]. Specifically, it should be noted that the synchrotron radiation spectrum, if normalized to the critical frequency, does not depend on the particle energy and is represented by the universal function shown in Fig. 9.12. The energy dependence is contained in the cubic dependence of the critical frequency acting as a scaling factor for the real spectral distribution.

The mathematical function is properly normalized as we can see by integrating over all frequencies.

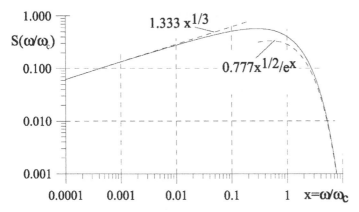

**Fig. 9.12.** Universal function: $S(\xi) = \frac{9\sqrt{3}}{8\pi}\xi\int_\xi^\infty K_{5/3}(x)\,\mathrm{d}x$, with $\xi = \omega/\omega_c$

$$\int_0^\infty \frac{\mathrm{d}P_\gamma}{\mathrm{d}\omega}\mathrm{d}\omega = \frac{9\sqrt{3}}{8\pi}P_\gamma\int_0^\infty\int_{x_0}^\infty K_{5/3}(x)\mathrm{d}x\,\mathrm{d}x_0\,. \tag{9.152}$$

After integration by parts, the result can be derived from GR[6.561.16] [54]

$$\int_0^\infty \frac{\mathrm{d}P_\gamma}{\mathrm{d}\omega}\mathrm{d}\omega = \frac{9\sqrt{3}}{16\pi}P_\gamma\int_0^\infty x_0^2\, K_{5/3}(x_0)\mathrm{d}x_0 = \Gamma\left(4/3\right)\Gamma\left(2/3\right)\,. \tag{9.153}$$

Using the triplication formula AS(6.1.19) [24] the product of the gamma functions becomes

$$\Gamma\left(4/3\right)\Gamma\left(2/3\right) = \frac{4}{9}\frac{2\pi}{\sqrt{3}}\,. \tag{9.154}$$

With this equation the proper normalization of (9.152) is demonstrated

$$\int_0^\infty \frac{\mathrm{d}P_\gamma}{\mathrm{d}\omega}\mathrm{d}\omega = P_\gamma\,. \tag{9.155}$$

Of more practical use is the spectral photon flux per unit angle of deflection in the bending magnet. With the photon flux $\mathrm{d}\dot{N}_{\mathrm{ph}} = \mathrm{d}P/\hbar\omega$ we get from (9.150)

$$\frac{\mathrm{d}\dot{N}_{\mathrm{ph}}}{\mathrm{d}\psi} = \frac{P_\gamma}{2\pi\hbar\omega_c}\frac{\Delta\omega}{\omega}S\left(\frac{\omega}{\omega_c}\right) \tag{9.156}$$

and with (9.59) and (9.78)

$$\frac{\mathrm{d}\dot{N}_{\mathrm{ph}}}{\mathrm{d}\psi} = \frac{4\alpha}{9}\gamma\frac{I}{e}\frac{\Delta\omega}{\omega}S\left(\frac{\omega}{\omega_c}\right)\,, \tag{9.157}$$

where $\psi$ is the deflection angle in the bending magnet and $\alpha$ the fine structure constant. In practical units, this becomes

$$\frac{\mathrm{d}\dot{N}_{\mathrm{ph}}}{\mathrm{d}\psi} = C_\psi \, E \, I \, \frac{\Delta\omega}{\omega} \, S\left(\frac{\omega}{\omega_c}\right) \tag{9.158}$$

with

$$C_\psi = \frac{4\alpha}{9 e \, mc^2} = 3.967 \cdot 10^{16} \, \frac{\text{photons}}{\text{s mrad A GeV}} \, . \tag{9.159}$$

The synchrotron radiation spectrum in Fig. 9.12 is rather uniform up to the critical frequency beyond which the intensity falls off rapidly. Equation (9.150) is not well suited for quick calculation of the radiation intensity at a particular frequency. We may, however, express (9.150) in much simpler form for very low and very large frequencies making use of approximate expressions of Bessel's functions for large and small arguments.

For small arguments $\left(x = \frac{\omega}{\omega_c} \ll 1\right)$ we find with AS(9.6.9) [24]

$$K_{5/3}(x \longrightarrow 0) \approx \Gamma\left(\tfrac{5}{3}\right) \frac{2^{2/3}}{x^{5/3}} \, , \tag{9.160}$$

which allows us to integrate (9.153) readily and get instead of (9.150)

$$\frac{\mathrm{d}P_\gamma}{\mathrm{d}\omega} \approx \frac{9\sqrt{3}}{8\pi} \frac{P_\gamma}{\omega_c} 2^{2/3} \, \Gamma(2/3) \left(\frac{\omega}{\omega_c}\right)^{1/3} \approx 1.333 \left(\frac{\omega}{\omega_c}\right)^{1/3} \frac{P_\gamma}{\omega_c} \, . \tag{9.161}$$

For high photon frequencies $\left(x = \frac{\omega}{\omega_c} \gg 1\right)$ the modified Bessel's function becomes from AS(9.7.2) [24]

$$K_{5/3}(x \gg 1) \approx \sqrt{\frac{\pi}{2}} \frac{\mathrm{e}^{-x}}{\sqrt{x}} \tag{9.162}$$

and after integration with GR(3.361.1) and GR(3.361.2), [54] (9.150) becomes

$$\frac{\mathrm{d}P_\gamma}{\mathrm{d}\omega} \approx \frac{9\sqrt{3}}{\sqrt{2\pi}} \frac{P_\gamma}{\omega_c} \sqrt{\frac{\omega}{\omega_c}} \mathrm{e}^{\omega/\omega_c} \approx 0.77736 \frac{P_\gamma}{\omega_c} \sqrt{\frac{\omega}{\omega_c}} \mathrm{e}^{\omega/\omega_c} \, . \tag{9.163}$$

Both approximations are included in Fig. 9.12 and display actually a rather good representation of the real spectral radiation distribution. Specifically, we note the slow increase in the radiation intensity at low frequencies and the exponential drop off above the critical frequency.

### 9.7.1 Statistical Radiation Parameters

The emission of synchrotron radiation is a classical phenomenon. For some applications it is, however, useful to express some parameters in statistical form. Knowing the spectral radiation distribution, we may follow Sands [14] and express some quantities in the photon picture. We have used such parameters in Chapter 7 to derive expressions for the equilibrium beam size and energy spread. Equilibrium beam parameters are determined by the statistical emission of photons and its recoil on the particle motion. For this purpose, we are mainly interested in an expression for $\varepsilon_{ph}^2$ and the photon flux at energy $\varepsilon_{ph}$. From these quantities, we may derive an expression for the average photon energy $\langle \varepsilon_{ph}^2 \rangle_s$ emitted along the circumference of the storage ring.

With $\Pi (\varepsilon_{ph})$ being the probability to emit a photon with energy $\varepsilon_{ph}$ we have

$$\langle \varepsilon_{ph}^2 \rangle = \int_0^\infty \varepsilon_{ph}^2 \, \Pi (\varepsilon_{ph}) \, \mathrm{d}\varepsilon_{ph} . \tag{9.164}$$

The probability $\Pi (\varepsilon_{ph})$ is defined by the ratio of the photon flux $\dot{n}(\varepsilon_{ph})$ emitted at energy $\varepsilon_{ph}$ to the total photon flux $\dot{N}_{ph}$

$$\Pi (\varepsilon_{ph}) = \frac{\dot{n}(\varepsilon_{ph})}{\dot{N}_{ph}} . \tag{9.165}$$

The photon flux at $\varepsilon_{ph}$ can be derived from $\varepsilon_{ph} \dot{n}(\varepsilon_{ph}) \, \mathrm{d}\varepsilon_{ph} = P (\varepsilon_{ph}) \, \mathrm{d}\varepsilon_{ph}$. Integrating (9.156) over all angles $\psi$ and multiplying by $\hbar\omega = \varepsilon_{ph}$ we get for the spectral radiation power

$$P (\varepsilon_{ph}) \, \mathrm{d}\varepsilon_{ph} = \varepsilon_{ph} \frac{\mathrm{d}\dot{N}}{\mathrm{d}\varepsilon_{ph}} \, \mathrm{d}\varepsilon_{ph} = \frac{P_\gamma}{\varepsilon_c} S \left( \frac{\varepsilon_{ph}}{\varepsilon_c} \right) \, \mathrm{d}\varepsilon_{ph} ,$$

and

$$\dot{n}(\varepsilon_{ph}) = \frac{P_\gamma}{\varepsilon_c^2} \frac{S (x)}{x} , \qquad \text{where} \qquad x = \frac{\varepsilon_{ph}}{\varepsilon_c} . \tag{9.166}$$

The total number of emitted photons per unit time is just the integral

$$\dot{N}_{ph} = \int_0^\infty \dot{n}(\varepsilon_{ph}) \, \mathrm{d}\varepsilon_{ph} = \frac{P_\gamma}{\varepsilon_c} \int_0^\infty \frac{S (x)}{x} \mathrm{d}x = \frac{15\sqrt{3}}{8} \frac{P_\gamma}{\varepsilon_c} . \tag{9.167}$$

With this, the probability to emit a photon of energy $\varepsilon_{ph}$ is finally

$$\Pi (\varepsilon_{ph}) = \frac{8}{15\sqrt{3}} \frac{1}{\varepsilon_c} \frac{S (x)}{x} , \tag{9.168}$$

and

$$\langle \varepsilon_{ph}^2 \rangle_s = \frac{8\varepsilon_c^2}{15\sqrt{3}} \int_0^\infty x \, S(x) \, \mathrm{d}x = \frac{11}{27} \varepsilon_c^2 . \tag{9.169}$$

To calculate equilibrium beam parameters in Section 7.4.1 and thereafter, we need to know the quantity $\left\langle \dot{N}_{\mathrm{ph}}\langle \varepsilon_{\mathrm{ph}}^2\rangle \right\rangle_s$ which is now from (9.167, 9.169)

$$\left\langle \dot{N}_{\mathrm{ph}}\langle \varepsilon^2\rangle \right\rangle_s = \frac{55}{24\sqrt{3}}\langle \varepsilon_c P_\gamma\rangle_s , \tag{9.170}$$

where the average is to be taken along the orbit and around the storage ring through all magnets. Expressing the critical photon energy by (9.78) and the radiation power by (9.59) and we get finally

$$\left\langle \dot{N}_{\mathrm{ph}}\langle \varepsilon^2\rangle \right\rangle_s = \frac{55}{24\sqrt{3}} r_c c\, mc^2\, \hbar c\, \gamma^7 \left\langle \frac{1}{\rho^3}\right\rangle_s . \tag{9.171}$$

# Exercises *

**Exercise 9.1 (S).** Integrate the radiation power distribution (9.76) over all solid angles and prove that the total radiation power is equal to (9.59).

**Exercise 9.2 (S).** In the ESRF (European Synchrotron Radiation Facility) synchrotron radiation source in Grenoble (France) an electron beam of 200 mA circulates at an energy of 6 GeV. The bending magnet field is 1.0 T. Calculate and plot the spectral photon flux into a band width of 0.1% and an acceptance angle of 10 mrad as a function of photon energy.

**Exercise 9.3 (S).** Derive an expression identifying the angle at which the spectral intensity has dropped to $p\%$ from the maximum intensity. Derive approximate expressions for very low or very large photon energies. Find the angle at which the total radiation intensity has dropped to 10%.

**Exercise 9.4.** Derive the wave equations (9.3) and (9.4).

**Exercise 9.5.** Derive (9.17).

**Exercise 9.6.** Derive (9.28) from (9.27) .Show that the electrical field in the radiation regime is purely orthogonal to the direction of observation. Is the field also parallel to the acceleration?

**Exercise 9.7.** Design a synchrotron radiation source for a photon energy of your choice. Use a simple FODO lattice and specify the minimum beam energy, beam current, and bending radius which will produce a bending magnet photon flux of $10^{14}$ photons/sec/mrad at the desired photon energy and into a band width of $\Delta\omega/\omega = 1\%$. What is the minimum and maximum photon energy for which the photon flux is at least $10^{11}$ photons/sec/mrad? How big is your ring assuming a 30% fill factor for bending magnets?

---

* The argument (S) indicates an exercise for which a solution is given in Appendix A.

# 10. Insertion Device Radiation

Synchrotron radiation from bending magnets is characterized by a wide spectrum from microwaves up to soft or hard x-rays as determined by the critical photon energy. To optimally meet the needs of basic research with synchrotron radiation, it is desirable to provide specific radiation characteristics that cannot be obtained from ring bending magnets but require special magnets. The field strength of bending magnets and the maximum particle beam energy in circular accelerators like a storage ring is fixed leaving no adjustments to optimize the synchrotron radiation spectrum for particular experiments. To generate specific synchrotron radiation characteristics, radiation is often produced from special insertion devices installed along the particle beam path. Such insertion devices introduce no net deflection of the beam and can therefore be incorporated in a beam line without changing its geometry. Motz [56] proposed first the use of undulators or wiggler magnets to optimize characteristics of synchrotron radiation. By now, such magnets have become the most common insertion devices consisting of a series of alternating magnet poles deflecting the beam periodically in opposite directions as shown in Fig. 10.1.

In Chap. 4 the properties of wiggler radiation were discussed shortly in an introductory way. Here we concentrate on more detailed and formal derivations of radiation characteristics from relativistic electrons passing through undulator and wiggler magnets.

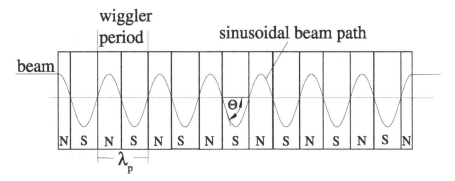

**Fig. 10.1.** Trajectory of a particle beam in a flat wiggler magnet

There is no fundamental difference between wiggler and undulator radiation. An undulator is basically a weak wiggler magnet. The deflection in an undulator is weak and the transverse particle momentum remains nonrelativistic. The motion is purely sinusoidal in a sinusoidal field, and the emitted radiation is monochromatic at the particle oscillation frequency which is the Lorentz-contracted periodicity of the undulator period. Since the radiation is emitted from a moving source the observer in the laboratory frame of reference then sees a Doppler shifted frequency. We call this monochromatic radiation the fundamental radiation or radiation at the fundamental frequency of the undulator.

As the undulator field is increased, the transverse motion becomes stronger and the transverse momentum starts to become relativistic. As a consequence, the so far purely sinusoidal motion becomes periodically distorted causing the appearance of harmonics of the fundamental monochromatic radiation. These harmonics increase in number and density with further increase of the magnetic field and, at higher frequencies, eventually merge into one broad spectrum characteristic for wiggler or bending magnet radiation. At very low frequencies, the theoretical spectrum is still a line spectrum showing the harmonics of the revolution frequency. Of course, there is a low frequency cutoff at a wavelength comparable or longer than vacuum chamber dimensions which therefore do not show-up as radiation.

An insertion device does not introduce a net deflection of the beam and we may therefore choose any arbitrary field strength which is technically feasible to adjust the radiation spectrum to experimental needs. The radiation intensity from a wiggler magnet also can be made much higher compared to that from a single bending magnet. A wiggler magnet with say ten poles acts like a string of ten bending magnets or radiation sources aligned in a straight line along the photon beam direction. The effective photon source is therefore ten times more intense than the radiation from a single bending magnet with the same field strength.

Wiggler magnets come in a variety of types with the flat wiggler magnet being the most common. In this wiggler type only the component $B_y$ is nonzero deflecting the beam in the horizontal plane. To generate circularly or elliptically polarized radiation, a helical wiggler magnet [36] may be used or a combination of several flat wiggler magnets deflecting the beam in orthogonal planes which will be discussed in more detail in Sect. 10.3.2.

## 10.1 Periodic Magnetic Field

Wiggler magnets are generally designed as flat wiggler magnets [56] with field components only in one plane or as helical wiggler magnets [36][57][58], where the transverse field component rotates along the magnetic axis. In this discussion, we concentrate on flat wigglers which are used widely to generate, for example, intense beams of synchrotron radiation from electron beams, to manipulate beam parameters or to pump a free electron laser.

Whatever the application may be, the wiggler magnet deflects the electron beam transversely in an alternating fashion without introducing an overall net deflection on the beam. Wiggler magnets are generally considered to be insertion devices, meaning they are not part of the basic magnet lattice of the accelerator but are installed in a magnet free straight section of the lattice. They can be turned on or off without affecting the functioning of the accelerator.

### 10.1.1 Periodic Field Configuration

To eliminate an overall effect of wiggler fields on the particle beam trajectory, the integrated magnetic field along the axis of the whole magnet must be zero

$$\int_{\text{wiggler}} B_\perp \, \mathrm{d}z = 0 \,. \tag{10.1}$$

Within this boundary condition we derive the general field configuration. Since a wiggler magnet is a straight device, we use a fixed Cartesian coordinate system $(x, y, z)$ with the $z$-axis parallel to the wiggler axis to describe the wiggler field. The origin of the coordinate system is placed in the middle of one of the wiggler magnet poles. The whole magnet may be composed of $N$ equal and symmetric pole pieces placed along the $z$-axis at a distance $\lambda_\mathrm{p}/2$ from pole center to pole center as depicted in Fig. 10.1. Each pair of adjacent wiggler poles forms one wiggler period with a length $\lambda_\mathrm{p}$ and the whole magnet is composed of $N_\mathrm{p} = \frac{1}{2} N$ periods. Since all periods are assumed to be equal and the beam deflection is compensated within each period no net beam deflection occurs for the complete magnet. At either end of the wiggler magnet, we must have "half poles" or "half fields" to match the external beam path.

We consider only periodic fields which can be expanded into a Fourier series along the axis including a strong fundamental component with a period length $\lambda_\mathrm{p}$ and higher harmonics expressed by the ansatz

$$B_y = B_\mathrm{o} \sum_{n=0}^{\infty} b_{2n+1}(x, y) \, \cos[(2n+1) \, k_\mathrm{p} z] \,, \tag{10.2}$$

where the wave number $k_\mathrm{p} = 2\pi/\lambda_\mathrm{p}$. The ideal configuration depends on the application. For the production of high brightness photon beams from an undulator one would choose a pure sinusoidal variation of the field with period $\lambda_\mathrm{p}$ and no higher harmonics. In applications where only flux or high photon energies are desired one would look for a magnet which exhibits some flat field profile along $z$ in each pole. Such a configuration would include many harmonics as reflected in (10.2).

The functions $b_{2n+1}(x,y)$ describe the variation of the field amplitude orthogonal to the beam axis for the harmonic $(2n+1)$. The content of higher harmonics is greatly influenced by the particular design of the wiggler magnet and the ratio of the period length to the pole gap aperture. For very long periods relative to the pole aperture the field profile approaches that of a hard-edge dipole field with a square field profile along the $z$-axis. For very short periods compared to the pole aperture, on the other hand, we find only a significant amplitude for the fundamental period and very small perturbations due to higher harmonics.

We may derive the magnetic field from Maxwell's equations based on a sinusoidal field along the axis. Each field harmonic may be determined separately due to the linear superposition of fields. To eliminate a dependence of the magnetic field on the horizontal variable $x$ we assume a pole width which is large compared to the pole aperture in which case we may set $b_{2n+1}(x,y) = b_{2n+1}(y)$. For the same reason and from symmetry $B_x \equiv 0$. The fundamental field component $(n=0)$ can then be expressed by

$$B_y(y,z) = B_0\, b_1(y)\, \cos k_{\mathrm{p}} z.\tag{10.3}$$

From Maxwell's curl equation $\nabla \times \mathbf{B} = 0$ we get $\frac{\partial B_z}{\partial y} = \frac{\partial B_y}{\partial z}$ and with (10.3) we have

$$\frac{\partial B_z}{\partial y} = \frac{\partial B_y}{\partial z} = -B_0\, b_1(y)\, k_{\mathrm{p}}\, \sin k_{\mathrm{p}} z.\tag{10.4}$$

We have not yet determined the $y$-dependence of the amplitude function $b_1(y)$. From $\nabla \mathbf{B} = 0$ and the independence of the field on the horizontal position we get with (10.3)

$$\frac{\partial B_z}{\partial z} = -B_0\, \frac{\partial b_1(y)}{\partial y}\, \cos k_{\mathrm{p}} z.\tag{10.5}$$

Forming the second derivatives $\partial^2 B_z/(\partial y\, \partial z)$ from (10.4) and (10.5) we get for the amplitude function the differential equation

$$\frac{\partial^2 b_1(y)}{\partial y^2} = k_{\mathrm{p}}^2\, b_1(y),\tag{10.6}$$

which can be solved by the hyperbolic functions

$$b_1(y) = a\, \cosh k_{\mathrm{p}} y + b\, \sinh k_{\mathrm{p}} y.\tag{10.7}$$

Since $b_1(0) = 1$ and the magnetic field is symmetric with respect to $y=0$ the coefficients are $a=1$ and $b=0$. Collecting all partial results, the wiggler magnet field is finally determined by the components

$$\begin{aligned} B_x &= 0,\\ B_y &= B_0 \cosh k_{\mathrm{p}} y\, \cos k_{\mathrm{p}} z,\\ B_z &= B_0 \sinh k_{\mathrm{p}} y\, \sin k_{\mathrm{p}} z, \end{aligned}\tag{10.8}$$

where $B_z$ is obtained by integration of (10.4) with respect to $y$.

The hyperbolic dependence of the field amplitude on the vertical position introduces higher-order field-errors which we determine by expanding the hyperbolic functions

$$\cosh k_{\mathrm{p}} y = 1 + \frac{(k_{\mathrm{p}} y)^2}{2!} + \frac{(k_{\mathrm{p}} y)^4}{4!} + \frac{(k_{\mathrm{p}} y)^6}{6!} + \frac{(k_{\mathrm{p}} y)^8}{8!} + \dots \tag{10.9a}$$

$$\sinh k_{\mathrm{p}} y = k_{\mathrm{p}} y + \frac{(k_{\mathrm{p}} y)^3}{3!} + \frac{(k_{\mathrm{p}} y)^5}{5!} + \frac{(k_{\mathrm{p}} y)\,7}{7!} + \dots \tag{10.9b}$$

Typically, the vertical gap in a wiggler magnet is smaller than the period length or $y < \lambda_{\mathrm{p}}$. For larger apertures the field strength reduces drastically. Due to the fast convergence of the series expansions only a few terms are required to obtain an accurate expression for the hyperbolic function within the wiggler aperture. The expansion displays the higher-order field components explicitly which, however, do not have the form of higher-order multipole fields and we cannot treat these fields just like any other multipole perturbation but must consider them separately.

To determine the path distortion due to the wiggler fields, we follow the reference trajectory through one quarter period starting at a symmetry plane in the middle of a pole. At the starting point $z = 0$ in the middle of a wiggler pole, the beam direction is parallel to the reference trajectory and the deflection angle at a downstream point $z$ is given by

$$\vartheta(z) = \frac{[c]\,e}{cp} \int_0^z B_y \, \mathrm{d}\bar{z} = \frac{[c]\,e}{cp} B_0 \cosh k_{\mathrm{p}} y \int_0^z \cos k_{\mathrm{p}} \bar{z} \, \mathrm{d}\bar{z}$$

$$= -\frac{[c]\,e}{cp} B_0 \frac{1}{k_{\mathrm{p}}} \cosh k_{\mathrm{p}} y \, \sin k_{\mathrm{p}} z \,. \tag{10.10}$$

The maximum deflection angle is equal to the deflection angle for a quarter period or half a wiggler pole and is from (10.10) for $y = 0$ and $k_{\mathrm{p}} z = \pi/2$

$$\theta = -\frac{[c]\,e}{cp} B_0 \frac{\lambda_{\mathrm{p}}}{2\,\pi} \,. \tag{10.11}$$

This deflection angle is used to define the wiggler strength parameter

$$K = \beta\gamma\,\theta = \frac{[c]\,e}{2\,\pi\,m\,c^2} B_0 \, \lambda_{\mathrm{p}} \,, \tag{10.12}$$

where $m\,c^2$ is the particle rest energy and $\gamma$ the particle energy in units of the rest energy. In more practical units this strength parameter is

$$K = C_K\, B_0 \,(\mathrm{T})\, \lambda_{\mathrm{p}}\,(\mathrm{cm}) \approx B_0\,(\mathrm{T})\, \lambda_{\mathrm{p}}\,(\mathrm{cm}) \,, \tag{10.13}$$

where

$$C_K = \frac{[c]\,e}{2\,\pi\,m\,c^2} = 0.93373 \ \mathrm{T}^{-1}\,\mathrm{cm}^{-1} \,. \tag{10.14}$$

## 10.1.2 Particle Dynamics in a Periodic Field Magnet

Particle dynamics and resulting radiation characteristics for an undulator have been derived first by Motz [56] and later in more detail by other authors [59]-[60]. A sinusoidally varying vertical field causes a periodic deflection of particles in the $(x, z)$-plane shown in Fig. 10.1. To describe the particle trajectory, we use the equation of motion

$$\frac{\mathbf{n}}{\rho} = [c] \frac{e}{mc^2\gamma\beta^2} [\boldsymbol{\beta} \times \boldsymbol{B}] , \tag{10.15}$$

where $\beta$ is the particle velocity and get with (10.3) the equations of motion in component form

$$\frac{d^2 x}{d t^2} = -\frac{eB_0}{\gamma mc} \frac{d z}{d t} \cos{(k_{\mathrm{p}} z)} , \tag{10.16a}$$

$$\frac{d^2 z}{dt^2} = +\frac{eB_0}{\gamma mc} \frac{d x}{d t} \cos{(k_{\mathrm{p}} z)} , \tag{10.16b}$$

where we have set $k_{\mathrm{p}} = 2\pi/\lambda_{\mathrm{p}}$ and $d z = \beta c d t$ with $\beta = v/c$.

Equations (10.16) describe the coupled motion of a particle in the sinusoidal field of a flat wiggler magnet. This coupling is common to the particle motion in any magnetic field but generally in beam dynamics we set $d z/d t \approx v$ and $d x/d t \approx 0$ because $d x/d t \ll d z/d t$. This approximation is justified in most beam transport applications for relativistic particles, but here we have to be cautious not to neglect effects that might be of relevance on a very short time or small geometric scale comparable to the oscillation period and wavelength of synchrotron radiation.

We will keep the $d x/d t$-term and get from (10.16a) with $d z/d t \approx v$ and after integrating twice that the particle trajectory follows the magnetic field in the sense that the oscillatory motion reaches a maximum where the magnetic field reaches a maximum and crosses the beam axis where the field is zero. We start at the time $t = 0$ in the middle of a magnet pole where the transverse velocity $\dot{x}_0 = 0$ while the longitudinal velocity $\dot{z}_0 = \beta c$ and integrate both equations (10.16) utilizing the integral of the first equation in the second to get

$$\frac{d x}{d t} = -\beta c \frac{K}{\gamma} \sin{(k_{\mathrm{p}} z)} , \tag{10.17a}$$

$$\frac{d z}{d t} = \beta c \left[ 1 - \frac{K^2}{2\gamma^2} \sin^2{(k_{\mathrm{p}} z)} \right] . \tag{10.17b}$$

The transverse motion describes the expected oscillatory motion and the longitudinal velocity $v$ exhibits a periodic modulation reflecting the varying projection of the velocity vector to the $z$-axis. Closer inspection of this velocity modulation shows that its frequency is twice that of the periodic motion.

It is convenient to describe the longitudinal particle motion with respect to a Cartesian reference frame moving uniformly along the $z$-axis with the average longitudinal particle velocity $\bar{\beta} c = \langle \dot{z} \rangle$ which can be derived from (10.17b)

$$\bar{\beta} = \beta \left( 1 - \frac{K^2}{4\gamma^2} \right) . \tag{10.18}$$

In this reference frame the particle follows a figure-of-eight trajectory composed of the transverse oscillation and a longitudinal oscillation with twice the frequency. We will come back to this point since both oscillations contribute to the radiation spectrum. A second integration of (10.17b) results finally in the equation of motion in component representation

$$x(t) = \frac{K}{\gamma k_p} \cos \left( k_p \bar{\beta} ct \right) , \tag{10.19a}$$

$$z(t) = \bar{\beta} ct + \frac{K^2}{8\gamma^2 k_p} \sin^2 \left( 2k_p \bar{\beta} ct \right) , \tag{10.19b}$$

where we set $z = \bar{\beta} ct$. The maximum amplitude $a$ of the transverse particle oscillation is finally

$$a = \frac{K}{\gamma k_p} = \frac{\lambda_p K}{2\pi \gamma} . \tag{10.20}$$

This last expression gives another simple relationship between the wiggler strength parameter and the transverse displacement of the beam trajectory

$$a \, (\mu m) = 0.8133 \frac{\lambda_p \, (\text{cm}) \; K}{E \, (\text{GeV})} . \tag{10.21}$$

For general cases, this beam displacement is very small.

### 10.1.3 Focusing in a Wiggler Magnet

As mentioned earlier, a wiggler magnet should be transparent to the electron beam which can be achieved only approximately. Every pole end generates some fringe fields which cause a focusing effect on the particle beam. In low energy storage rings with strong superconducting wavelength shifters this effect can be a major perturbation which requires significant compensation in the ring lattice proper. A detailed derivation of such fringe fields and their effect on the beam can be found in [45][46]. Here, we will only repeat some of the more salient features and results.

The beam path in a wiggler magnet is generally not parallel to the reference trajectory $z$ because of the transverse deflection in the wiggler field following a periodic sinusoidal form along the reference path. For this reason, the fringe field component $B_z$ appears to the particle partially as a transverse field which varies linearly with $y$. Such a field term constitutes focusing similar to that in a quadrupole with a strength for each wiggler pole end of

$$k_y \ell = -\frac{1}{f_y} = -\frac{\lambda_p}{8\rho_0^2}. \tag{10.22}$$

The focusing occurs in the vertical plane only assuming that the wiggler magnet deflects the beam in the horizontal plane, and is positive and independent of the sign of the deflection. For $N$ wiggler poles, we have $2N$ times the focusing strength of each individual pole end and the focal length of the total wiggler magnet of length $L_w = \frac{1}{2}N \lambda_p$ expressed in units of the wiggler strength parameter $K$ becomes

$$\frac{1}{f_y} = \frac{K^2}{2\gamma^2} k_p^2 L_w. \tag{10.23}$$

Tacitly, a rectangular form of the wiggler poles has been assumed (Fig. 10.2) and consistent with our sign convention we find that wiggler fringe fields cause focusing in the nondeflecting plane. Within the approximation used, there is no corresponding focusing effect in the deflecting plane. This is the situation for most wiggler magnet poles except for the first and last half pole where the beam enters the magnetic field normal to the pole face and no focusing occurs.

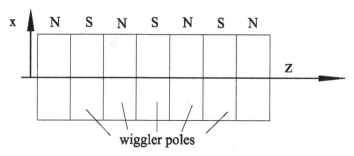

**Fig. 10.2.** Wiggler magnet with parallel pole end-faces

A reason to possibly use wiggler magnets with rotated pole faces like wedge magnets originates from the fact that the wiggler focusing is asymmetric, not part of the lattice focusing and may therefore need to be compensated. For moderately strong wiggler fields the asymmetric focusing in both planes can be compensated by small adjustments of lattice quadrupoles. The focusing effect of strong wiggler magnets, however, may generate a significant perturbation of the lattice focusing structure or create a situation where no stable solution for betatron functions exist. The severity of this problem can be reduced by designing the wiggler poles as wedge magnets in such a way as to split the focusing equally between both the horizontal and vertical plane (Fig. 10.3). In this case, local correction can be applied efficiently in nearby lattice or separate quadrupoles.

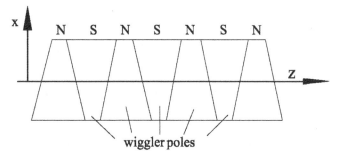

**Fig. 10.3.** Wiggler magnet with wedge shaped poles

The focal length of one half pole in the horizontal deflecting plane is from
[45], Sect. 5.3.2

$$\frac{1}{f_x} = \frac{\eta + \theta}{\rho_x}, \tag{10.24}$$

where the pole face rotation angle $\eta$ has been assumed to be small and of the
order of the wiggler deflection angle per pole and $\theta = K/\gamma$ is the deflection
angle of a half pole. In the case of a rectangular wiggler pole $\eta = -\theta$ and the
focusing in the deflecting plane vanishes as we would expect.

In the nondeflecting plane, equation (5.66) in [45] applies and the focal length
for small angles $\eta$ and $\theta$ is

$$\frac{1}{f_y} = -\frac{\eta + \theta}{\rho_x} - \frac{\pi\theta}{4\rho_x}. \tag{10.25}$$

The focusing in each single wiggler pole is rather weak and we may apply
thin lens approximation to derive the transformation matrices. For this we
consider the focusing to occur in the middle of each wiggler pole with drift
spaces of length $\lambda_p/4$ on each side. With $2/f$ being the focal length of a full
pole in either the horizontal plane (10.24) or the vertical plane (10.25) the
transformation matrix for each wiggler pole is finally

$$\mathcal{M}_{\text{pole}} = \begin{pmatrix} 1 & \lambda_p/4 \\ 0 & 1 \end{pmatrix} \begin{pmatrix} 1 & 0 \\ -2/f & 1 \end{pmatrix} \begin{pmatrix} 1 & \lambda_p/4 \\ 0 & 1 \end{pmatrix}$$

$$= \begin{pmatrix} 1 - \frac{\lambda_p}{2f} & \frac{\lambda_p}{2}\left(1 - \frac{\lambda_p}{4f}\right) \\ -\frac{2}{f} & 1 - \frac{\lambda_p}{2f} \end{pmatrix} \approx \begin{pmatrix} 1 & \frac{1}{2}\lambda_p \\ -2/f & 1 \end{pmatrix}, \tag{10.26}$$

where the approximation $\lambda_p \ll f$ was used. For a wiggler magnet of length
$L_w = \frac{1}{2}N\lambda_p$ we have $N$ poles and the total transformation matrix is

$$\mathcal{M}_{\text{wiggler}} = \mathcal{M}_{\text{pole}}^N. \tag{10.27}$$

This transformation matrix can be applied to each plane and any pole
rotation angle $\eta$. Specifically, we set $\eta = -K/\gamma$ for a rectangular pole cross

section and $\eta = 0$ for pole rotations orthogonal to the path like in sector magnets.

### 10.1.4 Hard Edge Wiggler Model

Although the magnetic properties of wiggler magnets are well understood and easy to apply it is nonetheless often desirable to derive the focusing effects from hard-edge wiggler magnets. This is particularly true when special numerical programs are to be used which are not designed to properly model a sinusoidal wiggler field. We would like therefore to represent a sinusoidal wiggler magnet by a constant field magnet, called a hard-edge model. On the other hand, accurate field representation is important since frequently strong wiggler magnets are to be inserted into a beam transport lattice.

For the proper representation of linear focusing properties of wiggler magnets by a hard-edge model we require three conditions to be fulfilled. First, the deflection angle for each hard-edge pole should be the same as that for the real wiggler magnet. Second, the edge focusing must be the same. Third, like any other bending magnet in an electron circular accelerator, a wiggler magnet also contributes to quantum excitation and damping of beam emittance and energy spread. The quantum excitation is in first approximation proportional to the third power of the curvature while the damping scales like the square of the curvature similar to focusing.

We consider now a wiggler field

$$B(z) = B_0 \cos k_{\mathrm{p}} z \tag{10.28}$$

and try to model the field for a half pole with parallel endpoles by a hard-edge magnet. The deflection angle of the hard-edge model of length $\ell$ and field $B$ must be the same as that for a wiggler half pole, or

$$\theta = \frac{\ell_{\mathrm{h}}}{\rho_{\mathrm{h}}} = \frac{[c]e}{cp_0} \int_{\text{halfpole}} B_y(z)\,\mathrm{d}z = \frac{\lambda_{\mathrm{p}}}{2\pi\,\rho_0}. \tag{10.29}$$

Here we use $\rho_{\mathrm{h}}$ for the bending radius of the hard-edge model and $\rho_0$ for the bending radius at the peak wiggler field $B_o$. The edge focusing condition can be expressed by

$$\frac{1}{f} = \frac{\ell_{\mathrm{h}}}{\rho_{\mathrm{h}}^2} = \frac{1}{\rho_0^2} \int_{\text{halfpole}} \cos^2 k_{\mathrm{p}} z\,\mathrm{d}z = \frac{\lambda_{\mathrm{p}}}{8\rho_0^2}. \tag{10.30}$$

Modeling a wiggler field by a single hard-edge magnet requires in linear beam optics only two conditions to be met which can be done with the two available parameters $B(z)$ and $\ell$. From (10.29, 10.30) we get therefore the hard-edge magnet parameters (Fig. 10.4)

$$\rho_{\mathrm{h}} = \frac{4}{\pi}\,\rho_0 \qquad \text{and} \qquad \ell_{\mathrm{h}} = \frac{2}{\pi^2}\,\lambda_{\mathrm{p}}. \tag{10.31}$$

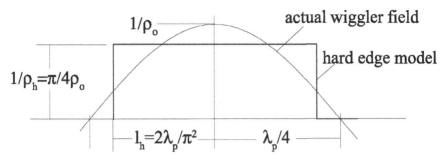

**Fig. 10.4.** Hard edge model for a wiggler magnet pole

For a perfect modeling of the equilibrium energy spread and emittance due to quantum excitation in electron storage rings we would also like the cubic term to be the same

$$\frac{\ell_h}{\rho_h^3} \stackrel{?}{=} \frac{1}{\rho_0^3} \int_{\text{halfpole}} \cos^3 k_p z \, dz = \frac{\lambda_p}{3\pi \, \rho_0^3} .\qquad(10.32)$$

Since we have no more free parameters available, we can at this point only estimate the mismatch. With (10.30,10.31) we get from (10.32) the inequality

$$\frac{1}{3\pi} \neq \frac{\pi}{32},$$

which indicates that the quantum excitation from wiggler magnets is not correctly treated although the error is only about 8%. Alternatively to the choice of modeling conditions just made, one could decide that the quadratic and cubic terms must be equal while the deflection angle is not constrained. This would be a reasonable assumption since the total deflection angle of a wiggler is compensated anyway. In this case the deflection angle would be underestimated by about 8%. Where these mismatches are not significant, the simple hard-edge model (10.32) can be applied. For more accuracy the sinusoidal wiggler field can be represented more accurately by splitting each half-pole into a series of hard-edge magnets with.

## 10.2 Undulator Radiation

The physical process of undulator radiation is not different from the radiation produced from a single bending magnet. However, the radiation received at great distances from the undulator exhibits special features which we will discus in more detail. Basically, we observe an electron performing $N_p$ oscillations while passing through an undulator, where $N_p$ is the number of undulator periods. The observed radiation spectrum is the Fourier transform of the electron motion and therefore quasi-monochromatic with a finite line width inversely proportional to the number of oscillations performed.

### 10.2.1 Fundamental Wavelength

Undulator radiation can also be viewed as a superposition of radiation fields from $N_p$ sources yielding quasi-monochromatic radiation as a consequence of interference. To see that, we observe the radiation at an angle $\vartheta$ with respect to the path of the electron as shown in Fig. 10.5.

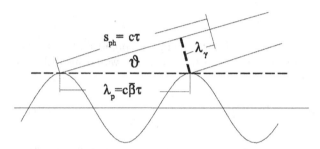

**Fig. 10.5.** Interference of undulator radiation

The electron travels on its path at an average velocity given by (10.18) and it takes the time

$$\tau = \frac{\lambda_p}{c\bar{\beta}} = \frac{\lambda_p}{c\beta\,[1 - K^2/(4\gamma^2)]} \tag{10.33}$$

to move along one undulator period. During that same time, the radiation front proceeds a distance

$$s_{\mathrm{ph}} = \tau c = \frac{\lambda_p}{\beta\,[1 - K^2/(4\gamma^2)]} \tag{10.34}$$

moving ahead of the particle since $s_{\mathrm{ph}} > \tau c\bar{\beta}$. For constructive superposition of radiation from all undulator periods, we require that the difference $s_{\mathrm{ph}} - \lambda_p \cos\vartheta$ be equal to an integer multiple of the wavelength $\lambda_k$ or for small observation angles $\vartheta \ll 1$

$$k\,\lambda_k = \frac{\lambda_p}{\beta\,[1 - K^2/(4\gamma^2)]} - \lambda_p\,(1 - \tfrac{1}{2}\vartheta^2). \tag{10.35}$$

After some manipulations, we get with $K^2/\gamma^2 \ll 1$ and $\beta \approx 1$ for the $k^{\mathrm{th}}$ harmonic of the fundamental wavelength of radiation into an angle $\vartheta$

$$\lambda_k = \frac{\lambda_p}{2\gamma^2\,k}\left(1 + \tfrac{1}{2}K^2 + \gamma^2\vartheta^2\right). \tag{10.36}$$

From an infinitely long undulator, the radiation spectrum consists of spectral lines at a wavelength determined by (10.36). In particular, we note that the shortest wavelength is emitted into the forward direction while the radiation at a finite angle $\vartheta$ appears red shifted by the Doppler effect. For an undulator with a finite number of periods, the spectral lines are widened to a width of about $1/N_p$ or less as we will discuss in the next section.

### 10.2.2 Radiation Power

The radiation power is from (9.41)

$$P = \tfrac{2}{3} r_c mc \, |\dot{\beta}^*|_r^2 \,, \tag{10.37}$$

where $^*$ indicates quantities to be evaluated in the particle reference system. We may use this expression in the particle system to calculate the total radiated energy from an electron passing through an undulator. The transverse particle acceleration is expressed by $m\dot{\mathbf{v}}^* = \mathbf{dp}_\perp/dt^* = \gamma \mathbf{dp}_\perp/dt$ where we used $t^* = t/\gamma$ and inserting into (10.37) we get

$$P = \tfrac{2}{3} \frac{r_c \gamma^2}{mc} \left( \frac{\mathbf{dp}_\perp}{dt} \right)^2 . \tag{10.38}$$

The transverse momentum is determined by the particle deflection in the undulator with a period length $\lambda_p$ and is for a particle of momentum $cp_0$

$$p_\perp = \hat{p} \sin \omega_p t \,, \tag{10.39}$$

where $\hat{p} = p_0 \theta$ and $\omega_p = ck_p = 2\pi c/\lambda_p$. The angle $\theta = K/\gamma$ is the maximum deflection angle defined in (10.12). With these expressions and averaging over one period, we get from (10.38) for the instantaneous radiation power from a charge $e$ traveling through an undulator

$$P_{inst} = \tfrac{1}{3} c r_c mc^2 \gamma^2 K^2 k_p^2 \,, \tag{10.40}$$

where $r_c$ is the classical electron radius. The duration of the radiation pulse is equal to the travel time through an undulator of length $L_u = \lambda_p N_p$ and the total radiated energy per electron is therefore

$$\Delta E = \tfrac{1}{3} r_c mc^2 \, \gamma^2 \, K^2 \, k_p^2 \, L_u \,. \tag{10.41}$$

In more practical units

$$\Delta E(eV) = C_u \frac{E^2 K^2}{\lambda_p^2} L_u = 725.69 \frac{E^2(\text{GeV}) K^2}{\lambda_p^2(\text{cm})} L_u(\text{m}) \tag{10.42}$$

with

$$C_u = \frac{4\pi^2 r_c}{3\,mc^2} = 7.2569 \times 10^{-20} \frac{m}{eV}. \tag{10.43}$$

The average total undulator radiation power for an electron beam circulating in a storage ring is then just the radiated energy (10.41) multiplied by the number of particles $N_b$ in the beam and the revolution frequency or

$$P_{avg} = \tfrac{1}{3} r_c \, c \, mc^2 \, \gamma^2 \, K^2 \, k_p^2 \, N_b \, \frac{L_u}{2\pi\,\bar{R}} \tag{10.44}$$

or

$$P_{avg}(W) = 6.336\,E^2(\text{GeV})\,B_0^2(\text{kG})\,I(\text{A})\,L_u(\text{m}), \tag{10.45}$$

where $I$ is the circulating electron beam current. The total angle integrated radiation power from an undulator in a storage ring is proportional to the square of the beam energy and maximum undulator field $B_0$ and proportional to the beam current and undulator length.

### 10.2.3 Spatial and Spectral Distribution

For bending magnet radiation, the particle dynamics is relatively simple being determined only by the particle velocity and the bending radius of the magnet. In a wiggler magnet, the magnetic field parameters are different from those in a constant field magnet and we will therefore derive again the synchrotron radiation spectrum for the beam dynamics in a general wiggler magnet. No special assumptions on magnetic field configurations have been made to derive the radiation spectrum (9.89) and we can therefore use this expression together with the appropriate beam dynamics to derive the radiation spectrum from a wiggler magnet

$$\frac{d^2W}{d\omega\,d\Omega} = \frac{r_c\,mc\,\omega^2}{4\pi^2}\left|\int_{-\infty}^{\infty} n \times [n \times \beta]\,e^{-i\omega\left(t_r + \frac{R}{c}\right)}\,dt_r\right|^2. \tag{10.46}$$

The integrand in (10.46) can be evaluated from known particle dynamics in a wiggler magnet noting that all quantities are to be taken at the retarded time $t_r$. The unit vector from the observer to the radiating particle is from Fig. 10.6

$$n = -x \cos\varphi \sin\vartheta - y \sin\varphi \sin\vartheta - z \cos\vartheta. \tag{10.47}$$

The exponent in (10.46) includes the term $R/c = nR/c$. We express again the vector $R$ from the observer to the particle by the constant vector $r$ from the origin of the coordinate system to the observer and the vector $r_p$ from the coordinate origin to the particle for $R = -r + r_p$ as shown in Fig. 10.6.

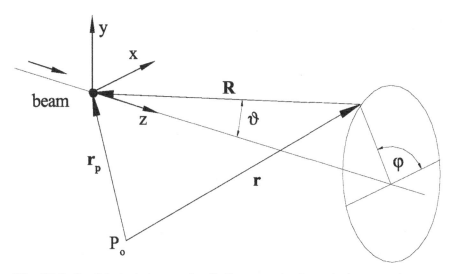

**Fig. 10.6.** Particle trajectory and radiation geometry in a wiggler magnet

The $r$-term gives only a constant phase shift and can therefore be ignored. The location vector $\boldsymbol{r}_\mathrm{p}$ of the particle with respect to the origin of the coordinate system is

$$\boldsymbol{r}_\mathrm{p}(t_\mathrm{r}) = x(t_\mathrm{r})\,\boldsymbol{x} + z(t_\mathrm{r})\,\boldsymbol{z}$$

and with the solutions (10.19) we have

$$\boldsymbol{r}_\mathrm{p}(t_\mathrm{r}) = \frac{K}{k_\mathrm{p}\,\gamma}\,\cos(\omega_\mathrm{p} t_\mathrm{r})\,\boldsymbol{x} + \left[\bar{\beta}\,c\,t_\mathrm{r} + \frac{K^2}{8\pi\,k_\mathrm{p}}\,\sin(2\omega_\mathrm{p}\,t_\mathrm{r})\right]\boldsymbol{z}\,, \qquad (10.48)$$

where

$$\omega_\mathrm{p} = k_\mathrm{p}\,\bar{\beta}c\,. \qquad (10.49)$$

The velocity vector finally is just the time derivative of (10.48)

$$\boldsymbol{\beta}(t_\mathrm{r}) = -\frac{K}{\gamma}\,\bar{\beta}\,\sin(\omega_\mathrm{p} t_\mathrm{r})\,\boldsymbol{x} + \bar{\beta}\left[1 + \frac{K^2}{4\gamma^2}\,\cos(2\omega_\mathrm{p} t_\mathrm{r})\right]\boldsymbol{z}\,. \qquad (10.50)$$

We use these vector relations to evaluate the integrand in (10.46). First, we express the triple vector product $\boldsymbol{n}\times[\boldsymbol{n}\times\boldsymbol{\beta}]$ by its components and get with (10.47, 10.50)

$$\boldsymbol{n} \times [\boldsymbol{n} \times \boldsymbol{\beta}] = +\boldsymbol{x} \left[ -\frac{K}{\gamma}\bar{\beta}\sin^2\vartheta\,\cos^2\varphi\,\cos\omega_\mathrm{p}t_\mathrm{r} + \frac{K}{\gamma}\bar{\beta}\sin\omega_\mathrm{p}t_\mathrm{r} \right.$$

$$\left. +\bar{\beta}\left(1+\frac{K^2}{4\gamma^2}\cos 2\omega_\mathrm{p}t_\mathrm{r}\right)\sin\vartheta\,\cos\vartheta\,\cos\varphi \right]$$

$$+\boldsymbol{y}\left[ -\frac{K}{\gamma}\bar{\beta}\sin^2\vartheta\,\sin\varphi\cos\varphi\,\sin\omega_\mathrm{p}t_\mathrm{r} \right.$$

$$\left. +\bar{\beta}\left(1+\frac{K^2}{4\gamma^2}\cos 2\omega_\mathrm{p}t_\mathrm{r}\right)\sin\vartheta\,\cos\vartheta\,\sin\varphi \right] \qquad (10.51)$$

$$+\boldsymbol{z}\left[ -\frac{K}{\gamma}\bar{\beta}\sin\vartheta\,\cos\vartheta\,\cos\varphi\,\cos\omega_\mathrm{p}t_\mathrm{r} \right.$$

$$\left. +\bar{\beta}\left(1+\frac{K^2}{4\gamma^2}\cos 2\omega_\mathrm{p}t_\mathrm{r}\right)(\cos^2\vartheta-1) \right].$$

This expression can be greatly simplified considering that the radiation is emitted into only a very small angle $\vartheta \ll 1$. Furthermore, we note that the deflection due to the wiggler field is in most practical cases very small and therefore $K \ll \gamma$ and $\bar{\beta} = \beta\left(1-\frac{K^2}{4\gamma^2}\right) \approx \beta$. Finally, we carefully set $\beta \approx 1$ where this term does not appear as a difference to unity. With this and ignoring second order terms in $\vartheta$ and $K/\gamma$ we get from (10.51)

$$\boldsymbol{n} \times [\boldsymbol{n} \times \boldsymbol{\beta}] = \left(\bar{\beta}\vartheta\cos\varphi + \bar{\beta}\frac{K}{\gamma}\sin\omega_\mathrm{p}t_\mathrm{r}\right)\boldsymbol{x} + \bar{\beta}\vartheta\sin\varphi\;\boldsymbol{y}. \qquad (10.52)$$

The vector product in the exponent of the exponential function is just the product of (10.47) and (10.48)

$$\frac{1}{c}\boldsymbol{n}\boldsymbol{r}_\mathrm{p}(t_\mathrm{r}) = -\frac{K\bar{\beta}}{\gamma\omega_\mathrm{p}}\sin\vartheta\,\cos\varphi\,\cos\omega_\mathrm{p}t_\mathrm{r} - \left(\bar{\beta}\,t_\mathrm{r} + \frac{K^2\bar{\beta}}{8\gamma^2\omega_\mathrm{p}}\sin 2\omega_\mathrm{p}t_\mathrm{r}\right)\cos\vartheta. \qquad (10.53)$$

Employing again the approximation $\vartheta \ll 1$ and keeping only linear terms we get from (10.53)

$$t_\mathrm{r} + \frac{1}{c}\boldsymbol{n}\boldsymbol{r}_\mathrm{p}(t_\mathrm{r}) = t_\mathrm{r}(1-\bar{\beta}\cos\vartheta) - \frac{K\bar{\beta}\vartheta}{\gamma\omega_\mathrm{p}}\cos\varphi\,\cos\omega_\mathrm{p}t_\mathrm{r} - \frac{K^2\bar{\beta}}{8\gamma^2\omega_\mathrm{p}}\sin 2\omega_\mathrm{p}\,t_\mathrm{r}. \qquad (10.54)$$

With (10.18) and $\cos\vartheta \approx 1 - \frac{1}{2}\vartheta^2$, the first term becomes

$$1 - \bar{\beta}\cos\vartheta = \frac{1}{2\gamma^2}\left(1+\tfrac{1}{2}K^2+\gamma^2\vartheta^2\right) = \frac{\omega_\mathrm{p}}{\omega_1}, \qquad (10.55)$$

where we have defined the fundamental wiggler frequency $\omega_1$ by

$$\omega_1 = \omega_\mathrm{p}\frac{2\gamma^2}{1+\tfrac{1}{2}K^2+\gamma^2\vartheta^2} \qquad (10.56)$$

or the fundamental wavelength of the radiation

$$\lambda_1 = \frac{\lambda_p}{2\gamma^2} \left(1 + \tfrac{1}{2} K^2 + \gamma^2 \vartheta^2\right) \tag{10.57}$$

in full agreement with (10.36). At this point, it is worth to remember that the term $\tfrac{1}{2} K^2$ becomes $K^2$ for a helical wiggler [36]. With (10.55), the complete exponential term in (10.46) can be evaluated

$$-\mathrm{i}\,\omega \left[t_r + \tfrac{1}{c} \boldsymbol{n} \boldsymbol{r}_p(t_r)\right] =$$
$$-\mathrm{i}\,\frac{\omega}{\omega_1} \left(\omega_p t_r - \frac{K\bar{\beta}\vartheta}{\gamma}\frac{\omega_1}{\omega_p}\cos\varphi\,\cos\omega_p t_r - \frac{K^2\bar{\beta}}{8\gamma^2}\frac{\omega_1}{\omega_p}\sin 2\omega_p t_r\right).$$

Equation (10.46) can be modified with these expressions into a form suitable for integration by inserting (10.52) and (10.55) into (10.46) for

$$\frac{\mathrm{d}^2 W}{\mathrm{d}\omega\,\mathrm{d}\Omega} = \frac{r_c\,mc\,\omega^2}{4\pi^2}\bar{\beta}^2 \tag{10.58}$$

$$\times \left| \int_{-\infty}^{\infty} \left[\left(\vartheta\cos\varphi + \frac{K}{\gamma}\sin\omega_p t_r\right)\boldsymbol{x} + \vartheta\sin\varphi\,\boldsymbol{y}\right]\mathrm{e}^X\,\mathrm{d}t_r\right|^2,$$

where

$$X = \left[-\mathrm{i}\,\frac{\omega}{\omega_1}\left(\omega_p t_r - \frac{K\vartheta}{\gamma}\frac{\omega_1}{\omega_p}\cos\varphi\,\cos\omega_p t_r - \frac{K^2}{8\gamma^2}\frac{\omega_1}{\omega_p}\sin 2\omega_p\,t_r\right)\right].$$

We are now ready to perform the integration of (10.58) noticing that the integration over all times can be simplified by separation into an integral along the wiggler magnet alone and an integration over the rest of the time while the particle is traveling in a field free space. We write symbolically

$$\int_{-\infty}^{\infty} = \int_{-\pi N_p/\omega_p}^{\pi N_p/\omega_p} (K \neq 0) + \int_{-\infty}^{\infty} (K = 0) - \int_{-\pi N_p/\omega_p}^{\pi N_p/\omega_p} (K = 0). \tag{10.59}$$

First we evaluate the second integral for $K = 0$ which is of the form

$$\int_{-\infty}^{\infty} \mathrm{e}^{\mathrm{i}\kappa\omega t}\,\mathrm{d}t = \frac{2\pi}{|\kappa|}\,\delta(\omega),$$

where $\delta(\omega)$ is the Dirac $\delta$-function. The value of the integral is nonzero only for $\omega = 0$ in which case the factor $\omega^2$ in (10.58) causes the whole expression to vanish. The second integral is therefore zero.

The third integral has the same form as the second integral, but since the integration is conducted only over the length of the wiggler magnet we get

$$\int\limits_{-\pi N_p/\omega_p}^{\pi N_p/\omega_p} e^{-i\frac{\omega}{2\gamma^2}t_r}\, dt_r = \frac{2\pi N_p}{\omega_p}\frac{\sin\frac{\pi N_p}{2\gamma^2}\frac{\omega}{\omega_p}}{\frac{\pi N_p}{2\gamma^2}\frac{\omega}{\omega_p}}. \tag{10.60}$$

The value of this integral reaches a maximum of $2\pi\frac{N_p}{\omega_p}$ for $\omega \to 0$. From (10.58) we note the coefficient of this integral to include the angle $\vartheta \gtrsim 1/\gamma$ and the whole integral is therefore of the order or less than $L_u/(c\gamma)$, where $L_u = N_p\lambda_p$ is the total length of the wiggler magnet. This value is in general very small compared to the first integral and can therefore be neglected. Actually, this statement is only partially true since the first integral, as we will see, is a fast varying function of the radiation frequency with a distinct line spectrum. Being, however, primarily interested in the peak intensities of the spectrum we may indeed neglect the third integral. Only between the spectral lines does the radiation intensity from the first integral become so small that the third integral would be a relatively significant although absolutely a small contribution.

To evaluate the first integral in (10.59) with $K \neq 0$ we follow Alferov [59] and introduce with (10.56) the abbreviations

$$C = \frac{2K\,\bar{\beta}\,\gamma\vartheta\cos\varphi}{1 + \frac{1}{2}K^2 + \gamma^2\vartheta^2}, \tag{10.61a}$$

$$S = \frac{K^2\,\bar{\beta}}{4\left(1 + \frac{1}{2}K^2 + \gamma^2\vartheta^2\right)} \tag{10.61b}$$

and get from (10.58) the exponential functions in the form

$$e^{-i\frac{\omega}{\omega_1}\omega_p t_r}\, e^{i\frac{\omega}{\omega_1}C\cos\omega_p t_r}\, e^{i\frac{\omega}{\omega_1}S\sin 2\omega_p t_r}. \tag{10.62}$$

The integral in the radiation power spectrum (10.58) has two distinct forms, one where the integrand is just the exponential function multiplied by a time independent factor while the other includes the sine function $\sin\omega_p t_r$ as a factor of the exponential function. To proceed further we replace the exponential functions by an infinite sum of Bessel's functions

$$e^{i\kappa\sin\psi} = \sum_{p=-\infty}^{p=\infty} J_p(\kappa)\,e^{ip\psi} \tag{10.63}$$

and apply this identity to the first integral type in (10.58). Applying the identity (10.63) also to the second and third exponential factors in (10.62), we get with $e^{a\cos x} = e^{a\sin(x+\pi/2)}$ the product of the exponential functions

$$\mathrm{e}^{-\mathrm{i}\left(\frac{\omega}{\omega_1}\omega_\mathrm{p}t_\mathrm{r}-\frac{\omega}{\omega_1}C\cos\omega_\mathrm{p}t_\mathrm{r}-\frac{\omega}{\omega_1}S\sin 2\omega_\mathrm{p}t_\mathrm{r}\right)}$$

$$= \sum_{m=-\infty}^{\infty}\sum_{n=-\infty}^{\infty} J_m(u)\,J_n(v)\,\mathrm{e}^{\mathrm{i}\frac{1}{2}\pi n}\mathrm{e}^{-\mathrm{i}R_\omega\omega_\mathrm{p}t_\mathrm{r}}, \tag{10.64}$$

where

$$R_\omega = \frac{\omega}{\omega_1} - n - 2m\,,$$

$$u = \frac{\omega}{\omega_1}S, \qquad\text{and} \tag{10.65}$$

$$v = \frac{\omega}{\omega_1}C\,.$$

The time integration along the length of the wiggler magnet is straight forward for this term since no other time dependent factors are involved and we get

$$\int_{-\pi N_\mathrm{p}/\omega_\mathrm{p}}^{\pi N_\mathrm{p}/\omega_\mathrm{p}} \mathrm{e}^{-\mathrm{i}\left(\frac{\omega}{\omega_1}-n-2m\right)\omega_\mathrm{p}t_\mathrm{r}}\,\mathrm{d}t_\mathrm{r} = \frac{2\pi N_\mathrm{p}}{\omega_\mathrm{p}}\frac{\sin\pi N_\mathrm{p}\,R_\omega}{\pi N_\mathrm{p}\,R_\omega}\,. \tag{10.66}$$

In the second form of the integrand, we replace the trigonometric factor, $\sin\omega_\mathrm{p}t_\mathrm{r}$, by exponential functions and get with (10.66) integrals of the form

$$\int_{-\pi N_\mathrm{p}/\omega_\mathrm{p}}^{\pi N_\mathrm{p}/\omega_\mathrm{p}} \sin\omega_\mathrm{p}t_\mathrm{r}\,\mathrm{e}^{-\mathrm{i}R_\omega\,\omega_\mathrm{p}t_\mathrm{r}}\,\mathrm{d}t_\mathrm{r}$$

$$= -\mathrm{i}\frac{1}{2}\int_{-\pi N_\mathrm{p}/\omega_\mathrm{p}}^{\pi N_\mathrm{p}/\omega_\mathrm{p}} \left(\mathrm{e}^{\mathrm{i}\,\omega_\mathrm{p}t_\mathrm{r}}-\mathrm{e}^{-\mathrm{i}\,\omega_\mathrm{p}t_\mathrm{r}}\right)\mathrm{e}^{-\mathrm{i}R_\omega\,\omega_\mathrm{p}t_\mathrm{r}}\mathrm{d}t_\mathrm{r} \tag{10.67}$$

$$= \mathrm{i}\frac{\pi N_\mathrm{p}}{\omega_\mathrm{p}}\frac{\sin\pi N_\mathrm{p}\,(R_\omega+1)}{\pi N_\mathrm{p}(R_\omega+1)} - \mathrm{i}\frac{\pi N_\mathrm{p}}{\omega_\mathrm{p}}\frac{\sin\pi N_\mathrm{p}\,(R_\omega-1)}{\pi N_\mathrm{p}\,(R_\omega-1)}\,.$$

Both integrals (10.66) and (10.67) exhibit the character of multibeam interference spectra well known from optical interference theory. The physical interpretation here is that the radiation from the $N_\mathrm{p}$ wiggler periods consists of $N_\mathrm{p}$ photon beamlets which have a specific phase relationship such that the intensities are strongly reduced for all frequencies but a few specific frequencies as determined by the $\frac{\sin x}{x}$-factors. The resulting line spectrum, characteristic for undulator radiation, is the more pronounced the more periods or beamlets are available for interference. To get a more complete picture of the interference pattern, we collect now all terms derived separately so far and use them in (10.58) which becomes with (10.62)

$$\frac{\mathrm{d}^2 W}{\mathrm{d}\omega\,\mathrm{d}\Omega} = a \left| \int_{-\pi N_{\mathrm{p}}/\omega_{\mathrm{p}}}^{\pi N_{\mathrm{p}}/\omega_{\mathrm{p}}} [(A_0 + A_1 \sin \omega_{\mathrm{p}} t_{\mathrm{r}})\,\boldsymbol{x} + B_0\,\boldsymbol{y}] \right.$$

$$\left. \times e^{-\mathrm{i}\frac{\omega}{\omega_1}\omega_{\mathrm{p}} t_{\mathrm{r}}}\, e^{\mathrm{i}\,v\,\cos\omega_{\mathrm{p}} t_{\mathrm{r}}}\, e^{\mathrm{i}\,u\,\sin 2\omega_{\mathrm{p}} t_{\mathrm{r}}}\, \mathrm{d} t_{\mathrm{r}} \right|^2 ,$$

where $a = \frac{r_{\mathrm{e}} mc\,\beta^2}{4\pi^2}\omega^2$, $A_0 = \vartheta\cos\varphi$, $A_1 = \frac{K}{\gamma}$, and $B_0 = \vartheta\sin\varphi$. Introducing the identity (10.62), the photon energy spectrum becomes

$$\frac{\mathrm{d}^2 W}{\mathrm{d}\omega\,\mathrm{d}\Omega} = a \left| \int_{-\pi N_{\mathrm{p}}/\omega_{\mathrm{p}}}^{\pi N_{\mathrm{p}}/\omega_{\mathrm{p}}} [(A_0 + A_1 \sin \omega_{\mathrm{p}} t_{\mathrm{r}})\,\boldsymbol{x} + B_0\,\boldsymbol{y}] \right.$$

$$\left. \times \sum_{m=-\infty}^{\infty}\sum_{n=-\infty}^{\infty} J_m(u)\,J_n(v)\,e^{\mathrm{i}\frac{1}{2}\pi n}\,e^{-\mathrm{i}R_\omega \omega_{\mathrm{p}} t_{\mathrm{r}}}\, \mathrm{d} t_{\mathrm{r}} \right|^2$$

and after integration with (10.66) and (10.67)

$$\frac{\mathrm{d}^2 W}{\mathrm{d}\omega\,\mathrm{d}\Omega} = a \left| \boldsymbol{x}\, A_0 \sum_{m=-\infty}^{\infty}\sum_{n=-\infty}^{\infty} J_m(u)\,J_n(v)\,e^{\mathrm{i}\frac{1}{2}\pi n}\frac{2\pi N_{\mathrm{p}}}{\omega_{\mathrm{p}}}\frac{\sin\pi N_{\mathrm{p}} R_\omega}{\pi N_{\mathrm{p}} R_\omega} \right.$$

$$+ \boldsymbol{x}\, A_1 \sum_{m=-\infty}^{\infty}\sum_{n=-\infty}^{\infty} J_m(u)\,J_n(v)\,e^{\mathrm{i}\frac{1}{2}\pi n} \qquad (10.68)$$

$$\times\, \mathrm{i}\frac{\pi N_{\mathrm{p}}}{2\omega_{\mathrm{p}}}\left[\frac{\sin\pi N_{\mathrm{p}}(R_\omega+1)}{\pi N_{\mathrm{p}}(R_\omega+1)} - \mathrm{i}\frac{\pi N_{\mathrm{p}}}{\omega_{\mathrm{p}}}\frac{\sin\pi N_{\mathrm{p}}(R_\omega-1)}{\pi N_{\mathrm{p}}(R_\omega-1)}\right]$$

$$\left. + \boldsymbol{y}\, B_0 \sum_{m=-\infty}^{\infty}\sum_{n=-\infty}^{\infty} J_m(u)\,J_n(v)\,e^{\mathrm{i}\frac{1}{2}\pi n}\frac{2\pi N_{\mathrm{p}}}{\omega_{\mathrm{p}}}\frac{\sin\pi N_{\mathrm{p}} R_\omega}{\pi N_{\mathrm{p}} R_\omega} \right|^2 .$$

To determine the frequency and radiation intensity of the line maxima, we simplify the double sum of Bessel's functions by selecting only the most dominant terms. The first and third sums in (10.68) show an intensity maximum for $R_\omega = 0$ at frequencies

$$\omega = (n + 2m)\,\omega_1 , \qquad (10.69)$$

and intensity maxima appear therefore at the frequency $\omega_1$ and harmonics thereof. The transformation of a lower frequency to very high values has two physical components. In the system of relativistic particles, the static magnetic field of the wiggler magnet appears Lorentz contracted by the factor $\gamma$, and particles passing through the wiggler magnet oscillate with the frequency $\gamma\omega_{\mathrm{p}}$ in its own system emitting radiation at that frequency. The observer in the laboratory system receives this radiation from a source moving with relativistic velocity and experiences therefore a Doppler shift by the factor $2\gamma$. The wavelength of the radiation emitted in the forward direction, $\vartheta = 0$, from a weak wiggler magnet, $K \ll 1$, with the period length $\lambda_\mathrm{p}$ is therefore reduced by the factor $2\gamma^2$. In cases of a stronger wiggler magnet or when observing at a finite angle $\vartheta$, the wavelength is somewhat longer as one would expect from higher order terms of the Doppler effect.

From (10.68) we determine two more dominant terms originating from the second term for $R_\omega \pm 1 = 0$ at frequencies

$$\omega = (n + 2m - 1)\,\omega_1 \tag{10.70a}$$

$$\omega = (n + 2m + 1)\,\omega_1\,, \tag{10.70b}$$

respectively. The summation indices $n$ and $m$ are arbitrary integers between $-\infty$ and $\infty$. Among all possible resonant terms we collect such terms which contribute to the same harmonic $k$ of the fundamental frequency $\omega_1$. To collect these dominant terms for the same harmonic we set $\omega = \omega_k = k\,\omega_1$ where $k$ is the harmonic number of the fundamental and express the index $n$ by $k$ and $m$ to get

$$\begin{aligned} \text{from (10.69):} \quad & n = k - 2m, \\ \text{and from (10.70a):} \quad & n = k - 2m + 1 \\ \text{and (10.70b):} \quad & n = k - 2m - 1\,. \end{aligned} \tag{10.71}$$

Introducing these conditions into (10.68) all trigonometric factors assume the form $\frac{\sin(\pi N_\mathrm{p}\,\Delta\omega_k/\omega_1)}{\pi N_\mathrm{p}\,\Delta\omega_k/\omega_1}$, where

$$\frac{\Delta\omega_k}{\omega_1} = \frac{\omega}{\omega_1} - k \tag{10.72}$$

and we get the photon energy spectrum of the $k$-th harmonic

$$\begin{aligned}
\frac{\mathrm{d}^2 W_k(\omega)}{\mathrm{d}\omega\,\mathrm{d}\Omega} = {} & \frac{r_\mathrm{c}\,mc\,\bar{\beta}^2 N_\mathrm{p}^2}{\gamma^2}\,\frac{\omega^2}{\omega_\mathrm{p}^2}\left(\frac{\sin\pi N_\mathrm{p}\,\Delta\omega_k/\omega_1}{\pi N_\mathrm{p}\,\Delta\omega_k/\omega_1}\right)^2 \\[2mm]
& \times \Bigg|+\boldsymbol{x}\,A_0 \sum_{m=-\infty}^{\infty} J_m(u)\,J_{k-2m}(v)\,\mathrm{e}^{\mathrm{i}\frac{1}{2}\pi(k-2m)} \\[2mm]
& \quad +\boldsymbol{y}\,B_0 \sum_{m=-\infty}^{\infty} J_m(u)\,J_{k-2m}(v)\,\mathrm{e}^{\mathrm{i}\frac{1}{2}\pi(k-2m)} \\[2mm]
& \quad +\mathrm{i}\tfrac{1}{2}\,\boldsymbol{x}\,A_1 \sum_{m=-\infty}^{\infty} J_m(u)\,J_{k-2m+1}(v)\,\mathrm{e}^{\mathrm{i}\frac{1}{2}\pi(k-2m+1)} \\[2mm]
& \quad -\mathrm{i}\tfrac{1}{2}\,\boldsymbol{x}\,A_1 \sum_{m=-\infty}^{\infty} J_m(u)\,J_{k-2m-1}(v)\,\mathrm{e}^{\mathrm{i}\frac{1}{2}\pi(k-2m-1)}\Bigg|^2\,.
\end{aligned} \tag{10.73}$$

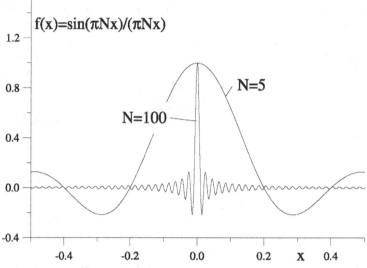

**Fig. 10.7.** $\frac{\sin \pi N_p x}{\pi N_p x}$ distribution for $N_p = 5$ and $N_p = 100$

All integrals exhibit the resonance character defining the locations of the spectral lines. The $(\sin x/x)$-terms are known from interference theory and represents the line spectrum of the radiation. Specifically, the number $N_p$ of beamlets, here source points, determines the spectral purity of the radiation. In Fig. 10.7 the $(\sin x/x)$-function is shown for $N_p = 5$ and $N_p = 100$. It is clear that the spectral purity improves greatly as the number of undulator periods is increased. This is one of the key features of undulator magnets to gain spectral purity by maximizing the number of undulator periods.

The spectral purity or line width is determined by the shape of the $(\sin x/x)$-function. We define the line width by the frequency at which $\sin x/x = 0$ or where $\pi N_p \, \Delta \omega_k / \omega_1 = \pi$ defining the line width for the $k^{\text{th}}$ harmonic

$$\frac{\Delta \omega_k}{\omega_k} = \pm \frac{1}{k \, N_p} . \tag{10.74}$$

The spectral width of the undulator radiation is reduced proportional to the number of undulator periods, but reduces also proportional to the harmonic number.

The Bessel functions $J_m(u)$ etc. determine mainly the intensity of the line spectrum. For an undulator with $K \ll 1$, the argument $u \propto K^2 \ll 1$ and the contributions of higher order Bessel's functions are very small. The radiation spectrum consists therefore only of the fundamental line. For stronger undulators with $K > 1$, higher order Bessel's functions grow and higher harmonic radiation appears in the line spectrum of the radiation.

Summing over all harmonics of interest, one gets the total power spectrum. In the third and fourth terms of (10.73) we use the identities $i e^{\pm i\pi/2} = \mp 1$, $J_m(u)\, e^{i\pi m} = J_{-m}(u)$ and abbreviate the sums of Bessel's functions by the symbols

$$\sum_1 = \sum_{m=-\infty}^{\infty} J_{-m}(u)\, J_{k-2m}(v) \tag{10.75a}$$

$$\sum_2 = \sum_{m=-\infty}^{\infty} J_{-m}(u)\, [J_{k-2m-1}(v) + J_{k-2m+1}(v)]\,. \tag{10.75b}$$

The total number of photons $N_{\mathrm{ph}}$ emitted into a spectral band width $\Delta\omega/\omega$ by a single electron moving through a wiggler magnet is finally with $N_{\mathrm{ph}}(\omega) = W(\omega)/(\hbar\omega)$

$$\frac{\mathrm{d}N_{\mathrm{ph}}(\omega)}{\mathrm{d}\Omega} = \alpha\gamma^2\bar{\beta}^2 N_{\mathrm{p}}^2 \frac{\Delta\omega}{\omega} \sum_{k=1}^{\infty} k^2 \left( \frac{\sin \pi N_{\mathrm{p}}\, \Delta\omega_k/\omega_1}{\pi N_{\mathrm{p}}\, \Delta\omega_k/\omega_1} \right)^2 \tag{10.76}$$

$$\times\ \frac{\left(2\gamma\vartheta \sum_1 \cos\varphi - K \sum_2\right)^2 x^2 + \left(2\gamma\vartheta \sum_1 \sin\varphi\right)^2 y^2}{\left(1 + \frac{1}{2}K^2 + \gamma^2\vartheta^2\right)^2}\,,$$

where $\alpha$ is the fine structure constant and where we have kept the coordinate unit vectors to keep track of the polarization modes. The vectors $x$ and $y$ are orthogonal unit vectors indicating the directions of the electric field or the polarization of the radiation. Performing the squares does therefore not produce cross terms and the two terms in (10.76) with the expressions (10.75) represent the amplitude factors for both polarization directions, the $\sigma$-mode and $\pi$-mode respectively.

We also made use of (10.72) and the resonance condition

$$\frac{\omega}{\omega_{\mathrm{p}}} = \frac{k\,\omega_1 + \Delta\omega_k}{\omega_{\mathrm{p}}} \approx k\,\frac{\omega_1}{\omega_{\mathrm{p}}} = \frac{2\gamma^2\, k}{1 + \frac{1}{2}K^2 + \gamma^2\vartheta^2}\,, \tag{10.77}$$

realizing that the photon spectrum is determined by the $(\sin x/x)^2$-function. For not too few periods, this function is very small for frequencies away from the resonance conditions.

Storage rings optimized for very small beam emittance are being used as modern synchrotron radiation sources to reduce the line width of undulator radiation and concentrate all radiation to the frequency desired. The progress in this direction is demonstrated in the spectrum of Fig. 10.8 derived from the first electron storage ring operated at a beam emittance below 10 nm at 7.1 GeV [61]. In Fig. 10.8 a measured undulator spectrum is shown as a function of the undulator strength $K$ [37]. For a strength parameter $K \ll 1$ there is only one line at the fundamental frequency. As the strength parameter increases, additional lines appear in addition to being shifted to lower frequencies. The spectral lines from a real synchrotron radiation source

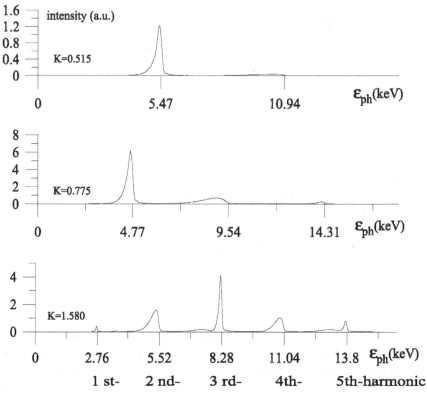

**Fig. 10.8.** Measured frequency spectrum from an undulator for different strength parameters $K$ [37]

are not infinitely narrow as (10.88) would suggest. Because of the finite size of the pinhole opening, some light at small angles with respect to the axis passes through, and we observe therefore also some signal of the even order harmonic radiation.

Even for an extremely small pin hole, we would observe a similar spectrum as shown in Fig. 10.8 because of the finite beam divergence of the electron beam. The electrons follow oscillatory trajectories due not only to the undulator field but also due to betatron oscillations. We observe therefore always some radiation at a finite angle given by the particle trajectory with respect to the undulator axis. Fig. 10.8 also demonstrates the fact that all experimental circumstances must be included to meet theoretical expectations. The amplitudes of the measured low energy spectrum is significantly suppressed compared to theoretical expectations which is due to a Be-window being used to extract the radiation from the ultra high vacuum chamber of the accelerator. This material absorbs radiation significantly below a photon energy of about 3 keV.

While we observe a line spectrum expressed by the $(\sin x/x)^2$-function, we also notice that this line spectrum is red shifted as we increase the observation angle $\vartheta$. Only, when we observe the radiation though a very small aperture, pin hole, do we actually see this line spectrum. Viewing the undulator radiation through a large aperture integrates the linespectra over a finite range of angles $\vartheta$ producing an almost continuous spectrum with small spikes at the locations of the harmonic lines.

The difference between a pin hole undulator spectrum and an angle-integrated spectrum becomes apparent from the experimental spectra shown in Fig. 10.9 [61]. While the pin hole spectrum demonstrates well the line character of undulator radiation, much radiation appears between these spectral lines as the pin hole is removed and radiation over a large solid angle is collected by the detector. The pin hole undulator line spectrum shows up as mere spikes on top of a broad continuous spectrum.

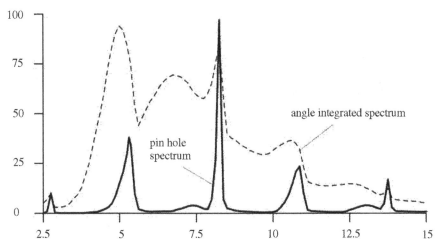

**Fig. 10.9.** Actual radiation spectra from an undulator with a maximum field of 0.2 T and a beam energy of 7.1 GeV through a pin hole and angle-integrated after removal of the pin hole [61]

The overall spatial intensity distribution includes a complex set of different radiation lobes depending on frequency, emission angle and polarization. In Fig. 10.10 the radiation intensity distributions described by the last factor in (10.76)

$$I_{\sigma,k} = \frac{(2\gamma\vartheta\,\Sigma_1\,\cos\varphi - K\,\Sigma_2)^2}{(1+\tfrac{1}{2}K^2 + \gamma^2\vartheta^2)^2}$$

for the $\sigma$-mode polarization and

σ-mode                                    π-mode

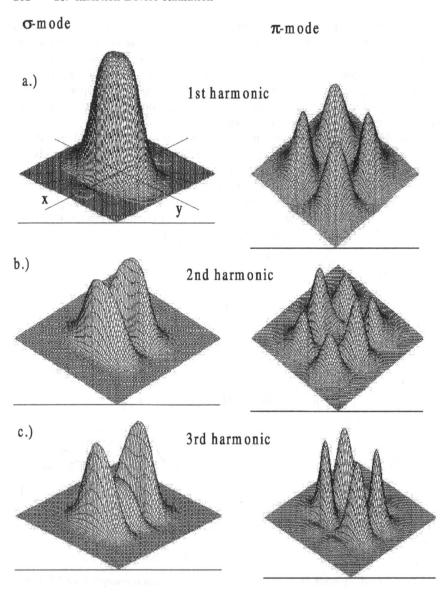

**Fig. 10.10.** Undulator radiation distribution in σ- and π-mode for the lowest order harmonics

$$I_{\pi,k} = \frac{(2\gamma\vartheta\,\Sigma_1\,\sin\varphi)^2}{(1+\frac{1}{2}K^2+\gamma^2\vartheta^2)^2}$$

for the π-mode polarization are shown for the lowest order harmonics.

We note clearly the strong forward lobe at the fundamental frequency in $\sigma$-mode while there is no emission in $\pi$-mode along the path of the particle. The second harmonic radiation vanishes in the forward direction, an observation that is true for all even harmonics. By inspection of (10.76), we note that $v = 0$ for $\vartheta = 0$ and the square bracket in (10.75b) vanishes for all odd indices or for all even harmonics $k$. There is therefore no forward radiation for even harmonics of the fundamental undulator frequency.

A contour plot of the first harmonic $\sigma$-and $\pi$-mode radiation is shown in Fig. 10.11. There is an slight asymmetry in the radiation distribution between the deflecting and nondeflecting plane as one might expect. It is obvious that the pin hole radiation is surrounded by many radiation lobes not only from the first harmonics but also from higher harmonics compromising the pure line spectrum for large apertures.

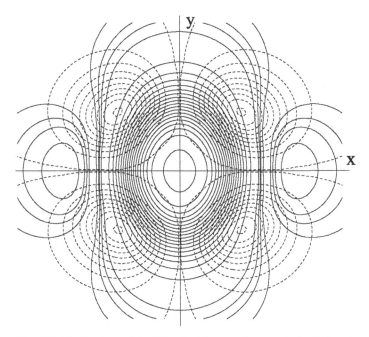

**Fig. 10.11.** Contour plot of the first harmonic $\sigma$-mode (*solid*) and $\pi$-mode (*dashed*) undulator radiation distribution

### 10.2.4 Line Spectrum

To exhibit other important and desirable features of the radiation spectrum (10.76), we ignore the actual frequency distribution in the vicinity of the harmonics and set $\Delta\omega_k = 0$ because the spectral lines are narrow for large numbers of wiggler periods $N_p$. Further, we are interested for now only in the forward radiation where $\vartheta = 0$, keeping in mind that the radiation is mostly emitted into a small angle $\langle \vartheta \rangle = 1/\gamma$.

There is no radiation for the $\pi$-mode in the forward direction and the only contribution to the forward radiation comes from the second term in (10.76) of the $\sigma$-mode. From (10.65), we get for this case with $\omega /\omega_1 = k$

$$u_0 = \frac{k\,K^2}{4 + 2K^2} \qquad \text{and} \qquad v_0 = 0. \tag{10.78}$$

The sums of Bessel's functions simplify in this case greatly because only the lowest order Bessel's function has a nonvanishing value for $v_0 = 0$. In the expression for $\Sigma_2$, all summation terms vanish except for the two terms for which the index is zero or for which

$$k - 2m - 1 = 0, \qquad \text{or} \qquad k - 2m + 1 = 0 \tag{10.79}$$

and

$$\Sigma_2 = \sum_{m=-\infty}^{\infty} J_{-m}(u)\,[J_{k-2m-1}(0) + J_{k-2m+1}(0)]$$
$$= J_{-\frac{1}{2}(k-1)}(u_0) + J_{-\frac{1}{2}(k+1)}(u_0). \tag{10.80}$$

The harmonic condition (10.79) implies that $k$ is an odd integer. For even integers, the condition cannot be met as we would expect from earlier discussions on harmonic radiation in the forward direction. Using the identity $J_{-n} = (-1)^n\,J_n$ and (10.78), we get finally with $N_{\text{ph}} = W / \hbar\omega$ the photon flux per unit solid angle from a highly relativistic particle passing through an undulator

$$\frac{dN_{\text{ph}}(\omega)}{d\Omega}\bigg|_{\theta=0} = \alpha\gamma^2 N_{\text{p}}^2 \frac{\Delta\omega}{\omega}\frac{K^2}{\left(1+\frac{1}{2}K^2\right)^2}\sum_{k=1}^{\infty} k^2 \left(\frac{\sin \pi N_{\text{p}}\,\Delta\omega_k/\omega_1}{\pi N_{\text{p}}\,\Delta\omega_k/\omega_1}\right)^2 JJ^2, \tag{10.81}$$

where the $JJ$−function is defined by

$$JJ = \left[J_{\frac{1}{2}(k-1)}\left(\frac{kK^2}{4+2K^2}\right) + J_{\frac{1}{2}(k+1)}\left(\frac{kK^2}{4+2K^2}\right)\right]. \tag{10.82}$$

The amplitudes of the harmonics are given by

$$A_k(K) = \frac{k^2\,K^2}{(1+\frac{1}{2}K^2)^2}\,JJ^2. \tag{10.83}$$

The strength parameter greatly determines the radiation intensity as shown in Fig. 10.12 for the lowest order harmonics. For the convenience of numerical calculations, $A_k(K)$ is tabulated for odd harmonics in Table 10.1. For weak magnets, $K \ll 1$, the intensity increases with the square of the

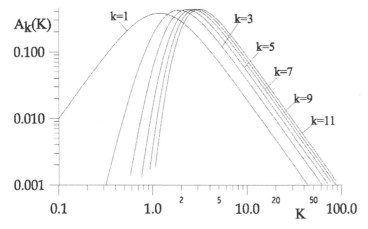

**Fig. 10.12.** Undulator radiation intensity $A_k(K)$ in the forward direction as a function of the strength parameter $K$ for the six lowest order odd harmonics

magnet field or undulator strength parameter. There is an optimum value for the strength parameter for maximum photon flux depending on the harmonic under consideration. In particular, radiation in the forward direction at the fundamental frequency reaches a maximum photon flux for strength parameters $K \approx 1.3$. The photon flux per unit solid angle increases like the square of the number of wiggler periods $N_p$, which is a result of the interference effect of many beams concentrating the radiation more and more into one frequency and its harmonics as the number of interfering beams is increased.

The radiation opening angle is primarily determined by the $(\sin x/x)^2$-term. We define the rms opening angle for the $k^{\text{th}}$ harmonic radiation by $\vartheta_k$ being the angle for which $\sin x/x = 0$ for the first time. In this case,

**Table 10.1.** Amplitudes $A_k(K)$ for $k = 1, 3, 5, 7, 9, 11$

| $K$ | $A_1$ | $A_3$ | $A_5$ | $A_7$ | $A_9$ | $A_{11}$ |
|------|-------|-------|-------|-------|-------|----------|
| 0.1  | 0.010 | 0     | 0     | 0     | 0     | 0        |
| 0.2  | 0.038 | 0     | 0     | 0     | 0     | 0        |
| 0.4  | 0.132 | 0.004 | 0     | 0     | 0     | 0        |
| 0.6  | 0.238 | 0.027 | 0.002 | 0     | 0     | 0        |
| 0.8  | 0.322 | 0.087 | 0.015 | 0.002 | 0     | 0        |
| 1.0  | 0.368 | 0.179 | 0.055 | 0.015 | 0.004 | 0.001    |
| 1.2  | 0.381 | 0.276 | 0.128 | 0.051 | 0.019 | 0.007    |
| 1.4  | 0.371 | 0.354 | 0.219 | 0.118 | 0.059 | 0.028    |
| 1.8  | 0.320 | 0.423 | 0.371 | 0.286 | 0.206 | 0.142    |
| 2.0  | 0.290 | 0.423 | 0.413 | 0.354 | 0.285 | 0.220    |
| 5.0  | 0.071 | 0.071 | 0.139 | 0.188 | 0.228 | 0.290    |
| 10.0 | 0.019 | 0.037 | 0.051 | 0.064 | 0.075 | 0.085    |
| 20.0 | 0.005 | 0.010 | 0.013 | 0.016 | 0.019 | 0.022    |

$x = \pi$ or $N_{\mathrm{p}} \, \Delta\omega_k/\omega_1 = 1$. With $\omega_1 = \omega_{\mathrm{p}} \frac{2\gamma^2}{1+\frac{1}{2}K^2}$, $\omega_k = k\,\omega_{\mathrm{p}} \frac{2\gamma^2}{1+\frac{1}{2}K^2+\gamma^2\vartheta_k^2}$ and $\frac{\Delta\omega_k}{\omega_1} = \left| \frac{\omega_k}{\omega_1} - k \right|$, we get $\frac{N_{\mathrm{p}}\,k\,\gamma^2\,\vartheta_k^2}{1+\frac{1}{2}K^2+\gamma^2\vartheta_k^2} = 1$ or after solving for $\vartheta_k$

$$\vartheta_k^2 = \frac{1+\frac{1}{2}K^2}{\gamma^2\,(kN_{\mathrm{p}} - 1)} \,. \tag{10.84}$$

Assuming an undulator with many periods $k\,N_{\mathrm{p}} \gg 1$, the rms opening angle of undulator radiation is finally

$$\sigma_{r'} = \tfrac{1}{\sqrt{2}}\vartheta_k \approx \frac{1}{\gamma}\sqrt{\frac{1+\frac{1}{2}K^2}{2\,k\,N_{\mathrm{p}}}} \,. \tag{10.85}$$

Radiation emitted into a solid angle defined by this opening angle

$$\mathrm{d}\Omega = 2\pi\,\sigma_{r'}^2, \tag{10.86}$$

is referred to as the forward radiation cone. The opening angle of undulator radiation becomes more collimated as the number of periods and the order of the harmonic increases. On the other hand, the radiation cone opens up as the undulator strength $K$ is increased. We may use this opening angle to calculate the total photon intensity of the $k^{\mathrm{th}}$ harmonic within a bandwidth $\frac{\Delta\omega}{\omega}$ into the forward cone

$$N_{\mathrm{ph}}(\omega_k)|_{\vartheta=0} = \pi\,\alpha\,N_{\mathrm{p}}\,\frac{\Delta\omega}{\omega_k}\,k\,\frac{K^2}{1+\frac{1}{2}K^2}\,JJ^2\,, \tag{10.87}$$

where $\omega_k = k\,\omega_1$. The radiation spectrum from an undulator magnet into the forward direction has been reduced to a simple form exhibiting the most important characteristic parameters. Utilizing (10.83), the number of photons emitted into a band width $\frac{\Delta\omega}{\omega_k}$ from a single electron passing through an undulator in the $k$–th harmonic is

$$N_{\mathrm{ph}}(\omega_k)|_{\vartheta=0} = \pi\,\alpha\,N_{\mathrm{p}}\,\frac{\Delta\omega}{\omega_k}\,\frac{1+\frac{1}{2}K^2}{k}\,A(K)\,. \tag{10.88}$$

Equation (10.88) is to be multiplied by the number of particles in the electron beam to get the total photon intensity. In case of a storage ring, particles circulate with a high revolution frequency and we get from (10.88) by multiplication with $I/e$, where $I$ is the circulating beam current, the photon flux

$$\left.\frac{\mathrm{d}\,N_{\mathrm{ph}}(\omega_k)}{\mathrm{d}\,t}\right|_{\vartheta=0} = \pi\,\alpha\,N_{\mathrm{p}}\,\frac{I}{e}\,\frac{\Delta\omega}{\omega_k}\,\frac{1+\frac{1}{2}K^2}{k}A(K)\,. \tag{10.89}$$

The spectrum includes only odd harmonic since all even harmonics are suppressed through the cancellation of Bessel's functions.

### 10.2.5 Spectral Undulator Brightness

The spectral brightness of undulator radiation is defined as the photon density in six-dimensional phase space

$$\mathcal{B}\left(\omega\right) = \frac{\dot{N}_{\mathrm{ph}}(\omega)}{4\pi^2\,\sigma_x\sigma_{x'}\sigma_y\sigma_{y'}\left(\mathrm{d}\omega/\omega\right)}. \tag{10.90}$$

In the laser community, this quantity is called the radiance while the term spectral brightness is common in the synchrotron radiation community. The maximum value of the brightness is limited by diffraction to

$$\mathcal{B}_{\mathrm{max}} = \dot{N}_{\mathrm{ph}}\,\frac{(4\,/\,\lambda^2)}{\mathrm{d}\omega/\omega}. \tag{10.91}$$

The actual photon brightness is reduced from the diffraction limit due to betatron motion of the particles, transverse beam oscillation in the undulator, apparent source size on axis and under an oblique angle. All of these effects tend to increase the source size and reduce brightness.

The particle beam cross section varies in general along the undulator. We assume here for simplicity that the beam size varies symmetrically along the undulator with a waist in its center. From beam dynamics it is then known that, for example, the horizontal beam size varies like $\sigma_b^2 = \sigma_{b0}^2 + \sigma_{b0}'^2\,s^2$, where $\sigma_{b0}$ is the beam size at the waist, $\sigma_{b0}'$ the divergence of the beam at the waist and $-\frac{1}{2}L \leqq s \leqq \frac{1}{2}L$ the distance from the waist. The average beam size along the undulator length $L$ is then

$$\langle\sigma_b^2\rangle = \sigma_{b0}^2 + \tfrac{1}{12}\sigma_{b0}'^2\,L^2. \tag{10.92}$$

Similarly, due to an oblique observation angle $\vartheta$ with respect to the $(y, z)$-plane or $\psi$ with respect to the $(x, z)$-plane we get a further additive contribution $\frac{1}{6}\vartheta L$ to the apparent beam size. Finally, the apparent source size is widened by the transverse beam wiggle in the periodic undulator field. This oscillation amplitude is from (10.20) $a = \lambda_{\mathrm{p}}K\,/\,(2\pi\gamma)$.

Collecting all contributions and adding them in quadrature, the total effective beam-size parameters are given by

$$\sigma_{t,x}^2 = \tfrac{1}{2}\sigma_r^2 + \sigma_{b0,x}^2 + \left(\frac{\lambda_{\mathrm{p}}K}{2\pi\gamma}\right)^2 + \tfrac{1}{12}\sigma_{b0,x'}^2 L^2 + \tfrac{1}{36}\vartheta^2 L^2, \tag{10.93a}$$

$$\sigma_{t,x'}^2 = \tfrac{1}{2}\sigma_{r'}^2 + \sigma_{b0,x'}^2, \tag{10.93b}$$

$$\sigma_{t,y}^2 = \tfrac{1}{2}\sigma_r^2 + \sigma_{b0,y}^2 + \left(\frac{\lambda_{\mathrm{p}}K}{2\pi\gamma}\right)^2 + \tfrac{1}{12}\sigma_{b0,y'}^2 L^2 + \tfrac{1}{36}\psi^2 L^2, \tag{10.93c}$$

$$\sigma_{t,y'}^2 = \tfrac{1}{2}\sigma_{r'}^2 + \sigma_{b0,y'}^2, \tag{10.93d}$$

where the particle beam sizes can be expressed by the beam emittance and betatron function as $\sigma_b^2 = \epsilon\beta$, $\sigma_b'^2 = \epsilon/\beta$, and the diffraction limited beam parameters are $\sigma_r = \sqrt{\lambda/L}$, and $\sigma_{r'} = \sqrt{\lambda L}/(2\pi)$.

## 10.3 Elliptical Polarization

During the discussion of bending magnet radiation in Chap. 9 and insertion radiation in this chapter we noticed the appearance of two orthogonal components of the radiation field which we identified with the $\sigma$-mode and $\pi$-mode polarization. The $\pi$-mode radiation is observable only at a finite angle with the plane defined by the particle trajectory and the acceleration force vector, which is in general the horizontal plane. As we will see, both polarization modes can, under certain circumstances, be out of phase giving rise to elliptical polarization. In this section, we will shortly discuss such conditions.

### 10.3.1 Elliptical Polarization from Bending Magnet Radiation

The direction of the electric component of the radiation field is parallel to the particle acceleration. Since radiation is the perturbation of electric field lines from the charge at the retarded time to the observer, we must take into account all apparent acceleration. To see this more clear, we assume an electron to travel counter clockwise on an orbit travelling from say a 12-o'clock position to 9-o'clock and then 6-o'clock. Watching the particle in the plane of deflection, the midplane, we notice only a horizontal acceleration which is maximum at 9-o'clock. Radiation observed in the midplane is therefore linearly polarized in the plane of deflection.

Now we observe the same electron at a small angle above the midplane. Apart from the horizontal motion, we notice now also a vertical motion. Since the electron follows pieces of a circle this vertical motion is not uniform but exhibits acceleration. Specifically, at 12-o'clock the particle seems to be accelerated only in the vertical direction (downward), horizontally it is in uniform motion; at 9-o'clock the acceleration is only horizontal (towards 3-o'clock) and the vertical motion is uniform; finally, at 6-o'clock the electron is accelerated only in the vertical plane again (upward). Because light travels faster than the electron, we observe radiation first coming from the 12-o'clock position, then from 9-o'clock and finally from 6-o'clock. The polarization of this radiation pulse changes from downward to horizontal (left-right) to upward which is what we call elliptical polarization where the polarization vector rotates with time. Of course, in reality we do not observe radiation from half the orbit, but only from a very short arc segment of angle $\pm 1/\gamma$. Yet, even this short piece of the orbit has all the features just used to explain elliptical polarization in a bending magnet.

If we observe the radiation at a small angle from below the midplane, the sequence of accelerations is opposite, upward-horizontal (left-right)-downward. The helicity of the polarization is therefore opposite for an observer below or above the midplane. This qualitative discussion of elliptical polarization must become obvious also in the formal derivation of the radiation field. Closer inspection of the radiation field (9.105) from a bending magnet

$$\boldsymbol{E}_{\mathrm{r}}(\omega) = \frac{-1}{[4\pi\epsilon_0]}\frac{\sqrt{3}e}{cR}\frac{\omega}{\omega_{\mathrm{c}}}\gamma(1+\gamma^2\vartheta^2)\left[\mathrm{sign}(1/\rho)\,K_{2/3}(\xi)\,\boldsymbol{u}_\sigma - \mathrm{i}\,\frac{\gamma\vartheta\,K_{1/3}(\xi)}{\sqrt{1+\gamma^2\vartheta^2}}\,\boldsymbol{u}_\pi\right]$$

$$(10.94)$$

shows that both polarization terms are $90°$ out of phase. As a consequence, the combination of both terms does not just introduce a rotation of the polarization direction but generates a time dependent rotation of the polarization vector which we identify with circular or elliptical polarization. In this particular case, the polarization is elliptical since the $\pi$-mode radiation is always weaker than the $\sigma$-mode radiation. The field rotates in time just as expected from the qualitative discussion above.

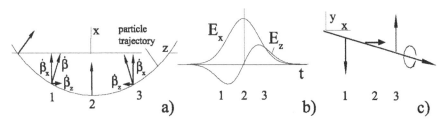

**Fig. 10.13.** Acceleration along an arc-segment of the particle trajectory in (**a**) a bending magnet, (**b**) polarization as a function of time, and (**c**) radiation field components as a function of time

We may quantify the polarization property considering that the electrical field is proportional to the acceleration vector $\dot{\boldsymbol{\beta}}$. Observing radiation at an angle with the horizontal plane, we note that the acceleration being normal to the trajectory and in the midplane can be decomposed into two components $\dot{\beta}_x$ and $\dot{\beta}_z$ as shown in Fig. 10.13a.

The longitudinal acceleration component together with a finite observation angle $\vartheta$ gives rise to an apparent vertical acceleration with respect to the observation direction and the associated vertical electric field component is

$$\boldsymbol{E}_y \propto \dot{\beta}_y = n_y\,\dot{\beta}_z + n_x\,n_y\,\dot{\beta}_x\,.$$

An additional component appears, if we observe the radiation also at an angle with respect to the $(x,y)$-plane which we, however, ignore here for this discussion. The components $n_x, n_y$ are components of the observation unit vector from the observer to the source with $n_y = -\sin\vartheta$. We observe radiation first from an angle $\vartheta > 0$. The horizontal and vertical radiation field components as a function of time are shown in Fig. 10.13b. Both being proportional to the acceleration (Fig. 10.13a), we observe a symmetric horizontal field $E_x$ and an antisymmetric vertical field $E_y$. The polarization vector (Fig. 10.13c) therefore rotates with time in a counter clockwise direction giving

210    10. Insertion Device Radiation

rise to elliptical polarization with lefthanded helicity. Observing the radiation from below with $\vartheta < 0$, the antisymmetric field switches sign and the helicity becomes righthanded. The visual discussion of the origin of elliptical polarization of bending magnet radiation is in agreement with the mathematical result (10.94) displaying the sign dependence of the $\pi$-mode component with $\vartheta$.

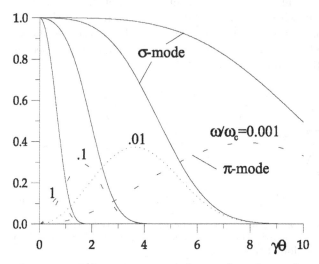

**Fig. 10.14.** Relative intensities of $\sigma$-mode and $\pi$-mode radiation as a function of vertical observation angle $\theta$ for different photon energies

The intensities for both polarization modes are shown in Fig. 10.14 as a function of the vertical observation angle $\vartheta$ for different photon energies. Both intensities are normalized to the forward intensity of the $\sigma$-mode radiation. From Fig. 10.14 it becomes obvious that circular polarization is approached for large observation angles. At high photon energies both radiation lobes are confined to very small angles but expand to larger angle distributions for photon energies much lower than the critical photon energy.

The elliptical polarization is left or right handed depending on whether we observe the radiation from above or below the horizontal mid plane. Furthermore, the helicity depends on the direction of deflection in the bending magnet or the sign of the curvature, sign$(1/\rho)$. By changing the sign of the bending magnet field the helicity of the elliptical polarization can be reversed. This is of no importance for radiation from a bending magnet since we cannot change the field without loss of the particle beam but is of specific importance for elliptical polarization state of radiation from wiggler and undulator magnets.

## 10.3.2 Elliptical Polarization from Periodic Insertion Devices

We apply the visual picture for the formation of elliptically polarized radiation in a bending magnet to the periodic magnetic field of wiggler and undulator magnets. The acceleration vectors and associated field vectors are shown in Fig. 10.15a and b for one period, and, similar to the situation in bending magnets, we do not expect any elliptical polarization in the mid plane where $\vartheta = 0$. Off the mid-plane, we observe now the radiation from a positive and a negative pole. From each pole we get elliptical polarization but the combination of lefthanded polarization from one pole with righthanded polarization from the next pole leads to a cancellation of elliptical polarization from periodic magnets (Fig. 10.15c). In bending magnets, this cancellation did not occur for lack of alternating deflection. Since there are generally an equal number of positive and negative poles in a wiggler or undulator magnet the elliptical polarization is completely suppressed. Ordinary wiggler and undulator magnets do not produce elliptically polarized radiation.

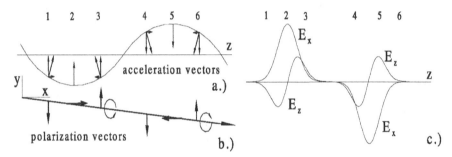

**Fig. 10.15.** Acceleration vectors along one period of (**a**) a wiggler magnet, (**b**) associated polarization vectors, and (**c**) corresponding radiation fields

**Asymmetric wiggler magnet.** The elimination of elliptical polarization in periodic magnets results from a compensation of left and righthanded helicity and we may therefore look for an insertion device in which this symmetry is broken. Such an insertion device is the asymmetric wiggler magnet which is designed similar to a wavelength shifter with one strong central pole and two weaker poles on either side such that the total integrated field vanishes or $\int B_y \, ds = 0$. A series of such magnets may be aligned to produce an insertion device with many poles to enhance the intensity. The compensation of both helicities does not work anymore since the radiation depends on the magnetic field and not on the total deflection angle. A permanent magnet rendition of an asymmetric wiggler magnet is shown in Fig. 10.16.

The degree of polarization from an asymmetric wiggler depends on the desired photon energy. The critical photon energy is high for radiation from the high field pole, $\epsilon_c^+$, and lower for radiation from the low field pole, $\epsilon_c^-$.

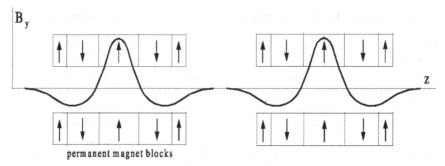

**Fig. 10.16.** Asymmetric wiggler magnet

For high photon energies $\epsilon_{ph} \approx \epsilon_c^+$ the radiation from the low field poles is negligible and the radiation is essentially the same as from a series of bending magnets with its particular polarization characteristics. For lower photon energies $\epsilon_c^- < \epsilon_{ph} < \epsilon_c^+$ the radiation intensity from high and low field pole become similar and cancellation of the elliptical polarization occurs. At low photon energies $\epsilon_{ph} < \epsilon_c^-$ the intensity from the low field poles exceeds that from the high field poles and we observe again elliptical polarization although with reversed helicity.

**Elliptically polarizing undulator.** The creation of elliptically and circularly polarized radiation is important for a large class of experiments using synchrotron radiation and special insertion devices have therefore been developed to meet such needs in an optimal way. Different approaches have been suggested and realized as sources for elliptically polarized radiation, among them for example, those described in refs. [62][63]. All methods are based on

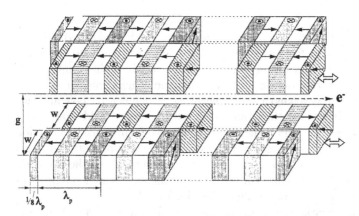

**Fig. 10.17.** Permanent magnet arrangement to produce elliptically polarized undulator radiation [64]

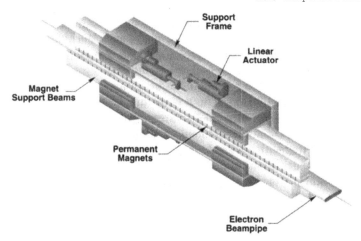

**Fig. 10.18.** 3-D view of an elliptically polarizing undulator, EPU [64]

**Fig. 10.19.** Undulator for elliptically polarized radiation [63]

permanent magnet technology, sometimes combined with electromagnets, to produce vertical and horizontal fields shifted in phase such that elliptically polarized radiation can be produced. Utilizing four rows of permanent magnets which are movable with respect to each other and magnetized as shown in Fig. 10.17, elliptically polarized radiation can be obtained.

Figure 10.18 shows the arrangement in a three dimensional rendition to visualize the relative movement of the magnet rows [62][64].

The top as well as the bottom row of magnet poles are split into two rows, each of which can be shifted with respect to each other. This way, a continuous variation of elliptical polarization from left to linear to right handed helicity can be obtained. By shifting the top magnet arrays with respect to the bottom magnets the fundamental frequency of the undulator radiation can be varied as well. Figure 10.19 shows a photo of such a magnet [63].

## Exercises *

**Exercise 10.1 (S).** Consider an undulator magnet with a period length of $\lambda_p = 5$ cm in a 7 GeV storage ring. The strength parameter be $K = 1$. What is the maximum oscillation amplitude of an electron passing through this undulator? What is the maximum longitudinal oscillation amplitude with respect to the reference system moving with velocity $\bar{\beta}$?

**Exercise 10.2 (S).** An undulator with 50 poles, a period length of $\lambda_p = 5$ cm and a strength parameter of $K = 1$ is to be installed into a 1 GeV storage ring. Calculate the focal length of the undulator magnet. Does the installation of this undulator require compensation of its focusing properties? How about a wiggler magnet with $K = 5$ ?

**Exercise 10.3 (S).** Consider the expression (10.89) for the photon flux into the forward cone. We also know that the band width of undulator radiation scales like $\Delta\omega/\omega_k \propto 1/N_p$. With this, the photon flux (10.89) becomes independent of the number of undulator periods!? Explain in words, why this expression for the photon flux is indeed a correct scaling law.

**Exercise 10.4 (S).** A hybrid undulator is to be installed into a 7 GeV storage ring to produce undulator radiation in a photon energy range of 4 keV to 15 keV. The maximum undulator field shall not exceed a value of $B_0 \leq 2$ T at a gap aperture of 10 mm. The available photon flux in the forward cone shall be at least 10% of the maximum flux within the whole spectral range. Specify the undulator parameters and show that the required photon energy range can be covered by changing the magnet gap only.

---

* The argument (S) indicates an exercise for which a solution is given in
  Appendix A.

**Exercise 10.5 (S).** Consider an electron colliding head-on with a laser beam. What is the wavelength of the laser as seen from the electron system. Derive from this the wavelength of the "undulator" radiation in the laboratory system.

**Exercise 10.6 (S).** An electron of energy 2 GeV performs transverse oscillations in a wiggler magnet of strength $K = 1.5$ and period length $\lambda_p = 7.5$ cm. Calculate the maximum transverse oscillation amplitude. What is the maximum transverse velocity in units of $c$ during those oscillations. Define and calculate a transverse relativistic factor $\gamma_\perp$. Note, that for $K \gtrsim 1$ the transverse relativistic effect becomes significant in the generation of harmonic radiation..

**Exercise 10.7 (S).** Calculate for a 3 GeV electron beam the fundamental photon energy ($\vartheta = 0$) for a 100 period-undulator with $K = 1.0$ and a period length of $\lambda_p = 5$ cm. What is the maximum angular acceptance angle $\vartheta$ of the beam line, if the radiation spectrum is to be restricted to a bandwidth of 10%?

**Exercise 10.8.** Add to the purely sinusoidal field of an ideal undulator additional terms (say 3-5), which would become necessary for a symmetric perturbation of the fundamental field, for example due to relativistic effects in strong undulators or due to long poles. Solve the equations of motion in the moving reference system (10.16a, 10.16b). Which harmonics are involved in the perturbation of the purely sinusoidal motion? Can you relate them to the radiation spectrum in the laboratory system?

**Exercise 10.9.** The undulator radiation intensity is a function of the strength parameter $K$. Find the strength parameter $K$ for which the fundamental radiation intensity is a maximum. Determine the range of $K$-values where the intensity of the fundamental radiation is at least 10% of the maximum.

**Exercise 10.10.** Verify the relative intensities of $\sigma$-mode and $\pi$-mode radiation in Fig. 10.15 for two quantitatively different pairs of observation angles $\vartheta$ and photon energies $\epsilon/\epsilon_c$.

**Exercise 10.11.** Design an asymmetric wiggler magnet assuming hard edge fields and optimized for the production of elliptical polarized radiation at a photon energy of your choice. Calculate and plot the photon flux of polarized radiation in the vicinity of the optimum photon energy.

**Exercise 10.12.** Show from (10.76) that along the axis, $\vartheta = 0$, radiation is emitted only in odd harmonics.

**Exercise 10.13.** Show from (10.73) that undulator radiation does not produce elliptically polarized radiation.

**Exercise 10.14.** Design a hybrid undulator for a 3 GeV storage ring to produce 4 keV to 15 keV photon radiation. Optimize the undulator parameters such that this photon energy range can be covered with the highest flux possible and utilizing lower order harmonics (order 7 or less). Plot the radiation spectrum that can be covered by changing the gap height of the undulator.

**Exercise 10.15.** Calculate the total undulator ($N_p = 50$, $\lambda_p = 4.5$ cm, $K = 1.0$) radiation power from a 200 mA, 6 GeV electron beam. Pessimistically, assume all radiation to come from a point source and be contained within the central cone. Determine the power density at a distance of 15 m from the source. Compare this power density with the maximum acceptable of 10 W/mm$^2$. How can you reduce the power density, on say a mask, to the acceptable value or below?

**Exercise 10.16.** Use the beam and undulator from Exercise 10.15 and estimate the total radiation power into the forward cone alone. What percentage of all radiation falls within the forward cone?

# 11. Free Electron Lasers

Synchrotron radiation is emitted when electromagnetic fields exert a force on a charged particle. This opens the possibility to apply external fields with specific properties for the stimulation of electrons to emit even more radiation. Of course, not just any external electromagnetic field would be useful. Fields at some arbitrary frequency would not work because particles interacting with such fields would in general be periodically accelerated and decelerated without any net energy transfer. The external field must have a frequency and phase such that a particle may continuously lose energy into synchrotron radiation. Generally, it is most convenient to recycle and use spontaneous radiation emitted previously by the same emission process. In this part, we will discuss in some detail the process of stimulation as it applies to a free electron laser.

In a free electron laser (FEL) quasi-monochromatic, spontaneous radiation emitted from an undulator is recycled in an optical cavity to interact with the electron beam causing accelerations which are periodic with the frequency of the undulator radiation. In order to couple the particle motion to the strictly transverse electromagnetic radiation field, the path of the electrons is modulated by periodic deflections in a magnetic field to generate transverse velocity components. In a realistic setup, this magnetic field is provided in an undulator magnet serving both as the source of radiation and the means to couple to the electric field. The transverse motion of the particle results in a gain or loss of energy from/to the electromagnetic field depending on the location of the particle with respect to the phase of the external radiation field. The principle components of a FEL are shown in Fig. 11.1.

An electron beam is guided by a bending magnet unto the axis of an undulator. Upon exiting the undulator, the beam is again deflected away from the axis by a second bending magnet, both deflections to protect the mirrors of the optical cavity. Radiation that is emitted by the electron beam while travelling through the undulator is reflected by a mirror, travels to the mirror on the opposite side of the undulator and is reflected there again. Just as this radiation pulse enters the undulator again, another electron bunch joins to establish the emission of stimulated radiation. The electron beam pulse consists of a long train of many bunches, much longer than the length of the optical cavity such that many beam-radiation interactions can be established.

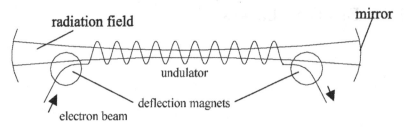

**Fig. 11.1.** Free electron laser setup (schematic)

We may follow this process in great detail observing an electron as it travels through say the positive half period of its oscillatory trajectory. During this phase, the electron experiences a negative acceleration from the undulator magnet field which is in phase with the oscillation amplitude. Acceleration causes a perturbation of the electric fields of the electron as was discussed in detail in Chap. 2. This perturbation travels away from the source at the speed of light, which is what we call electromagnetic radiation. For an electron, the electric radiation field points in the direction of the acceleration. As the electron travels through the positive half wave, it emits a radiation field made of half a wave. Simultaneously, this radiation field, being faster than the electron, travels ahead of the electron by precisely half a wavelength. This process tells us that the radiation wavelength is closely related to the electron motion and that it is quasi-monochromatic. Of course, for a strong undulator the sinusoidal motion becomes perturbed and higher harmonics appear, but the principle arguments made here are still true. Now, the electron starts performing the negative half of its oscillation and, experiencing a positive acceleration, emits the second halfwave of the radiation field matching perfectly the first halfwave. This happens in every period of the undulator and when the electron reaches the end of the last period a radiation wave composed of $N_p$ oscillations exists ahead of the electron. This process describes the spontaneous radiation emission from an electron in an undulator magnet.

The radiation pulse just created is recycled in the optical cavity to reenter the undulator again at a later time. The length of the optical cavity must be adjusted very precisely to an integer multiple of both the radiation wavelength and the distance between electron bunches. Under these conditions, electron bunches and radiation pulses enter the undulator synchronously. A complication arises from the fact that the electrons are contained in a bunch which is much longer than the wavelength of the radiation. The electrons are distributed for all practical purposes uniformly over many wavelengths. For the moment, we ignore this complication and note that there is an electron available whenever needed.

We pick now an electron starting to travel through a positive half wave of its oscillation exactly at the same time and location as the radiation wave starts its positive field halfperiod. The electron, experiences then a downward acceleration from the radiation field. During its motion the electron is continuously accelerated until it has completed its travel through the positive half oscillation. At the same time, the full positive have wave of the radiation field has moved over the electron. At this moment the electron and the radiation field are about to start their negative half periods. Continuing its motion now through the negative half period, the electron still keeps loosing energy because it now faces a negative radiation field. The fact that the radiation field "slides" over the electron just one wavelength per undulator period ensures a continuous energy transfer from electron to the radiation field. The electron emits radiation which is now exactly in synchronism with the existing radiation field and the new radiation intensity is proportional to the acceleration or the external radiation field. Multiple recycling and interaction of radiation field with electron bunches results therefore in an exponential increase in radiation intensity.

At this point, we must consider all electrons, not just the one for which the stimulation works as just described. This process does not work that perfect for all particles. An electron just half a wavelength behind the one discussed above would continuously gain energy from the radiation field and any other electron would loose or gain energy depending on its phase with respect to the radiation. It is not difficult to convince oneself that on average there may not be any net energy transfer one way or another and therefore no stimulation or acceleration. To get actual stimulation, some kind of asymmetry must be introduced.

To see this, we recollect the electron motion in a storage ring in the presence of the rf-field in the accelerating cavity. In Sect. 7.1 we discussed the phase space motion of particles under the influence of a radiation field. The radiation field of a FEL acts exactly the same although at a much shorter wavelength. The electron beam extends over many buckets as shown in Fig.11.2 and it is obvious that in its interaction with the field half of the electrons

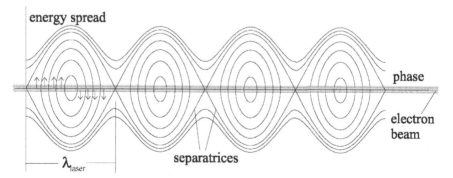

**Fig. 11.2.** Interaction of an electron beam (on-resosnace energy) with the radiation field of a FEL. The arrows in the first bucket indicate the direction of particle motion in its interaction with the electromagnetic field

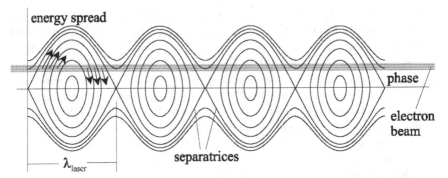

**Fig. 11.3.** Interaction of an electron beam (off-resonance energy) with the radiation field of a FEL

gain and the other half loose energy from/to the radiation field. The effect of the asymmetry required to make the FEL work is demonstrated in Fig. 11.3. Choosing an electron beam energy to be off-resonance by a small amount, the energy gain and losses for all electrons within a bucket becomes unbalanced and we can choose a case where all electrons on average loose energy into (FEL) or gain energy (particle acceleration by a radiation field) from the radiation field. The arrows in the first bucket of Fig. 11.3 show clearly the imbalance of energy gain or loss. What it means to choose an electron beam energy off-resonance will be discussed in more detail in the next section, where we formulate quantitatively the processes discussed so far only qualitatively.

## 11.1 Small Gain FEL

We concentrate on the case where only a small fraction of the particle energy is extracted such that we can neglect effects on particle parameters. This regime is called the "small-gain" regime. Specifically, we ignore changes in the particle energy and redistribution in space as a consequence of the periodic energy modulation.

### 11.1.1 Energy Transfer

Transfer of energy between a charged particle and an electromagnetic wave is effected by the electric field term of the Lorentz force equation and the amount of transferred energy is

$$\Delta W = e \int \boldsymbol{E}_{\mathrm{L}}\,\mathrm{d}\boldsymbol{s} = e \int \boldsymbol{E}_{\mathrm{L}}\boldsymbol{v}\mathrm{d}t\,, \qquad (11.1)$$

where $\boldsymbol{E}_{\mathrm{L}}$ is the external field or the Laser field in the optical cavity and $\boldsymbol{v}$ the particle velocity. In free space $\boldsymbol{v} \perp \boldsymbol{E}_{\mathrm{L}}$ and therefore there is no energy transfer possible, $\Delta W \equiv 0$. Generating some transverse velocity $\boldsymbol{v}_\perp$ through periodic deflection in an undulator, we get from (10.17a)

$$v_x = +\beta c \frac{K}{\gamma} \sin k_{\mathrm{p}} s \,, \tag{11.2}$$

where $k_{\mathrm{p}} = 2\pi/\lambda_{\mathrm{p}}$. The external radiation field can be expressed by

$$\boldsymbol{E}_{\mathrm{L}} = \boldsymbol{E}_{0\mathrm{L}} \cos\left(\omega_{\mathrm{L}} t - k_{\mathrm{L}} s + \varphi_0\right) \tag{11.3}$$

and the energy transfer is

$$
\begin{aligned}
\Delta W &= e \int \boldsymbol{v} \boldsymbol{E}_{\mathrm{L}} \mathrm{d}t = e \int v_x E_{\mathrm{L}} \mathrm{d}t \\
&= e\beta c \frac{K}{\gamma} E_{0\mathrm{L}} \int \cos\left(\omega_{\mathrm{L}} t - k_{\mathrm{L}} s + \varphi_0\right) \sin k_{\mathrm{p}} s \, \mathrm{d}t \\
&= \tfrac{1}{2} e\beta c \frac{K}{\gamma} E_{0\mathrm{L}} \int \left(\sin \Psi^+ - \sin \Psi^-\right) \mathrm{d}t \,,
\end{aligned}
\tag{11.4}
$$

where

$$\Psi^\pm = \omega_{\mathrm{L}} t - (k_{\mathrm{L}} \pm k_{\mathrm{p}})\, s + \varphi_0 \,. \tag{11.5}$$

The energy transfer appears to be oscillatory, but continuous energy transfer can be obtained if either $\Psi^+ = $ const. or $\Psi^- = $ const. In this case

$$\frac{\mathrm{d}\Psi^\pm}{\mathrm{d}t} = \omega_{\mathrm{L}} - (k_{\mathrm{L}} \pm k_{\mathrm{p}})\, \dot{s} = 0 \tag{11.6}$$

and we must derive conditions for this to be true. The velocity $\dot{s}$ is from (10.17b)

$$\dot{s} = \bar{\beta} c + \beta c \frac{K^2}{4\gamma^2} \cos\left(2k_{\mathrm{p}} z\right) \,, \tag{11.7}$$

where the average drift velocity $\bar{\beta} c$ is defined by

$$\frac{\mathrm{d}\bar{s}}{\mathrm{d}t} = \bar{\beta} c = \beta c \left(1 - \frac{K^2}{4\gamma^2}\right) \,. \tag{11.8}$$

We modify slightly the condition (11.6) and require that it be true only on average

$$\frac{\mathrm{d}\Psi^\pm}{\mathrm{d}t} = \omega_{\mathrm{L}} - (k_{\mathrm{L}} \pm k_{\mathrm{p}}) \frac{\mathrm{d}\bar{s}}{\mathrm{d}t} = 0 \,, \tag{11.9}$$

or

$$(k_{\mathrm{L}} \pm k_{\mathrm{p}})\, \beta \left(1 - \frac{K^2}{4\gamma^2}\right) - k_{\mathrm{L}} = 0 \,. \tag{11.10}$$

With $\beta \approx 1 - 1/2\gamma^2$ and $k_p \ll k_L$, (11.10) becomes

$$k_L \left[ \left( 1 - \frac{1}{2\gamma^2} \right) \left( 1 - \frac{K^2}{4\gamma^2} \right) - 1 \right] \pm k_p \approx 0 \,, \tag{11.11}$$

or for $\gamma \gg 1$

$$-\frac{k_L}{2\gamma^2} \left( 1 + \tfrac{1}{2}K^2 \right) \pm k_p = 0 \,. \tag{11.12}$$

Equation (11.12) can be met only for the +sign or for

$$k_p = \frac{k_L}{2\gamma^2} \left( 1 + \tfrac{1}{2}K^2 \right) \,, \tag{11.13}$$

which is identical to the definition of the fundamental undulator radiation wavelength

$$\lambda_L = \frac{\lambda_p}{2\gamma^2} \left( 1 + \tfrac{1}{2}K^2 \right) \,. \tag{11.14}$$

Radiation at the fundamental wavelength of undulator radiation guarantees a continuous energy transfer from the particles to the electromagnetic wave or stimulation of radiation emission by an external field. For this reason, it is most convenient to use spontaneous undulator radiation as the external field to start the build-up of the free electron laser.

### 11.1.2 Equation of Motion

The energy gain $dW$ of the electromagnetic field is related to the energy change $d\gamma$ of the electron by

$$\frac{d\gamma}{ds} = -\frac{1}{mc^2} \frac{dW}{\beta c dt} \tag{11.15}$$

or with (11.4)

$$\frac{d\gamma}{ds} = -\frac{eK E_{0L}}{2\gamma mc^2} \left( \sin \Psi^+ - \sin \Psi^- \right) \,. \tag{11.16}$$

With the substitution $\sin x = -\operatorname{Re} \left( i\, e^{ix} \right)$

$$\frac{d\gamma}{ds} = \frac{eK E_{0L}}{2\gamma mc^2} \operatorname{Re} \left( i\, e^{i\Psi^+} - i\, e^{i\Psi^-} \right) \,. \tag{11.17}$$

In $\Psi^{\pm} = \omega_{\mathrm{L}}t - (k_{\mathrm{L}} \pm k_{\mathrm{p}})\, s\,(t) + \varphi_0$, we replace the location function $s(t)$ by its expression (10.19b)

$$s\,(t) = \underbrace{\bar{\beta}ct}_{=\bar{s}} + \underbrace{\frac{K^2}{8\gamma^2 k_{\mathrm{p}}} \sin\left(2k_{\mathrm{p}}\bar{\beta}ct\right)}_{\ll \bar{\beta}ct}, \qquad (11.18)$$

composed of an average position $\bar{s}$ and an oscillatory term. With $k_{\mathrm{p}} \ll k_{\mathrm{L}}$

$$\frac{\mathrm{d}\gamma}{\mathrm{d}s} = \frac{e\beta K\,E_{0\mathrm{L}}}{2\gamma mc^2}\,\mathrm{Re}\left\{\mathrm{i}\exp\left[\mathrm{i}\,\frac{k_{\mathrm{L}}K^2}{8\gamma^2 k_{\mathrm{p}}}\sin\left(2k_{\mathrm{p}}\bar{s}\right)\right]\left[e^{\mathrm{i}\bar{\Psi}^{+}} - e^{\mathrm{i}\bar{\Psi}^{-}}\right]\right\} \qquad (11.19)$$

and the the phase

$$\bar{\Psi}^{\pm} = \omega_{\mathrm{L}}t - (k_{\mathrm{L}} \pm k_{\mathrm{p}})\,\bar{s} + \varphi_0. \qquad (11.20)$$

With the definition $\exp\left(\mathrm{i}x\sin\phi\right) = \sum_{n=-\infty}^{n=+\infty} J_n\,(x)e^{\mathrm{i}n\phi}$ we get finally

$$\frac{\mathrm{d}\gamma}{\mathrm{d}s} = \frac{e\beta K\,E_{0\mathrm{L}}}{2\gamma mc^2}\,\mathrm{Re}\left[\mathrm{i}\sum_{n=-\infty}^{n=+\infty} J_n\left(\frac{k_{\mathrm{L}}K^2}{8\gamma^2 k_{\mathrm{p}}}\right)e^{\mathrm{i}2nk_{\mathrm{p}}\bar{s}}\left(e^{\mathrm{i}\bar{\Psi}^{+}} - e^{\mathrm{i}\bar{\Psi}^{-}}\right)\right]. \qquad (11.21)$$

The infinite sum reflects the fact that the condition for continuous energy transfer can be met not only at one wavenumber but also at all harmonics of that frequency. Combining the exponential terms and sorting for equal wavenumbers $h\,k_{\mathrm{p}}$, where $h$ is an integer, we redefine the summation index by setting

$$2n\,k_{\mathrm{p}} + k_{\mathrm{p}} = h\,k_{\mathrm{p}} \quad\longrightarrow\quad n = \frac{h-1}{2} \qquad (11.22a)$$

$$2n\,k_{\mathrm{p}} - k_{\mathrm{p}} = h\,k_{\mathrm{p}} \quad\longrightarrow\quad n = \frac{h+1}{2} \qquad (11.22b)$$

and get

$$\frac{\mathrm{d}\gamma}{\mathrm{d}s} = \frac{e\beta K\,E_{0\mathrm{L}}}{2\gamma mc^2}\sum_{h=1}^{\infty}\left[J_{\frac{h-1}{2}}(x) - J_{\frac{h+1}{2}}(x)\right]\underbrace{\mathrm{Re}\left\{\mathrm{i}\,e^{\mathrm{i}[(k_{\mathrm{L}}+h\,k_{\mathrm{p}})\,\bar{s}-\omega_{\mathrm{L}}t+\varphi_0]}\right\}}_{=-\sin[(k_{\mathrm{L}}+h\,k_{\mathrm{p}})\,\bar{s}-\omega_{\mathrm{L}}t+\varphi_0]}, \qquad (11.23)$$

where $x = \frac{K^2}{4+2K^2}$. Introducing the abbreviation

$$[JJ] = J_{\frac{k-1}{2}}\left(\frac{K^2}{4+2K^2}\right) - J_{\frac{k+1}{2}}\left(\frac{K^2}{4+2K^2}\right), \qquad (11.24)$$

the energy transfer is

$$\frac{d\gamma}{ds} = -\frac{e\beta K E_{0L}}{2\gamma mc^2} \sum_{h=1}^{\infty} [JJ] \sin \Psi .$$ (11.25)

For maximum continuous energy transfer $\sin \Psi = \text{const.}$ or

$$\frac{d\Psi}{dt} = (k_L + h k_p) \bar{\dot{s}} - \omega_L$$ (11.26)

$$= (k_L + h k_p) \beta c \left(1 - \frac{K^2}{4\gamma^2}\right) - \omega_L$$

$$= (k_L + h k_p) \left(1 - \frac{1}{2\gamma^2}\right) c \left(1 - \frac{K^2}{4\gamma^2}\right) - ck_L$$

$$= -\frac{ck_L}{2\gamma^2} \left(1 + \tfrac{1}{2}K^2\right) + h k_p c = 0 ,$$

where we assumed that $k_L \gg h k_p$, which is true since $\lambda_p \gg \lambda_L$ and the harmonic number of interest is generally unity or a single digit number. This condition confirms our earlier finding (11.14) and extends the synchronicity condition to multiples $h$ of the fundamental radiation frequency

$$\lambda_L = \frac{\lambda_p}{2\gamma^2 h} \left(1 + \tfrac{1}{2}K^2\right) .$$ (11.27)

The integer $h$ therefore identifies the harmonic of the radiation frequency with respect to the fundamental radiation.

In a real particle beam with a finite energy spread we may not assume that all particles exactly meet the synchronicity condition. It is therefore useful to evaluate the tolerance for meeting this condition. To do this, we define a resonance energy

$$\gamma_r^2 = \frac{k_L}{2 h k_p} \left(1 + \tfrac{1}{2}K^2\right) ,$$ (11.28)

which is the energy at which the synchronicity condition is met exactly. For any other particle energy $\gamma = \gamma_r + \delta\gamma$ we get from (11.26) and (11.28)

$$\frac{d\Psi}{ds} = 2 h k_p \frac{\delta\gamma}{\gamma_r} .$$ (11.29)

With the variation of the energy deviation $\frac{d}{ds}\delta\gamma = \frac{d\gamma}{ds}\big|_{\gamma_r} - \frac{d\gamma_r}{ds} = \frac{d\gamma}{ds}\big|_{\gamma_r}$ and (11.25) we get from (11.29) after differentiating with respect to $s$

$$\frac{d^2\Psi}{ds^2} = 2 h k_p \frac{d}{ds}\frac{\delta\gamma}{\gamma_r} = -\frac{e h k_p K E_{0L}}{\gamma_r^2 mc^2} [JJ] \sin \Psi(s) ,$$ (11.30)

where, for simplicity, we use only one harmonic $h$. This equation can be written in the form

nnot be done, the volume $V$ is the overlap volume, or the larger of
With this the FEL-gain becomes

$$= -\frac{8\pi e^2 n_{\mathrm{b}} h K^2 k_{\mathrm{p}} [JJ]^2}{mc^2 \gamma_{\mathrm{r}}^3 \Omega_{\mathrm{L}}^4} \langle \Delta \Psi' \rangle_{n_{\mathrm{c}}} . \qquad (11.36)$$

merical evaluation of $\langle \Delta \Psi' \rangle_{n_{\mathrm{c}}}$ can be performed with the pendulum
on. Multiplying the pendulum equation $2\Psi'$ and integrating we get

$$^2 - 2\Omega_{\mathrm{L}}^2 \cos \Psi = \text{const.} \qquad (11.37)$$

ting this at the beginning of the undulator

$$^2 - \Psi_0'^2 = 2\Omega_{\mathrm{L}}^2 (\cos \Psi - \cos \Psi_0) , \qquad (11.38)$$

becomes with $\Psi_0' = 2N\, k_{\mathrm{p}} \frac{\gamma_0 - \gamma_{\mathrm{r}}}{\gamma_{\mathrm{r}}}$ and $\Psi_{\mathrm{b}}' = \Psi$ from (11.29)

$$^2 = \left( 2h k_{\mathrm{p}} \frac{\gamma_0 - \gamma_{\mathrm{r}}}{\gamma_{\mathrm{r}}} \right)^2 + 2\Omega_{\mathrm{L}}^2 (\cos \Psi - \cos \Psi_0) . \qquad (11.39)$$

',

$$(s) = 2h\, k_{\mathrm{p}} \frac{\gamma - \gamma_{\mathrm{r}}}{\gamma_{\mathrm{r}}} \sqrt{1 + \frac{\Omega_{\mathrm{L}}^2}{2k^2\, k_{\mathrm{p}}^2} \frac{\gamma_{\mathrm{r}}^2}{(\gamma - \gamma_{\mathrm{r}})^2} [\cos \Psi (s) - \cos \Psi_0]} , \quad (11.40)$$

h

$$= h\, k_{\mathrm{p}} L_{\mathrm{u}} \frac{\gamma - \gamma_{\mathrm{r}}}{\gamma_{\mathrm{r}}} , \qquad (11.41)$$

$L_{\mathrm{u}} = N_{\mathrm{p}} \lambda_{\mathrm{p}}$ is the undulator length,

$$(s) = \frac{2w}{L_{\mathrm{u}}} \sqrt{1 + \frac{L_{\mathrm{u}}^2 \Omega_{\mathrm{L}}^2}{2w^2} [\cos \Psi (s) - \cos \Psi_0]} . \qquad (11.42)$$

solve this by expansion and iteration. For a low gain FEL, the '
weak and does not influence the particle motion. Therefore $\Omega_{\mathrm{L}} \ll$
) becomes

$$\approx \frac{2w}{L} \left[ 1 + \frac{1}{2} \frac{L^2 \Omega_{\mathrm{L}}^2}{2w^2} (\cos \Psi - \cos \Psi_0) \right.$$

$$\left. - \frac{1}{8} \frac{L^4 \Omega_{\mathrm{L}}^4}{4w^4} (\cos \Psi - \cos \Psi_0)^2 + \dots \right] .$$

the lowest order of iteration $\Psi_0' = \frac{2w}{L}$ and $\Delta \Psi'_{(0'}$
means there is no energy transfer. For first o'
ate $\Psi_0' (s) = \frac{2w}{L_{\mathrm{u}}}$ to get $\Psi_{(1)}(s) = \frac{2w}{L_{\mathrm{u}}} s + \Psi_0$ ar

$$\frac{d^2\Psi}{ds^2} + \Omega_L^2 \sin\Psi = 0$$

exhibiting the dynamics of a harmonic oscillator. Eq
as the Pendulum equation [65] with the frequency

$$\Omega_L^2 = \frac{eh\,k_p K\,E_{0L}}{\gamma_r^2 mc^2}\,|[JJ]|\,.$$

While interacting with the external radiation fiel
harmonic oscillations in a potential generated by tl
is very similar to the synchrotron oscillation of par
interacting with the field of the rf-cavities as was di
phase space, the electron perform synchrotron oscilli
$\Omega_L$ while exchanging energy with the radiation field.

### 11.1.3 FEL-Gain

Having established the possibility of energy transfer
radiation field, we may evaluate the magnitude of this
gain in field energy per interaction process or per pa
by the interaction of an electron bunch with the radia
through the entire length of the undulator. The gain in
$-mc^2 n_e \Delta\gamma$, where $\Delta\gamma$ is the energy loss per electron an
field and $n_e$ the number of electrons per bunch. The c

$$W_L = \frac{1}{8\pi}\frac{1}{2}E_{0L}^2\,V\,,$$

where $V$ is the volume of the radiation field. With tl
FEL-gain for the $k$-th harmonic by

$$G_k = \frac{\Delta W_L}{W_L} = -\frac{mc^2 n_b \Delta\gamma}{\frac{1}{16\pi}E_{0L}^2 V} = -\frac{4\pi mc^2 \gamma_r n_e}{hk_p E_{0L}^2 V}\,\langle\Delta\Psi'\rangle$$

making use of (11.29). $\langle\Delta\Psi'\rangle_{n_e}$ is the average value o
electrons per bunch, where $\Psi_0'$ is defined at the begin
and $\Psi_f'$ at the end of the undulator. To further simpli
use (11.32), solve for the laser field

$$E_{0L} = \frac{mc^2\gamma_r^2 \Omega_L^2}{ehKk_p V[JJ]}\,,$$

define the electron density $n_b = n_e/V$. Here we have
olume of the radiation field perfectly overlaps the
This is not automatically the case and must be
ing the electron beam to the diffraction domina

this c
both.

$G$

N
equat

$\Psi$

Evalu

$\Psi$

which

$\Psi$

Final

$\Psi$

or wi

$u$

wher

$\Psi$

V
$E_{0L}$
(11.4

$\Psi$

I
whic
integ

$$\Delta\Psi'_{(1)} = \Psi'(L_u) - \Psi'_1(0) = \frac{L\,\Omega_L^2}{2w}\left[\cos\left(2w + \Psi_0\right) - \cos\Psi_0\right] + \mathcal{O}(2) \quad (11.44)$$

from (11.43). Averaging over all initial phases occupied by electrons $0 \leq \Psi_0 \leq 2\pi$

$$\langle\Delta\Psi'_1\rangle = \frac{L\,\Omega_L^2}{2w}\frac{1}{2\pi}\int_0^{2\pi}\left[\cos\left(2w + \Psi_0\right) - \cos\Psi_0\right]\,\mathrm{d}\Psi_0 = 0. \quad (11.45)$$

No energy transfer to the laser field occurs in this approximation either. We need a still higher order approximation. The higher order correction to $\Psi'_1(s) = \Psi'_0(s) + \delta\Psi'_1(s)$ is from (11.43)

$$\delta\Psi'_{(1)} = \frac{L\Omega_L^2}{2w}\left[\cos\Psi - \cos\Psi_0\right], \quad (11.46)$$

and the correction to $\Psi_1(s)$ is

$$\delta\Psi_{(1)} = \frac{L\Omega_L^2}{2w}\int_0^L\left[\cos\left(\frac{2w}{L}s + \Psi_0\right) - \cos\Psi_0\right]\,ds$$

$$= \frac{L\Omega_L^2}{4w^2}\left[\sin\left(2w + \Psi_0\right) - \sin\Psi_0 - 2w\cos\Psi_0\right]. \quad (11.47)$$

The second order approximation to the phase is then $\Psi_1(s) = \frac{2w}{L_u}s + \Psi_0 + \delta\Psi_{(1)}$ and using (11.43) in second order as well we get

$$\Delta\Psi'_{(2)} = \frac{L\,\Omega_L^2}{2w}\left[\cos\left(2w + \Psi_0 + \delta\Psi_{(1)}\right) - \cos\Psi_0\right]$$

$$-\frac{L^3\Omega_L^4}{4w^2}\left[\cos\left(2w + \Psi_0\right) - \cos\Psi_0\right]^2 + \ldots, \quad (11.48)$$

where in the second order term only the first order phase $\Psi_1(s) = \frac{2w}{L_u}s + \Psi_0$ is used. The first term becomes with $\delta\Psi_{(1)} \ll \Psi_0 + 2w$

$$\cos\left(2w + \Psi_0 + \delta\Psi_1\right) - \cos\Psi_0$$
$$\approx \cos\left(2w + \Psi_0\right) - \delta\Psi_1\sin\left(2w + \Psi_0\right) - \cos\Psi_0$$

and

$$\Delta\Psi'_2 = \frac{L_u^3\,\Omega_L^4}{16w^3}\left\{\frac{8w^2}{L_u^2\Omega^2}\left[\cos\left(2w + \Psi_0\right) - \cos\Psi_0\right]\right.$$

$$-2\sin\left(2w + \Psi_0\right)\left[\sin\left(2w + \Psi_0\right) - \sin\Psi_0 - 2w\cos\Psi_0\right]$$

$$\left.- \left[\cos\left(2w + \Psi_0\right) - \cos\Psi_0\right]^2 + \ldots\right\}. \quad (11.49)$$

Now, we average over all initial phases assuming a uniform distribution of particles in $s$ or in phase. The individual terms become then

$$\langle\cos\left(2w + \Psi_0\right) - \cos\Psi_0\rangle = 0$$
$$\langle\sin^2\left(2w + \Psi_0\right)\rangle = \tfrac{1}{2}$$
$$\langle\sin\left(2w + \Psi_0\right)\sin\Psi_0\rangle = \tfrac{1}{2}\cos\left(2w\right) \quad (11.50)$$
$$\langle\sin\left(2w + \Psi_0\right)\cos\Psi_0\rangle = \tfrac{1}{2}\sin\left(2w\right)$$
$$\langle\cos\left(2w + \Psi_0\right)\cos\Psi_0\rangle = \tfrac{1}{2}\cos\left(2w\right).$$

With this

$$\langle \Delta \Psi_2' \rangle = -\frac{L_u^3 \Omega_L^4}{16 w^3} [1 - \cos(2w) - w \sin(2w)] \tag{11.51}$$

and finally with $[1 - \cos(2w) - w \sin(2w)]/w^3 = -\frac{d}{dw} \left(\frac{\sin w}{w}\right)^2$

$$\langle \Delta \Psi_2' \rangle = \frac{L_u^3 \Omega_L^4}{8} \frac{d}{dw} \left(\frac{\sin w}{w}\right)^2. \tag{11.52}$$

The FEL-gain is finally from (11.36)

$$G_k = -\frac{\pi r_c n_b h\, K^2 L_u^3 k_p}{\gamma_r^3} [JJ]^2 \frac{d}{dw} \left(\frac{\sin w}{w}\right)^2, \tag{11.53}$$

where we may express the particle density $n_b$ by beam parameters as obtained from the electron beam source

$$n_b = \frac{n_c}{V} = \frac{n_c}{\pi^2 \sigma^2 \ell}, \tag{11.54}$$

where $\sigma$ is the radius of a round beam. With these definitions, and $\hat{I} = c e n_c/\ell$ the electron peak current the gain per pass becomes

$$G_k = -\frac{2^{2/3} \pi r_c h\, \lambda^{3/2} L_u^3}{c \sigma^2 \lambda_p^{5/2}} \frac{\hat{I}}{\ell} \frac{K^2 [JJ]^2}{\left(1 + \frac{1}{2} K^2\right)^{3/2}} \frac{d}{dw} \left(\frac{\sin w}{w}\right)^2. \tag{11.55}$$

The gain depends very much on the choice of the electron beam energy through the function (11.41), which is expressed by the gain curve as shown in Fig. 11.4.

There is no gain if the beam energy is equal to the resonance energy, $\gamma = \gamma_r$. As has been discussed in the introduction to this chapter, we must introduce an asymmetry to gain stimulation of radiation or gain and this asymmetry is generated by a shift in energy. For a monochromatic electron beam maximum gain can be reached for $w \approx 1.2$. A realistic beam, however, is not monochromatic and the narrow gain curve indicates that a beam with too large an energy spread may not produce any gain. There is no precise upper limit for the allowable energy spread but from Fig. 11.4 we see that gain is all but gone when $|w| \gtrsim 5$. We use this with (11.41) and (11.28) to formulate a condition for the maximum allowable energy spread

$$\left|\frac{\delta\gamma}{\gamma}\right| \ll \frac{2 \gamma_r^2 \lambda_L}{1 + \frac{1}{2} K^2}. \tag{11.56}$$

For efficient gain the geometric size of the electron beam and the radiation field must be matched. In (11.54) we have introduced a volume for the electron bunch. Actually, this volume is the overlap volume of radiation field

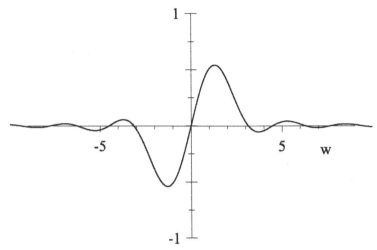

**Fig. 11.4.** Free electron laser gain curve $G \propto -\frac{\mathrm{d}}{\mathrm{d}w}\left(\frac{\sin w}{w}\right)^2$

and electron bunch. Ideally, one would try to get a perfect overlap by form-ing both beams to be equal. This is in fact possible and we will discuss the conditions for this to happen. First, we assume that the electron beam size varies symmetrically about the center of the undulator. From our discussion in Section 6.6.2 the beam size develops like

$$\sigma^2(s) = \sigma_0^2 + \left(\frac{\epsilon}{\sigma_0}\right)^2 s^2 \tag{11.57}$$

with distance $s$ from the beam waist. To maximize gain we look for the minimum average beam size within an undulator. This minimum demands a symmetric solution about the undulator center. Furthermore, we may select the optimum beam size at the center by looking for the minimum value of the maximum beam size within the undulator. From $\mathrm{d}\sigma^2/\mathrm{d}\sigma_0^2 = 0$, the optimum solution is obtained for $s = \frac{1}{2}L_\mathrm{u} = \sigma_0^2/\epsilon = \beta_0$. For $\beta_0 = \frac{1}{2}L_\mathrm{u}$ the beam cross section grows from a value of $\sigma_0^2$ in the middle of the undulator to a maximum value of $2\sigma_0^2$ at either end.

The radiation field is governed by diffraction. Starting at a beam waist the growth of the radiation field cross section due to diffraction is quantified by the Rayleigh length

$$s_\mathrm{R} = \pi \frac{w_0^2}{\lambda}, \tag{11.58}$$

where $w_0$ is the beam size at the waist and $\lambda$ the wavelength. This length is defined as the distance from the radiation source (waist) to the point at which the cross section of the radiation beam has grown by a factor of two. For a Gaussian beam we have for the beam size at a distance $s$ from the waist

$$w^2(s) = w_0^2 + \Theta^2 s^2 , \tag{11.59}$$

where $\Theta = \frac{\lambda}{\pi w_0}$ is the divergence angle of the radiation field. This is exactly the same condition as we have just discussed for the electron beam assuming the center of the undulator as the source of radiation.

## Exercises *

**Exercise 11.1 (S).** Consider an electron travelling through an undulator producing radiation. Show, that the radiation front travels faster than the electron by one fundamental radiation wavelength per undulator period.

**Exercise 11.2.** Choose an undulator with $N_p = 50$ and $K = 4.0$ . Specify the undulator period length $\lambda_p$ and electron beam energy from a linear accelerator such that the fundamental wavelength of the radiation is $1 \mu m$. What is the upper limit on the beam energy spread to make an FEL to function?

---

* The argument (S) indicates an exercise for which a solution is given in
  Appendix A.

# A. Solutions to Exercises

**Solution 1.1.** Solve $E^2 = (cp)^2 + (mc^2)^2$ for $(cp)^2 = E^2 - (mc^2)^2$ , extract $E$ , and replace $E/mc^2 = \gamma$, and $E = E_{\text{kin}} + mc^2$ to get with $\beta = \sqrt{1 - \gamma^{-2}}$ finally $cp = \beta(E_{\text{kin}} + mc^2)$. Replacing $\beta$ we get after some manipulation: $cp = mc^2\sqrt{(E_{\text{kin}}^2/mc^2 + 1)^2 - 1}$ , and finally $E_{\text{kin}} = mc^2(\gamma - 1)$.

**Solution 1.2.** For very large energies $\gamma \gg 1$ and $cp \approx E_{\text{kin}}$, and $E_{\text{kin}} \approx mc^2\gamma$ . For nonrelativistic particles, set $\gamma \approx 1+\delta$, where $\delta = E_{\text{kin}}/mc^2 = \frac{1}{2}\beta^2$, therefore $\beta \approx \sqrt{2\delta}$ . With $E_{\text{kin}} = \frac{1}{2}mv^2$ and keeping only terms linear in $\beta$, we get $cp = \sqrt{2\delta}(1 + \delta)mc^2 \approx cmv$ or the classical definition of the momentum $p = mv$.

**Solution 1.3.** For a kinetic energy of $E_{\text{kin}} = 200$ MeV and $m_p c^2 = 938.27$ MeV; the total energy $E = 1138.27$ MeV. The velocity $\beta = \sqrt{1 - \gamma^{-2}} = \sqrt{1 - (\frac{938.27}{1138.27})^2} = 0.566$ and the momentum $cp = \beta E = 644.44$ MeV or $p = 644.44$ MeV/c.

**Solution 1.4.** A length $ds$ of the linac in the laboratory system appears to the electron Lorentz contracted by the relativistic factor $\gamma$. Since the electron energy varies along the linac we must integrate the contraction along the full length of the linac. The linac length as seen by the electron is therefore $\int \frac{ds}{\gamma(s)}$, where $\gamma(s) = \gamma_0 + \alpha s$, and the acceleration $\alpha = 20/mc^2 = 39.14$ 1/m. From $\beta_0 = \frac{1}{2}$ we get $\gamma_0 = \frac{4}{3}$ and with $\gamma = \frac{4}{3} + 39.14 \cdot 3000 = 1.17 \times 10^5$ the integral is $\int \frac{ds}{\gamma(s)} = \frac{1}{\alpha} \ln \frac{\gamma}{\gamma_0}$ or numerically $\frac{1}{39.14} \ln \frac{1.17 \times 10^5}{4/3} = 0.291$ m. For an electron coasting with energy $\gamma$ along a 3000 m long tube, this tube appears to be $3000/\gamma = 0.0256$ m or 2.56 cm long.

**Solution 1.5.** The revolution frequency is $f_{\text{rev}} = c/C = 1.0 \times 10^6$ 1/s and the total number of particles orbiting $n_e = I/e/f_{\text{rev}} = 1.5604 \times 10^{12}$ electrons or $3.1208 \times 10^9$ electron/bunch. The photon pulses image exactly those of the electron bunches. Therefore, there is a 1 cm long photon pulse every 0.6 m, or in time one 30 ps photon pulses every 2 ns.

**Solution 1.6.** The mass of the pion is $m_\pi c^2 = 139.6$ MeV and $\gamma = 1.716$. From this the velocity is $\beta = 0.813$ or $v = 2.44 \times 10^8$ m/s. The travel time along a 15 m long beam line is $t = 6.15 \times 10^{-8}$ and the survival probability without time dilatation! is $P = \exp{(-6.15/2.6)} = 0.094$. That means, only 9.4 % of the pions would survive at the end of a 15 m long beam transport line. With time dilatation the survival probability for 100 MeV pions increases to $P = \exp{[-6.15/(2.6 \cdot 1.716)]} = 25.1\%$ and the available pion flux is greater by a factor of 2.7.

**Solution 1.7.** Travelling through the atmosphere, 70% of the muons survive or $t/\tau = 0.357$. With $\ell = 2000$ m, $t = \ell/v$ and $\tau = \gamma\tau_0 = \tau_0/\sqrt{1 - \beta^2}$ we get $t/\tau = \frac{\ell\sqrt{1-\beta^2}}{c\tau_0\beta} = 0.357$. Solving for $\beta$ we get $\beta = 0.993$, $\gamma = 8.546$ and the kinetic muon energy is therefore $E_{\mathrm{kin}} = 797.6$ MeV.

**Solution 1.8.** The geometry of the field lines in the particle system of reference can be expressed by $x = z \tan\alpha$, where $\alpha$ is an arbitrary angle defining the angle of the field line with respect to the $z$-axis. In the laboratory system, the $z$-coordinate is Lorentz contracted and the equation for the field lines becomes $x = \frac{\tan\alpha}{\gamma} z$. The distribution of radial field lines is compressed in the $z$-direction by a factor $\gamma$.

**Solution 1.9.** The circulating beam current is defined by $i = enf_{rev} = env/C$, where $n$ is the number of particles circulating, $f_{rev}$ is the revolution frequency, $v \approx c$ the particle velocity and $C$ the accelerator circumference. The number of particles representing a current of 1A are: $n = iC/ev = 2.08 \times 10^{12}$. The ejected beam resembles a pulse with a pulse current of 1A since particles are assumed to be distributed uniformly. The pulse length is given by the revolution time $\tau = C/c = 0.333$ $\mu s$. The synchrotron produces 100 pulses of $1\mu$ sec duration and at a pulse current of 1A. The average beam current is therefore $i_{\mathrm{avg}} = 100 \cdot \tau \cdot 1 = 33.3$ $\mu A$.

**Solution 2.1.** The Cherenkov condition is $\beta n_{\mathrm{air}} \cos\theta = 1$. For electrons $\beta(10 \text{ MeV}) = 0.99869$ and $\beta(50 \text{ MeV}) = 0.9999478$. The Cherenkov angle for 10 MeV electrons is ($\cos\theta = 1.001 > 1$) imaginary. In order to preserve energy and momentum, the electron energy must have a minimum energy such that $n\beta > 1$. For 50 MeV electrons this condition is met and the Cherenkov angle is $\theta_{\mathrm{Ch}} = \arccos{\left(\frac{1}{0.9999478 \cdot 1.0002769}\right)} = 1.214$ deg.

**Solution 2.2.** The Cherenkov angle is $\theta_{\mathrm{Ch}} = \arccos{\left(\frac{1}{0.99869 \cdot 1.7}\right)} = 53.9$ deg. This radiation will meet the other side of the plate at an this angle an will be totally reflected, because according to Snell's law the maximum angle is $\theta_{\mathrm{S}} = 41.8$ deg. The radiation continues to be reflected until it reaches the small side of the plate in which case the incident angle is now 36.1 deg and can escape the plate.

**Solution 2.3.** In the system $\mathcal{L}$ the two 4-vectors be $\tilde{a}(a_1, a_2, a_3, ia_4)$ and $\tilde{b}(b_1, b_2, b_3, ib_4)$ and their product is $a_1b_1 + a_2b_2 + a_3b_3 - a_4b_4$. In system $\mathcal{L}^*$

the same product would be $a_1^* b_1^* + a_2^* b_2^* + a_3^* b_3^* - a_4^* b_4^*$ and applying a Lorentz transformation we get $a_1^* b_1^* + a_2^* b_2^* + a_3^* b_3^* - a_4^* b_4^* = a_1 b_1 + a_2 b_2 + (\gamma a_3 - \beta \gamma a_4)(\gamma b_3 - \beta \gamma b_4) - (-\beta \gamma a_3 + \gamma a_4)(-\beta \gamma b_3 + \gamma b_4) = a_1 b_1 + a_2 b_2 + \gamma^2 a_3 b_3 - \beta \gamma^2 a_3 b_4 - \beta \gamma^2 a_4 b_3 + \beta^2 \gamma^2 a_4 b_4 - \beta^2 \gamma^2 a_3 b_3 + \beta \gamma^2 a_3 b_4 + \beta \gamma^2 a_4 b_3 - \gamma^2 a_4 b_4 = a_1 b_1 + a_2 b_2 + a_3 b_3 - a_4 b_4$ q.e.d.

**Solution 2.4.** The 4-momentum is $(c\hbar \boldsymbol{k}, \mathrm{i}\hbar\omega) = (c\hbar k \boldsymbol{n}, \mathrm{i}\hbar\omega)$, the 4-spacetime vector $(\boldsymbol{r}, \mathrm{i}ct)$ and with $ck = \omega$ their product is $c\hbar k n_x x + c\hbar k n_y y + c\hbar k n_z z - c\hbar \omega t = \hbar\omega (n_x x + n_y y + n_z z - ct)$ which is $\hbar c$-times the phase (2.18) of a plane wave.

**Solution 2.5.** The relativistic Doppler effect is $\omega^* \gamma \left(1 + \beta_z\, n_z^* \right) = \omega$ and for the classical case we set $\gamma = 1$, $n_z^* = \cos\vartheta$ and $\beta = v/v_0$, where $v_0$ is the velocity of the wave (light or acoustic). The relative Doppler shift is then $\frac{\Delta f}{f_s} = \frac{v}{v_0} \cos\vartheta$.

**Solution 2.6.** We use the uncertainty relation $\Delta x\, \Delta p = \Delta x\, \hbar k \geq \hbar$ or $\Delta x \geq 1/k$ and the "characteristic volume" of a photon is $V_{\mathrm{ph}} = \frac{\lambda^3}{8\pi^3}$. The average electric field within this volume is from $\varepsilon = \frac{1}{2\epsilon_0} E^2 V_{\mathrm{ph}} = \hbar\omega$ or $E = k^2 \sqrt{2\epsilon_0 \hbar c}$. For a 0.1238 eV photon ($CO_2$-laser) the wavelength is $\lambda = 10$ $\mu$m and the average electric field is $E = 2.96 \times 10^{-7}$ V/m. In case of a 10 keV x-ray photon the field is $E = 1.93$ kV/m

**Solution 2.7.** We describe EM-waves by $\boldsymbol{E} = \boldsymbol{E}_0 \exp\left[\mathrm{i}\left(\omega t - \boldsymbol{kr}\right)\right]$ and $\boldsymbol{B} = \boldsymbol{B}_0 \exp\left[\mathrm{i}\left(\omega t - \boldsymbol{kr}\right)\right]$ where $\boldsymbol{kr} = \boldsymbol{nr}k$ and $\boldsymbol{n}$ is a unit vector parallel to $\mathbf{k}$. Inserted into Maxwell's equation $\nabla \times \boldsymbol{E} = -\frac{[c]}{c}\dot{\boldsymbol{B}}$ we get with $k = \omega/c$ for the l.h.s.: $\nabla \times (\mathrm{i}\boldsymbol{n}rk)\boldsymbol{E} = \nabla(\mathrm{i}\boldsymbol{n}rk) \times \boldsymbol{E}$ and for the r.h.s. $\dot{\boldsymbol{B}} = \frac{\mathrm{i}\omega}{c}[c]\boldsymbol{B} = \mathrm{i}[c]k\boldsymbol{B}$. With $\nabla(\boldsymbol{nr}) = \nabla(n_x x + n_y y + n_z z) = \boldsymbol{n}$ we get finally $[c]\boldsymbol{B} = \boldsymbol{n} \times \boldsymbol{E}$. This equation tells us that the electric and magnetic wavefields are orthogonal.

**Solution 2.8.** With $[c]\boldsymbol{B} = \boldsymbol{n} \times \boldsymbol{E}$ from Exercise 2.7 and $(B.10)$ we get $\boldsymbol{E} \times (\boldsymbol{n} \times \boldsymbol{E}) = E^2 \boldsymbol{n}$, what was to be demonstrated.

**Solution 3.1.** The energy loss per turn is from (3.18) $U_0 = 20.32$ keV and the total radiation power $P = 20.32$ kW. In case of muons, we have the mass ratio $m_\mu/m_e = 206.8$ and the energy loss is reduced by the $4^{\mathrm{th}}$- power of this ratio to become $U_{0\mu} = 11.1$ $\mu$eV, which is completely negligible.

**Solution 3.2.** The maximum photon flux occurs at a photon energy of about $\varepsilon = 0.286\,\varepsilon_c$ and $S(0.286) \approx 0.569$. To find the 1% photon energy we use (3.38) to scale the photon flux and have $0.777\sqrt{x}/\exp x = 0.00569$, which is solved by $x = 5.795$. Appreciable radiation exists up to almost six times the critical photon energy.

**Solution 3.3.** From (3.28) we have $E = [0.4508\,\epsilon_c\,(\text{keV})\,\rho\,(\text{m})]^{1/3} = 8.0423^{1/3}$ $= 2.0035$ GeV. The magnetic field necessary for a bending radius of $\rho = 1.784$ m would be $B = 3.75$ T, which is way beyond conventional magnet technology. Either superconducting magnets must be used to preserve the ring geometry or a new ring must be constructed with bending magnets which must be longer by at least a factor of 2.5.

**Solution 3.4.** The bending radius is $\rho = 2887$ m, the energy loss is $U_0 = 88.5\dfrac{E^4}{\rho}\left(\dfrac{m_e}{m_p}\right)^4 = 399.4$ keV and the critical photon energy $\varepsilon_c = 2.2\dfrac{E^3}{\rho}\left(\dfrac{m_e}{m_p}\right)^3 = 929.3$ eV. The synchrotron radiation power is $P = 65.5$ kW.

**Solution 3.5.** The critical photon energy is $\varepsilon_c = 38.04$ keV and $\varepsilon/\varepsilon_c = 0.21$. The universal function is $S(0.21) = 0.5625$ and the photon flux $\dfrac{\mathrm{d}\dot{N}_{\text{ph}}}{\mathrm{d}\psi} = C_\psi\,E\,I\,\dfrac{\Delta\omega}{\omega}\,S\,(0.21) = 3.1185\times10^{12}$ photons/mrad. The vertical opening angle $\sqrt{2\pi}\sigma_\theta = 0.251$ mr resulting in an effective beam height at the experiment of $Y = 3.77$ mm. A beam size of 10 $\mu$m at 15 m corresponds to an angle of $0.667\ \mu$rad at the source. The total photon flux into the required sample cross section is then $\dot{N}_{\text{ph}} = 5.53 \times 10^6$ photons/s, which is more than required. For a still higher photon flux one might apply some photon focusing.

**Solution 3.6.** In the horizontal plane the radiation distribution is uniform and an angle of $\Delta\psi = 0.2$ mr will produce a photon beam width of 1 mm at a distance of 5 m. The critical photon energy is $\varepsilon_c = 563$ eV and $\varepsilon/\varepsilon_c = 0.124/563 = 0.00022$. For the IR radiation the vertical opening angle $\theta_{\text{rad}} = 11.3$ mr $(\gg 1/\gamma\,!)$ and the source length $L = 0.045$ m. The total source height is $\sigma_{\text{tot},y} = \sqrt{0.11^2 + 0.107^2} = 0.153$ mm and the vertical divergence $\sigma_{\text{tot},y'} = 14.9$ mr. The photon flux for $\lambda = 10\ \mu$m and $S(0.00022) = 0.0805$ is $\mathrm{d}\dot{N}_{\text{ph}}/\mathrm{d}\psi = 1.275\,10^{15}$ photons/s/mr/100%BW. The photon brightness is then $\mathcal{B} = \dfrac{(\mathrm{d}\dot{N}_{\text{ph}}/\mathrm{d}\psi)\Delta\psi}{2\pi\sigma_{\text{tot},y}\sigma_{\text{tot},y'}} = \dfrac{1.275\cdot10^{15}\,0.2}{2\pi\cdot0.153\cdot14.9} = 1.780 \times 10^{13}\ \dfrac{\text{photons}}{\text{s}\cdot\text{mm}^2\cdot\text{mr}^2\cdot100\%\text{BW}}$.

**Solution 3.7.** From Exercise 3.6 $L = 0.045$ m and the diffraction limited source size and divergence are $\sigma_r = \frac{1}{2\pi}\sqrt{\lambda L} = 0.107$ mm and $\sigma_{r'} = \sqrt{\dfrac{\lambda}{L}} = 14.9$ mr, respectively. This is to be compared with the electron beam parameters $(\sigma_{b,x}, \sigma_{b,y}) = (1.1, 0.11)$ mm and $\sigma\,(\sigma_{b,x'}, \sigma_{b,y'}) = (0.11, 0.011)$ mr. There is a considerable mismatch in the $x$-plane with $\sigma_r/\sqrt{2} = 0.076$ mm $\ll \sigma_{b,x}$ and $\sigma_{r'}/\sqrt{2} = 10.5$ mr $\gg \sigma_{b,x'}$. In the vertical plane the mismatch is small. In both planes the diffraction limited photon emittance is $\epsilon_{\text{ph},x,y} = 797$ nm, which is much larger than the electron beam emittances in both planes. The 10 $\mu$m IR radiation is therefore spatially coherent.

**Solution 4.1.** The magnetic field of 2 T would limit the proton energy to $E \leq 0.3B\rho = 90000$ GeV or 90 TeV. The energy loss at 90 TeV would be from (3.18) with (D.1) and (3.15) $\Delta E = 3.41$ keV and the total radiation power is 34.1 W. The total radiation power is less than the available rf-power and therefore the energy is limited by the magnetic field. The critical photon energy is from (3.26) and (3.27) corrected for protons $\epsilon_c = \hbar C_c \frac{E^3}{\rho} = 1.74 \times 10^{-6}$ GeV $= 1.74$ keV and the radiation is therefore mainly concentrated in the soft x-ray regime.

**Solution 4.2.** To keep the $66/99$ keV contamination below 1% we expect to work on the high energy end of the synchrotron radiation spectrum. In this regime the spectral intensity scales like $I(x) \propto 0.777\sqrt{x}/e^x$, where $x = \epsilon_{ph}/\epsilon_c$. The task requires that $I_{66}/I_{33} \leq 0.01$ or $\sqrt{2}e^{x_{33}-x_{66}} = \sqrt{2}e^{-x_{33}} \leq 0.01$. Solving, we get $x_{33} = \ln\left(100/\sqrt{2}\right) = 4.2586$ or $\epsilon_c \leq 33/4.2586 = 7.749$ keV. The spectral photon flux at 33 keV is then from (3.38) with $x = 4.2586$ given by $\dot{N}_{ph} \approx 2.25 \times 10^{13} n_p E I$ photons/s/0.1%BW, where we added a factor $n_p$ to account for the number of wiggler poles. The magnetic field and beam energy are related by the critical photon energy and we get from (3.26) $E^2\left(\text{GeV}^2\right) B\left(\text{T}\right) \leq 19.256$. A reasonable field level for a superconducting wiggler magnet is $B = 4-6$ T; we take $B = 6$ T. For economy, we would like to keep the beam energy low and the ring size small. Here we try $E = 1.5$ GeV. In this case $\epsilon_c = 8.9775$ keV and $x_{33} = 3.676$. For this decreased value of $x_{33}$ the flux is increased and we have now $\dot{N}_{ph} \approx 5.61 \times 10^{13} n_p I$ photons/s/0.1%BW. With a beam current of say 0.5 A and a wiggler magnet with $n_p = 6$ poles we have finally a flux of $\dot{N}_{ph}(33 \text{ keV}) \approx 1.68 \times 10^{14}$ photons/s/0.1%BW, which is not quite what we want. To get more flux, one would have to increase either the beam energy, the current or the number of wiggler poles. The final choice of parameters is now determined by technical limitations or economic considerations.

**Solution 4.3.** In first approximation, we assume that all the fields are contained within the two rows of poles and no field leaks out. Separating the poles by $dg$ requires to generate the additional field energy $d\varepsilon = F dg$, where $F$ is the force between poles. Since $d\varepsilon > 0$ for $dg > 0$, the force is attractive, meaning that the poles are attracted. The force is then $F = \frac{d\varepsilon_m}{dg} = \frac{w}{2\mu_0}\int_0^{15\lambda_p} B^2(z)\,dz = 20889$ N$= 2.13$ tons.

**Solution 4.4.** The instantaneous radiation power is given by (3.9) $P_\gamma\left(\text{GeV/s}\right) = 379.35\,B^2\,E^2$. The total energy loss of an electron due to wiggler radiation power can be obtained by integrating through the wiggler field for $\Delta E\left(\text{GeV}\right) = 189.67\,B_0^2\,E^2\frac{L_u}{\beta c}$ and the total radiation power for a beam current $I$ is $P_u\left(\text{W}\right) = 632.67 B_0^2 E^2 L_u I$.

**Solution 4.5.** In the electron rest frame energy conservation requires $\hbar\omega + mc^2 = \hbar\omega' + \sqrt{c^2 p^2 + (mc^2)^2}$, where $\hbar\omega$ and $\hbar\omega'$ are the incoming and outgoing photon energies, respectively and $cp$ the electron momentum after the scattering process. Solving for $cp$ we get $c^2 p^2 = \hbar^2\left(\omega - \omega'\right)^2 + 2\hbar mc^2\left(\omega - \omega'\right)$.

For momentum conservation, we require that $\hbar k = \hbar k' + p$ with the angle $\vartheta$ between $k$ and $p$. From this we get $c^2 p^2 = (\hbar \omega')^2 + (\hbar \omega)^2 - 2\hbar^2 \omega \omega' \cos \vartheta$. Comparing booth expressions for $cp$ we get $-2\hbar \omega \omega' + 2\hbar mc^2 (\omega - \omega') = -2\hbar^2 \omega \omega' \cos \vartheta$ or $\frac{\hbar}{mc^2}(1 - \cos \vartheta) = \frac{1}{\omega'} - \frac{1}{\omega} = \frac{\lambda' - \lambda}{2\pi c}$. We look for radiation emitted in the forward direction or for $\vartheta = 180°$ and get for the scattered wavelength $\lambda' = \lambda$, because $\frac{4\pi \hbar c}{mc^2} \approx 4.8\,10^{-12} \ll \lambda$. Note, that all quantities are still defined in the electron rest frame. The wavelength of the undulator field in the electron system is $\lambda = \frac{\lambda_p^*}{\gamma}$, where now $\mathcal{L}^*$ is the laboratory system of reference and the scattered radiation in the laboratory system due to the Doppler effect is $\lambda^* = \frac{\lambda_p^*}{2\gamma^2}\left(1 + \frac{1}{2}K^2\right)$, which is the expression for the fundamental wavelength of undulator radiation.

**Solution 4.6.** The fundamental wave length for a very weak undulator ($K \ll 1$, e.g. wide open gap) is $\lambda\,(800\ \text{MeV}) = 102$ Å and $\lambda\,(7\ \text{GeV}) = 1.33$ Å which are the shortest achievable wavelength. For a 10 mm gap the field is from (4.5) $B = 1.198$ T and the maximum value of the strength parameter is $K = 5.595$. With this the longest wavelength in the fundamental is $\lambda = 1698.5$ Å for the 800 MeV ring and $\lambda = 22.147$ Å for the 7 GeV ring.

**Solution 4.7.** The short wavelength limits are given for a wide open undulator, $K \ll 1$, and are $\lambda = 3.13$ Å for $\lambda_p = 15$ mm and $\lambda = 15.7$ Å for $\lambda_p = 75$ mm. The long wavelength limits are determined by the magnetic fields when the undulator gaps are closed to 10 mm. The fields are from (4.5) $B_0\,(\lambda_p = 15\ \text{mm}) = 0.19$ T and $B_0\,(\lambda_p = 75\ \text{mm}) = 1.66$ T, respectively. The undulator strengths are $K\,(\lambda_p = 15\ \text{mm}) = 0.270$ and $K\,(\lambda_p = 75\ \text{mm}) = 1.35$ and the wavelengths $\lambda\,(\lambda_p = 15\ \text{mm}) = 3.24$ Å and $\lambda\,(\lambda_p = 75\ \text{mm}) = 30.0$ Å. The tuning range is very small for the 15 mm undulator and about a factor of two for the long period undulator. The ranges are so different because the $K$-value can be varied much more for longer period undulators.

**Solution 5.1.** In each bunch there are $n_e = \tau I_b / e = 1.87 \times 10^7$ electrons and the circulating beam current per bunch is $i_b = e n_e f_{\text{rev}} = 3.0\ \mu\text{A}$. To reach a circulating beam current of 200 mA 66667 injection pulses are required. To deliver that many pulses an injector operating at 10 Hz would require 1.85 hours. To reduce the injection time to less than 5 min each injection pulse must deliver at least 23 bunches. Additional time and bunches are required if the injection process is less than 100% effective.

**Solution 6.1.** A uniform field $B_y$ in the laboratory system transforms into a field $\left(E_x^*, E_y^*, E_z^*\right) = \left([c]\,\beta\gamma B_y,\ 0,\ 0\right)$ and $\left(B_x^*, B_y^*, B_z^*\right) = \left(0,\ \gamma B_y,\ 0\right)$. The particle velocity is zero in it's own system and therefore the magnetic field $B_y^*$

is ineffective. There is a nonzero electric field $E_x^* = [c]\beta\gamma B_y$ which deflects an electron in the negative $x$-direction just like the magnetic field in the laboratory system does. The gain in transverse momentum is $\Delta p_x^* = \int e$ $E_x^* \, dt^* = \int \frac{eE_x^*}{\beta c\gamma} \, ds = \frac{eE_x^*}{\beta c\gamma}\ell_\mathrm{b}$ where we set $dt^* = \frac{dt}{\gamma} = \frac{ds}{\beta c\gamma}$ , and $\ell_\mathrm{b}$ is the length of the bending magnet. The deflection angle is then $\psi = \frac{\Delta p_x^*}{p_0^*} = [c]\frac{eB_y}{\beta E}\ell_\mathrm{b}$ , which is the same as (6.11) with (6.7).

**Solution 6.2.** From the Lorentz force equation we get $\boldsymbol{E} = [c]\beta\boldsymbol{B}$ . Solving for $\beta$ , we get for our case $\beta = \frac{E}{[c]\boldsymbol{B}} = 3.336 \times 10^{-3}$ or with $\gamma \approx 1 + \frac{1}{2}\beta^2$ the kinetic energy for force equivalence is $E_\mathrm{kin} = 852.35$ eV. Magnetic fields are more effective for electrons with a kinetic energy of more than 852 eV. That equivalence point is much higher for protons and ions.

**Solution 6.3.** For the total ring we have $0.3\frac{B}{E}n_\mathrm{b}\ell = 2\pi$ or $n_\mathrm{b} = \frac{2\pi E}{0.3B\ell} = 26.18$ . The number of magnets must be an integer and we set $n_\mathrm{b} = 28$ (it is customary to use an even number of magnets, although not necessary), the field $B = 1.122$ T, the bending radius $\rho = 8.913$ m, and the deflection angle is $\psi = 360/28 = 12.86$ degrees per magnet.

**Solution 6.4.** To generate a focusing magnet, we require a magnetic field $B_y\,(y=0) = gx$ , where $g$ is a constant. This field can be derived from a potential $V = -gxy$ . The surface of ferromagnetic pole is an equipotential surface and its cross section in the $xy$-plane is given by $xy = $ const. The shape of a pole is that of a hyperbola. The pole tip at the intersection of the $45^\circ$-line from the magnet axis and the pole profile is at the coordinates $\left(R/\sqrt{2},\ R/\sqrt{2}\right)$. With this the equation for the pole shape becomes $xy = \frac{1}{2}R^2$ which is (6.16).

**Solution 6.5.** The transformation matrix for quadrupole and 5 m drift space is $M = \begin{pmatrix} 1 & 5 \\ 0 & 1 \end{pmatrix}\begin{pmatrix} 1 & 0 \\ -f^{-1} & 1 \end{pmatrix} = \begin{pmatrix} 1 - \frac{5}{f} & 5 \\ -\frac{1}{f} & 1 \end{pmatrix}$ . To focus a parallel ray to a focal point at 5 m, we require $x_\mathrm{F} = \left(1 - \frac{5}{f}\right)x_0 = 0$ or $f = \frac{1}{kl} = 5$ m. From this $k = 1.0$ m$^{-2}$ and $g = \frac{1.5 \cdot 1}{0.3} = 5.0$ T/m. The relation between excitation current and field gradient in a quadrupole can be derived in a similar way as done for a bending magnet and is given by $\mu_0 I_\mathrm{coil} = \frac{1}{2}gR^2$ . For the case on hand, the excitation current is $I_\mathrm{coil} = 4973.6$ A·turns.

**Solution 6.6.** Starting from the waist in the center of the drift space the betatron function is $\beta\,(s) = \beta_\mathrm{w} + \frac{s^2}{\beta_\mathrm{w}}$ . If we choose a very small value for $\beta_\mathrm{w} = \beta\,(0)$ we get a very large value at $s = \frac{1}{2}L$ . Similarly, if $\beta_\mathrm{w}$ is very large $\beta\left(\frac{1}{2}L\right)$ is large too. There must be a minimum for $\beta\left(\frac{1}{2}L\right)$ . We calculate $\partial\beta/\partial\beta_\mathrm{w} = 0$ , solve for $\beta_\mathrm{w}$ and get for the optimum value $\beta_\mathrm{w} = \frac{1}{2}L$ . For this value at the waist we have the minimum variation of the betatron function or of the beam size along the drift space $L$ . If $\sigma_\mathrm{w} = \sqrt{\epsilon\beta_\mathrm{w}}$ is the beam size at the waist then the beam size at the ends of the drift space is $\sigma = \sqrt{2}\sigma_\mathrm{w}$. This is the same result as is known for a photon beam defining the Rayleigh length.

**Solution 6.7.** We form FODO cells which are each $2L = 8$ m long. That leaves between bending magnets 2 m of space for quadrupoles and drift spaces. For $\kappa = \sqrt{2}$ we get the minimum beam sizes along the FODO-cell, which is define as the minimum beam radius, or $R_\mathrm{min} = \left.\sqrt{\sigma_x^2 + \sigma_y^2}\right|_\mathrm{min}$. The horizontal and vertical betatron functions in the center of the QD-quadrupole are $\beta_x = 2.343$ m and $\beta_y = 13.657$ m. The focal length of the half-quadrupole is $f_\mathrm{hq} = \kappa L = 5.657$ m. Note, in FODO theory we use half quadrupoles. That means the focal length of the full quadrupole is $f_q = \frac{1}{2}f_\mathrm{hq} = 2.8285$ m. Each half FODO-cell now has the following structure: $\frac{1}{2}$QF(0.5 m) Drift(0.5 m) Bend(2 m) Drift(0.5 m) $\frac{1}{2}$QD(0.5 m).

**Solution 6.8.** In the center of say a QF the optimum betatron function is given by $\beta_0 = L\frac{\kappa(\kappa+1)}{\sqrt{\kappa^2-1}}$, and $\alpha_0 = 0$. The transformation matrix from the center to the exit of the thin quadrupole is $\begin{pmatrix} 1 & 0 \\ -f^{-1} & 1 \end{pmatrix}$ and the betatron functions at the quadrupole exit are therefore: $\beta = \beta_0$, $\alpha = -\frac{1}{f}\beta_0$ and $\gamma = \frac{1+\alpha^2}{\beta_0}$. These are the starting values for the drift space which spans the space between quadrupoles since the bending magnets are not included in the focusing scheme in this approximation. The expression for the betatron function between a QF and QD quadrupole is therefore $\beta(s) = \beta_0 - 2\alpha s + \gamma s^2$ and the phase advance $\Psi = \int_0^L \frac{ds}{\beta(s)} = \sqrt{\kappa^2-1}\int_0^1 \frac{ds}{\kappa(\kappa+1)-2(\kappa+1)s+2s^2} = -\arctan\frac{\kappa-1}{\sqrt{\kappa^2-1}} + \arctan\frac{\kappa+1}{\sqrt{\kappa^2-1}}$. For the optimum FODO-cell with $\kappa = \sqrt{2}$, the phase advance per half cell is $\Psi_{\frac{1}{2}} = 45°$, and because of symmetry for the full cell $\Psi_\mathrm{FODO} = 90°$.

**Solution 7.1.** In this case we do not expand the rf-voltage and keep $V_\mathrm{rf}(t) = V_\mathrm{rf}\sin\omega_\mathrm{rf}t$. We also ignore damping, because it is a very small effect. Equation (7.2) becomes then with (7.3)

$\ddot{\tau} + \frac{\eta_c e V_\mathrm{rf}}{T_0 E_0}(\sin\psi_\mathrm{s}\cos\omega_\mathrm{rf}\tau + \cos\psi_\mathrm{s}\sin\omega_\mathrm{rf}\tau - \sin\psi_\mathrm{s}) = 0$, where $\eta_c = \frac{1}{\gamma^2} - \alpha_c$. We multiply this equation with $\dot{\tau}$ and integrate to get

$\frac{1}{2}\dot{\tau}^2 + \frac{\eta_c e V_\mathrm{rf}}{\omega_\mathrm{rf}T_0 E_0}(\sin\psi_\mathrm{s}\sin\omega_\mathrm{rf}\tau - \cos\omega_\mathrm{rf}\tau\cos\psi_\mathrm{s} - \tau\sin\psi_\mathrm{s}) =$const. For simplicity, we assume that $\psi_\mathrm{s} \to 0$, which is not true for storage rings but describes very well the situation in FELs, and get $\dot{\tau}^2 - \frac{2\eta_c e V_\mathrm{rf}}{\omega_\mathrm{rf}T_0 E_0}\cos\omega_\mathrm{rf}\tau =$const. Plotting $\dot{\tau}$ as a function of $\tau$ results in the phase space diagram as shown in Fig. 11.2.

**Solution 7.2.** We use the storage ring of Exercise 6.3 with an energy of 3 GeV and a bending radius of $\rho = 8.913$ m. This ring has 14 FODO cells, each 8 m long. To make it more realistic we assume that all insertion straight sections will occupy the same length such that the ring circumference is $C = 224.0$ m. The synchrotron radiation power is from (3.14) $\langle P_\gamma \rangle = 1075.9$ GeV/s and the horizontal and vertical damping times are then $\tau_{x,y} = 2.788$ ms. The synchrotron damping time is half that or $\tau_s = 1.394$ ms. The beam energy spread is from (7.32) $\left(\frac{\sigma_\epsilon}{E}\right) = 0.0609$ %.

**Solution 7.3.** The probability to emit a photon of energy $\varepsilon$ in a unit time is $\dot{n}(\varepsilon_{\rm ph}) = \frac{P_\gamma}{\varepsilon_c^2} \frac{S(x)}{x}$ . We are looking for the case $\varepsilon = \sigma_\varepsilon = \frac{E^2}{mc^2} \sqrt{\frac{55\hbar c}{64\sqrt{3}mc^2 J_s \rho}} =$ 10.9 MeV. For $\varepsilon_c = \frac{3}{2}\hbar c \frac{\gamma^3}{\rho} = 19166$ eV, the ratio $x = \frac{1}{\gamma}\sqrt{\frac{55mc^2\rho}{144\sqrt{3}J_s\hbar c}} = 227.54$ $\gg 1$ and $\frac{P_\gamma}{\varepsilon_c^2} = 23826$ 1/eVs. The probability becomes with this $\dot{n}(\varepsilon_{\rm ph}) \approx 1.86 \times 10^{-96}$! We may, without calculation conclude that no second photon of this energy will be emitted within a damping time. Energy is emitted in very small fractions of the electron energy.

**Solution 7.4.** From (9.167) we get the number of photons emitted per unit time $\dot{N}_{\rm ph} = \frac{15\sqrt{3}}{8} \frac{P_\gamma}{\varepsilon_c} = 3.158 \times 10^6 \frac{\gamma}{\rho}$ and per radian $\dot{n} = 0.01063\gamma \approx \frac{\gamma}{100}$ .

**Solution 9.1.** Integration of (9.76) over $\varphi$ results in factors $2\pi$ and $\pi$ for the two terms in the nominator, respectively and we have the integrals $2\pi \int_0^\pi \frac{\sin\theta}{(1-\beta\cos\theta)^3}d\theta - \pi\left(1-\beta^2\right)\int_0^\pi \frac{\sin^3\theta}{(1-\beta\cos\theta)^5}d\theta = \frac{4\pi}{(1-\beta^2)^2} - \frac{\frac{4}{3}\pi}{(1-\beta^2)^2} = 4\pi\gamma^4\left(1-\frac{1}{3}\right) = 4\pi\gamma^4\frac{2}{3}$ . With this, the radiation power is $P_{\rm tot} = \frac{2}{3}r_c mc^2 c \frac{\beta^4\gamma^4}{\rho^2}$, which is (9.59).

**Solution 9.2.** The vertical opening angle is $1/\gamma = 0.085$ mr and therefore all radiation will be accepted. The spectral photon flux into an opening angle of $\Delta\psi = 10$ mr is therefore given from (9.156) by $\dot{N}_{\rm ph} = C_\psi EI\frac{\Delta\omega}{\omega}S\left(\frac{\omega}{\omega_c}\right)\Delta\psi$. With the critical photon energy is $\varepsilon_c = 23.94$ keV the spectral photon flux from an ESRF bending magnet is $\dot{N}_{\rm ph} = 4.75 \times 10^{14} S\left(\frac{\varepsilon_{\rm ph}({\rm keV})}{23.94}\right)$.

**Solution 9.3.** We use (9.106) and get with $\xi = \frac{1}{2}\frac{\omega}{\omega_c}(1+\gamma^2\theta^2)^{3/2}$ for the $p$% point $\frac{d^2W(10\%)}{d\Omega d\omega}\Big/\frac{d^2W}{d\Omega d\omega} = (1+\gamma^2\theta^2)^2\left[\frac{K_{2/3}^2(\xi)}{K_{2/3}^2(0)} + \frac{\gamma^2\theta^2}{1+\gamma^2\theta^2}\frac{K_{1/3}^2(\xi)}{K_{2/3}^2(0)}\right] = 0.1$. Solving for $\theta$ gives the angle at which the intensity has dropped to 10%. For low frequencies $\frac{d^2W(10\%)}{d\Omega d\omega}\Big/\frac{d^2W}{d\Omega d\omega} \xrightarrow{\xi\to 0} 1 + \frac{\gamma^2\theta^2}{1+\gamma^2\theta^2}\frac{\Gamma^4(1/3)}{2^{8/3}\Gamma^4(2/3)}\left(\frac{\omega}{\omega_c}\right)^{4/3} = p$, and for large arguments $\frac{d^2W(10\%)}{d\Omega d\omega}\Big/\frac{d^2W}{d\Omega d\omega} \xrightarrow{\xi\to\infty} \frac{\exp\left(\frac{\omega}{\omega_c}\right)}{\exp\left[\frac{\omega}{\omega_c}(1+\gamma^2\theta^2)^{3/2}\right]}\frac{1+2\gamma^2\theta^2}{\sqrt{1+\gamma^2\theta^2}} = p$. All expressions have to be evaluated numerically. The angle at which the total radiation intensity has dropped to 10% is from (9.115) given by $\frac{dW(10\%)}{d\Omega}\Big/\frac{dW}{d\Omega} = \frac{1}{(1+\gamma^2\theta^2)^{5/2}}\left(1 + \frac{5}{7}\frac{\gamma^2\theta^2}{1+\gamma^2\theta^2}\right) = p$, which can be solved by $\gamma\theta = 1.390$ for $p = 10\%$.

**Solution 10.1.** From (10.20) the amplitude of the oscillatory motion in an undulator is $a_\perp = \frac{\lambda_p K}{2\pi\gamma} = \frac{0.05 \cdot 1}{2\pi(7/0.000511)} = 0.581$ $\mu$m. The longitudinal oscillation amplitude is from (10.19$b$) $a_\parallel = \frac{K^2}{8\gamma^2 k_p} = 0.053$ Å. Both amplitudes are very small, yet are responsible for the high intensities of radiation.

**Solution 10.2.** The focal length for a single pole end is given by (10.23) $\frac{1}{f_{1,y}} = \frac{\pi^2}{2\gamma^2}\frac{K^2}{\lambda_p} = 2.58 \times 10^{-5}$ m$^{-1}$ and for the whole undulator $\frac{1}{f_y} = \frac{\pi^2}{2\gamma^2}\frac{K^2}{\lambda_p}2N = 0.00258$ m$^{-1}$ or $f_y = 387.60$ m. This focal length is very long compared to the focal lengths of the ring quadrupole, which are of the order of the distance between quadrupoles. Typically, the focal length of any insertion should be more than about 50 m to be negligible. The wiggler magnet with $K = 5$, on the other hand, produces a focal length of $f_y = 15.50$ m which is too strong to be ignored and must be compensated. The difference comes from the fact that its the deflection angle which is responsible for focusing and $\frac{1}{f_y} \propto \theta^2$. Focusing occurs only in the nondeflecting plane and $\frac{1}{f_x} = 0$.

**Solution 10.3.** This result appears nonphysical, yet it is correct, but requires some interpretation. The number of photons emitted into the forward cone is constant. Note, that the forward cone angle decreases with increasing number of periods. The constant number of photons is emitted into a smaller and smaller cone. Outside this forward cone there is still much radiation and integration of all radiation would give the more intuitive result that the total radiation power increases with number of undulator periods.

**Solution 10.4.** To solve this problem, we do not rely on exact calculations, but are satisfied with the precision of reading graphs in Chapter D. We also use iterations to get the solution we want. The fundamental flux drops below 10% for $K < 0.25$, and we use this value to get 15 keV radiation. From the definition of the fundamental photon energy we get the periodlength $\lambda_p = 3.0$ cm. To generate 4 keV radiation we need to change $K$ enough to raise the factor $\left(1 + \frac{1}{2}K^2\right)$ from a low value of 1.031 by a factor of 15/4 to a value of 3.87 or to a high of $K = 2.4$, which corresponds to a field of $B = 0.857$ T. Unfortunately, that field requires a gap of $g = 8.1$ mm which is less than allowed. We have to increase the periodlength to say $\lambda_p = 3.5$ cm, which gives a maximum photon energy for $K = 0.25$ of $\varepsilon_{ph} = 12.9$ keV. We plan to use the 3$^{rd}$ harmonic to reach 15 keV. To reach $\varepsilon_{ph} = 4$ keV, we need $K = 2.16$, a field of $B = 0.661$ T, which requires an allowable gap of $g = 11.7$ mm. We use the 3$^{rd}$ harmonic to reach $\varepsilon_{ph} = 15$ keV at $K = 1.82$. With this result we may even extend the spectral range on both ends.

**Solution 10.5.** In the electron system the wavelength of the laser beam is Lorentz contracted by a factor of $\frac{1}{2\gamma}$, where the factor of two is due to the fact that the relative velocity between both beams is $2c$. The wavelength in the laboratory system is therefore $\lambda = \frac{\lambda_L}{4\gamma^2}$, since $K \ll 1$ for the laser field.

**Solution 10.6.** The maximum transverse oscillation amplitude is 4.57 $\mu$m and the transverse velocity in units of $c$ is just equal to the maximum deflection angle $\beta_\perp = \theta = K/\gamma = 0.38$ mr. The transverse relativistic factor $\gamma_\perp \approx 1 + 7.22 \times 10^{-8}$ , indeed very small, yet enough to start generating relativistic perturbations in the transverse particle motion.

**Solution 10.7.** The fundamental wavelength is given by the expression $\lambda = \frac{\lambda_p}{2\gamma^2}\left(1 + \frac{1}{2}K^2 + \gamma^2\vartheta^2\right)$ and for $\vartheta = 0$, we have the fundamental wavelength $\lambda = 10.88$ Å. The natural bandwidth is $1/N_p = 1\%$ and we look therefore for an angle $\hat\vartheta$ such that the wavelength has increased by no more than 9%, or $\frac{\gamma^2\hat\vartheta^2}{1+\frac{1}{2}K^2} = 0.09$ and solving for $\hat\vartheta$, we get $\hat\vartheta = \pm 62.6$ $\mu$rad.

**Solution 11.1.** We may solve this problem two ways. First, we use the average drift velocity $\bar\beta = \beta\left(1 - \frac{K^2}{4\gamma^2}\right)$ and calculate the time it takes the electron to travel along one period, $t_c = \frac{\lambda_p}{c\bar\beta} \approx \frac{\lambda_p}{c\beta}\left(1 + \frac{K^2}{4\gamma^2}\right)$. During that same time the photon travels a distance $s_\gamma = ct_c = \frac{\lambda_p}{\beta}\left(1 + \frac{K^2}{4\gamma^2}\right)$ and the difference is
$$\delta s = s_\gamma - \lambda_p = \frac{\lambda_p}{\beta}\left(1 + \frac{K^2}{4\gamma^2}\right) - \lambda_p = \lambda_p\left(\frac{1}{\beta} - 1\right) + \frac{\lambda_p}{\beta}\frac{K^2}{4\gamma^2} \approx \frac{\lambda_p}{2\gamma^2}\left(1 + \frac{1}{2}K^2\right)$$
which is just equal to the fundamental radiation wave length. We may also integrate the path length along the sinusoidal trajectory and get for one quarter period $s_c = \frac{\lambda_p}{2\pi}\int_0^{\pi/2}\sqrt{1 + \theta^2\cos^2 x}\,dx = \frac{\lambda_p}{2\pi}$ EllipticE $\left(\sqrt{-\theta^2}\right)$ an elliptical function. Since the argument will always be very small we may expand EllipticE $\left(\sqrt{-\theta^2}\right) \approx \frac{\pi}{2} + 0.393\,\theta^2$. The electron travel time for one period is then $t_c = 4\frac{\lambda_p}{2\pi}\frac{1}{c\beta}\left(\frac{\pi}{2} + 0.393\,\theta^2\right)$ and the path length difference is

$$\delta s = ct_c - \lambda_p = \lambda_p\frac{1}{\beta}\left(1 + \frac{0.393\,\theta^2}{\pi/2}\right) - \lambda_p \approx \frac{\lambda_p}{2\gamma^2}\left(1 + \underbrace{\frac{8\cdot 0.393}{\pi}}_{\approx 1}\frac{1}{2}K^2\right) \text{ which is}$$

again the wavelength of the fundamental radiation.

# B. Mathematical Constants and Formulas

## B.1 Constants

$$\pi = 3.141592653589793238 \tag{B.1}$$
$$e = 2.718281828459045235 \tag{B.2}$$

$$\Gamma(1/3) = 2.6789385 \tag{B.3}$$
$$\Gamma(2/3) = 1.351179 \tag{B.4}$$

## B.2 Series Expansions

For $x \ll 1$

$$e^x \approx 1 + \frac{1}{1!}x + \frac{1}{2!}x^2 + \frac{1}{3!}x^3 + \dots \tag{B.5}$$

$$\sin x \approx \frac{x}{1!} - \frac{x^3}{3!} + \frac{x^5}{5!} - \dots \tag{B.6}$$

$$\cos x \approx 1 - \frac{x^2}{2!} + \frac{x^4}{4!} - \frac{x^6}{6!} + \dots \tag{B.7}$$

$$\frac{1}{1+x} \approx 1 - x + x^2 - x^3 + x^4 - \dots \tag{B.8}$$

$$\sqrt{1+x} \approx 1 + \tfrac{1}{2}x - \tfrac{1\cdot 1}{2\cdot 4}x^2 + \tfrac{1\cdot 1\cdot 3}{2\cdot 4\cdot 6}x^3 + \tfrac{1\cdot 1\cdot 3\cdot 5}{2\cdot 4\cdot 6\cdot 8}x^4 + \dots \tag{B.9}$$

## B.3 Multiple Vector Products

In a vector product $a \times b = c$, vectors $\{a, b, c\}$ form a right handed orthogonal system.

$a = (a_x, a_y, a_z)$;

$a \times b = (a_y b_z - a_z b_y,\ a_z b_x - a_x b_z,\ a_x b_y - a_y b_x)$

$$a \times (b \times c) = b(ac) - c(ab) \tag{B.10}$$

$$(a \times b) \times c = b(ac) - a(bc) \tag{B.11}$$

$$(a \times b) \times (c \times d) = c\left[(a \times b)\,d\right] - d\left[(a \times b)\,c\right] \tag{B.12}$$

$$a(b \times c) = b(c \times a) = c(a \times b) \tag{B.13}$$

$$(a \times b)(c \times d) = (ac)(bd) - (bc)(ad) \tag{B.14}$$

## B.4 Differential Vector Expressions

$a, b$ vectors; $\psi$ scalar;

$\nabla = \left( \frac{\partial}{\partial x}, \frac{\partial}{\partial y}, \frac{\partial}{\partial z} \right)$ ;

$\Delta = \left( \frac{\partial^2}{\partial x^2}, \frac{\partial^2}{\partial y^2}, \frac{\partial^2}{\partial z^2} \right)$

$$\nabla(a\psi) = \psi\,\nabla a + a\nabla\psi \tag{B.15}$$

$$\nabla \times (a\psi) = \psi\,(\nabla \times a) - (a \times \nabla\psi) \tag{B.16}$$

$$\nabla(a \times b) = b\,(\nabla \times a) - a(\nabla \times b) \tag{B.17}$$

$$\nabla \times (a \times b) = (b\nabla)\,a - (a\nabla)\,b + a(\nabla b) - b(\nabla a) \tag{B.18}$$

$$\nabla(ab) = (b\nabla)\,a + (a\nabla)\,b + a \times (\nabla \times b) + b \times (\nabla \times a) \tag{B.19}$$

$$\nabla(\nabla\psi) = \nabla^2\psi = \Delta\psi \tag{B.20}$$

$$\nabla \times (\nabla\psi) = 0 \tag{B.21}$$

$$\nabla \times (\nabla\phi) = 0 \tag{B.22}$$

$$\nabla(\nabla \times a) = 0 \tag{B.23}$$

$$\nabla \times (\nabla \times a) = \nabla(\nabla a) - \Delta a \tag{B.24}$$

$$\text{if }\ \nabla \times a = 0 \quad \text{a scalar function } \psi \text{ exists with }\ a = \nabla\psi \tag{B.25}$$

$$\text{if }\ \ \nabla a = 0 \quad \text{a vector function } b \text{ exists with }\ a = \nabla \times b \tag{B.26}$$

## B.5  Theorems

$$\int_V \boldsymbol{\nabla}\psi \, \mathrm{d}V = \oint \left(\psi \boldsymbol{n}\right) \mathrm{d}a\,. \tag{B.27}$$

**Gauss's theorem**

$$\int_V \boldsymbol{\nabla}\boldsymbol{a} \, \mathrm{d}V = \oint \left(\boldsymbol{a}\boldsymbol{n}\right) \mathrm{d}a\,. \tag{B.28}$$

**Stokes' theorem**

$$\int_S \left(\boldsymbol{\nabla}\times\boldsymbol{a}\right)_{\mathrm{n}} \mathrm{d}\boldsymbol{a} = \oint \boldsymbol{a}\,\mathrm{d}\boldsymbol{s}\,. \tag{B.29}$$

**Fourier transform**

$$f(\omega) = \int_{-\infty}^{\infty} f(t)\,\mathrm{e}^{\mathrm{i}\omega t}\mathrm{d}t\,, \tag{B.30}$$

$$f(t) = \frac{1}{2\pi}\int_{-\infty}^{\infty} f(\omega)\,\mathrm{e}^{-\mathrm{i}\omega t}\mathrm{d}\omega\,. \tag{B.31}$$

**Parseval's theorem**

$$\int_{-\infty}^{\infty} f^2(t)\,\mathrm{d}t = \frac{1}{2\pi}\int_{-\infty}^{\infty} f^2(\omega)\,\mathrm{d}\omega\,. \tag{B.32}$$

## B.6  Coordinate Systems

**Cartesian coordinates** $(x, y, z)$

$$\boldsymbol{\nabla}\phi = \left[\frac{\partial\phi}{\partial x}, \frac{\partial\phi}{\partial y}, \frac{\partial\phi}{\partial z}\right] \tag{B.33}$$

$$\boldsymbol{\nabla}\boldsymbol{a} = \frac{\partial a_x}{\partial x} + \frac{\partial a_y}{\partial y} + \frac{\partial a_z}{\partial z} \tag{B.34}$$

$$\boldsymbol{\nabla}\times\boldsymbol{a} = \left[\frac{\partial a_z}{\partial y} - \frac{\partial a_y}{\partial z}, \frac{\partial a_x}{\partial z} - \frac{\partial a_z}{\partial x}, \frac{\partial a_y}{\partial x} - \frac{\partial a_x}{\partial y}\right] \tag{B.35}$$

$$\Delta\phi = \frac{\partial^2\phi}{\partial x^2} + \frac{\partial^2\phi}{\partial y^2} + \frac{\partial^2\phi}{\partial z^2} \tag{B.36}$$

## Cylindrical coordinates $(\rho, \varphi, \zeta)$

$$\boldsymbol{\nabla}\phi = \left[\frac{\partial\phi}{\partial\rho}, \frac{1}{\rho}\frac{\partial\phi}{\partial\varphi}, \frac{\partial\phi}{\partial\zeta}\right] \tag{B.37}$$

$$\boldsymbol{\nabla}\boldsymbol{a} = \frac{1}{\rho}\frac{\partial}{\partial\rho}\left(\rho a_\rho\right) + \frac{1}{\rho}\frac{\partial a_\varphi}{\partial\varphi} + \frac{\partial a_\zeta}{\partial\zeta} \tag{B.38}$$

$$\boldsymbol{\nabla}\times\boldsymbol{a} = \left[\frac{1}{\rho}\frac{\partial a_\zeta}{\partial\varphi} - \frac{\partial a_\varphi}{\partial\zeta}, \frac{\partial a_\rho}{\partial\zeta} - \frac{\partial a_\zeta}{\partial\rho}, \frac{1}{\rho}\frac{\partial}{\partial\rho}\left(\rho a_\varphi\right) - \frac{1}{\rho}\frac{\partial a_\rho}{\partial\varphi}\right] \tag{B.39}$$

$$\Delta\phi = \frac{\partial^2\phi}{\partial\rho^2} + \frac{1}{\rho}\frac{\partial\phi}{\partial\rho} + \frac{1}{\rho^2}\frac{\partial^2\phi}{\partial\varphi^2} + \frac{\partial^2\phi}{\partial\zeta^2} \tag{B.40}$$

## Transformation to cylindrical coordinates $(\rho, \varphi, \zeta)$

$$(x, y, z) = (\rho\cos\varphi, \rho\sin\varphi, \zeta)$$
$$\mathrm{d}s^2 = \mathrm{d}\rho^2 + \rho^2\mathrm{d}\varphi^2 + \mathrm{d}\zeta^2 \tag{B.41}$$
$$\mathrm{d}V = \rho\,\mathrm{d}\rho\,\mathrm{d}\varphi\,\mathrm{d}\zeta$$

## Polar coordinates $(r, \varphi, \theta)$

$$\boldsymbol{\nabla}\phi = \left[\frac{\partial\phi}{\partial r}, \frac{1}{r}\frac{\partial\phi}{\partial\varphi}, \frac{1}{r\sin\theta}\frac{\partial\phi}{\partial\theta},\right] \tag{B.42}$$

$$\boldsymbol{\nabla}\boldsymbol{a} = \frac{1}{r^2}\frac{\partial}{\partial r}\left(r^2 a_r\right) + \frac{1}{r\sin\theta}\frac{\partial}{\partial\varphi}\left(\sin\varphi\, a_\varphi\right) + \frac{1}{r\sin\theta}\frac{\partial a_\theta}{\partial\theta} \tag{B.43}$$

$$\boldsymbol{\nabla}\times\boldsymbol{a} = \begin{bmatrix}\frac{1}{r\sin\theta}\left(\frac{\partial(\sin\theta\, a_\zeta)}{\partial\varphi} - \frac{\partial a_\varphi}{\partial\theta}\right), \\ \frac{1}{r\sin\theta}\left(\frac{\partial a_r}{\partial\theta} - \sin\theta\frac{\partial(r a_\theta)}{\partial r}\right), \\ \frac{1}{r}\left(\frac{\partial}{\partial r}\left(r a_\varphi\right) - \frac{\partial a_r}{\partial\varphi}\right).\end{bmatrix} \tag{B.44}$$

$$\Delta\phi = \frac{1}{r^2}\frac{\partial}{\partial r}\left(r^2\frac{\partial\phi}{\partial r}\right) + \frac{1}{r^2\sin^2\theta}\frac{\partial^2\phi}{\partial\varphi^2} + \frac{1}{r^2\sin\theta}\frac{\partial}{\partial\theta}\left(\sin\theta\frac{\partial\phi}{\partial\theta}\right) \tag{B.45}$$

## Transformation to polar coordinates $(r, \varphi, \theta)$

$$(x, y, z) = (r\cos\varphi\sin\theta, r\sin\varphi\sin\theta, r\cos\theta)$$
$$\mathrm{d}s^2 = \mathrm{d}r^2 + r^2\sin^2\theta\,\mathrm{d}\varphi^2 + r^2\mathrm{d}\theta^2 \tag{B.46}$$
$$\mathrm{d}V = r^2\sin\theta\,\mathrm{d}r\,\mathrm{d}\varphi\,\mathrm{d}\theta$$

## B.7 Gaussian Distribution

**1-dim Gaussian distribution** ($\sigma$: standard deviation) (Fig. 13.1)

$$\varphi(x) = \frac{1}{\sqrt{2\pi}\sigma} e^{-\frac{1}{2}\frac{x^2}{\sigma^2}} \qquad (B.47)$$

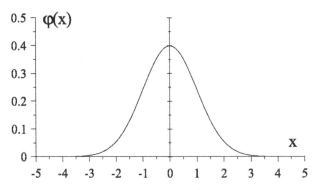

**Fig. B.1.** Gaussian error function

**Table B.1.** Gaussian error functions $\varphi(x)$ and $\Phi(x)$

| $x$ | $\varphi(x)$ | $\Phi(x)$ | $x$ | $\varphi(x)$ | $\Phi(x)$ | $x$ | $\varphi(x)$ | $\Phi(x)$ |
|---|---|---|---|---|---|---|---|---|
| 0.0 | 0.3989 | 0.0000 | 1.0 | 0.2420 | 0.6827 | 2.0 | 0.0540 | 0.9545 |
| 0.1 | 0.3970 | 0.0797 | 1.1 | 0.2179 | 0.7287 | 2.25 | 0.0317 | 0.9756 |
| 0.2 | 0.3910 | 0.1585 | 1.2 | 0.1942 | 0.7699 | 2.5 | 0.0175 | 0.9876 |
| 0.3 | 0.3814 | 0.2358 | 1.3 | 0.1714 | 0.8064 | 2.75 | 0.0091 | 0.9940 |
| 0.4 | 0.3683 | 0.3108 | 1.4 | 0.1497 | 0.8385 | 3.0 | 0.0044 | 0.9973 |
| 0.5 | 0.3521 | 0.3830 | 1.5 | 0.1295 | 0.8664 | 3.5 | 8.73e-4 | 0.9995 |
| 0.6 | 0.3332 | 0.4514 | 1.6 | 0.1109 | 0.8904 | 4.0 | 1.34e-4 | 0.9999 |
| 0.7 | 0.3123 | 0.5161 | 1.7 | 0.0941 | 0.9109 | 5.0 | 1.49e-6 | 1-6e-7 |
| 0.8 | 0.2897 | 0.5762 | 1.8 | 0.0790 | 0.9281 | 7.5 | 2.4e-13 | $\approx 1$ |
| 0.9 | 0.2661 | 0.6319 | 1.9 | 0.0656 | 0.9426 | 10.0 | 7.7e-23 | $\approx 1$ |

**Integral of Gaussian error function** (Fig. 13.2)

$$\Phi(x/\sigma) = 2 \int_0^{x/\sigma} \varphi(\bar{x}/\sigma) \, \mathrm{d}\frac{\bar{x}}{\sigma} \qquad (B.48)$$

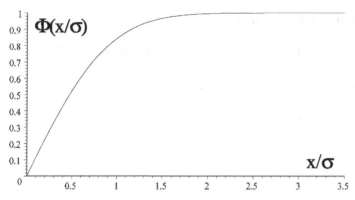

**Fig. B.2.** Integral of Gaussian error function

**2-dim Gaussian distribution**

$$\varphi(r) = \frac{1}{2\pi\,\sigma_x\sigma_y} \exp\left[-\frac{1}{2}\left(\frac{x^2}{\sigma_x^2} + \frac{y^2}{\sigma_y^2}\right)\right] \tag{B.49}$$

**2-dim Gaussian distribution (round)** $(\sigma_x = \sigma_y = \sigma_r)$

$$\varphi(r) = \frac{1}{2\pi\sigma_r^2}\mathrm{e}^{-\frac{1}{2}\frac{r^2}{\sigma_r^2}} \tag{B.50}$$

## B.8 Miscelaneous Mathematical Formulas

**Element of solid angle** ($\theta$ polar angle, $\psi$ azimuthal angle)

$$\mathrm{d}\,\Omega = \sin\theta\,\mathrm{d}\psi\,\mathrm{d}\,\theta \tag{B.51}$$

**Integrated solid angle within polar angle** $\theta$

$$\Delta\Omega = 2\pi\,(1 - \cos\theta) \tag{B.52}$$

**Modified Bessel's functions** (Fig. 13.3)

$$K_{1/3}(\xi) = \sqrt{3}\int_0^\infty \cos\left[\frac{1}{2}\,\xi(3x + x^3)\right]\,\mathrm{d}x\,, \tag{B.53}$$

$$K_{2/3}(\xi) = \sqrt{3}\int_0^\infty \sin\left[\frac{1}{2}\,\xi(3x + x^3)\right]\,\mathrm{d}x\,. \tag{B.54}$$

for small arguments $\xi \to 0$

$$K_{1/3}(\xi \to 0) \approx \frac{\Gamma^2\,(1/3)}{2^{2/3}}\left(\frac{\omega}{\omega_\mathrm{c}}\right)^{-2/3}\frac{1}{1 + \gamma^2\theta^2}\,, \tag{B.55}$$

$$K_{2/3}(\xi \to 0) \approx 2^{2/3}\Gamma^2\,(2/3)\left(\frac{\omega}{\omega_\mathrm{c}}\right)^{-4/3}\frac{1}{\left(1 + \gamma^2\theta^2\right)^2}\,, \tag{B.56}$$

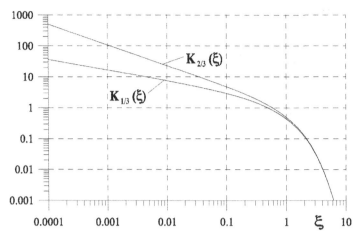

**Fig. B.3.** Modified Bessel's functions $K_{1/3}(\xi)$ and $K_{2/3}(\xi)$

and for large arguments $\xi \to \infty$

$$K_{1/3}^2(\xi \to \infty) \approx \frac{\pi}{2\xi\, e^{2\xi}} \,, \tag{B.57}$$

$$K_{2/3}^2(\xi \to \infty) \approx \frac{\pi}{2\xi\, e^{2\xi}} \,. \tag{B.58}$$

For $\nu = \frac{\omega}{\omega_L} \gg 1$, where the Larmor frequency $\omega_L = c/\rho$,

$$K_{1/3}(\xi) = \frac{\sqrt{3}\pi}{\sqrt{1 - \beta^2 \cos^2 \theta}} J_\nu(\nu\beta\cos\theta) \,, \tag{B.59}$$

$$K_{2/3}(\xi) = \frac{\sqrt{3}\pi}{\sqrt{\left(1 - \beta^2 \cos^2 \theta\right)}} J_\nu'(\nu\beta\cos\theta) \,. \tag{B.60}$$

**Airy's functions**

$$\mathcal{A}i\,(z) = \frac{\sqrt{z}}{\sqrt{3}\pi} K_{1/3}(\xi) \,, \tag{B.61}$$

$$\mathcal{A}'i\,(z) = -\frac{z}{\sqrt{3}\pi} K_{2/3}(\xi) \,. \tag{B.62}$$

# C. Physical Formulas and Parameters

## C.1 Constants

| | | | |
|---|---|---|---|
| velocity of light in vacuum | $c$ | $= 2.99792458 \times 10^8$ | m / s |
| electric charge unit | $e$ | $= 1.60217733 \times 10^{-19}$ | C |
| | $e^2$ | $= 14.399652$ | eV Å |
| electron rest energy | $m_{\mathrm{e}}c^2$ | $= 0.5110034$ | MeV |
| fine structure constant | $\alpha$ | $= 7.29735308 \times 10^{-3}$ | |
| | | $= 1/137.04$ | |
| Avogadro's number | $A$ | $= 6.0221367 \times 10^{23}$ | $1/\mathrm{mol}$ |
| molar volume at STP | | $22.41410 \times 10^{-3}$ | $\mathrm{m}^3/\mathrm{mol}$ |
| atomic mass unit | amu | $= 931.49432$ | MeV |
| classical electron radius | $r_{\mathrm{e}}$ | $= 2.81794092 \times 10^{-15}$ | m |
| proton/electron mass ratio | $m_{\mathrm{p}}/m_{\mathrm{e}}$ | $= 1836.2$ | |
| Planck's constant: | $h$ | $= 6.6260755 \times 10^{-34}$ | J s |
| | | $= 4.1356692 \times 10^{-15}$ | eV s |
| Planck's constant | $\hbar$ | $= 1.05457266 \times 10^{-34}$ | J s |
| | | $= 6.5821220 \times 10^{-16}$ | eV s |
| | $\hbar c$ | $= 197.327053$ | MeV s |
| electron Compton wavelength | $\lambda_{\mathrm{C}}$ | $= 2.42631058 \times 10^{-12}$ | m |
| wavelength for 1eV | $\hbar c/e$ | $= 12398.424$ | Å |
| el.cyclotron frequency/field | $\omega_{\mathrm{Cy}}/B = e/m_{\mathrm{e}}$ | | |
| | | $= 1.75881962 \times 10^{11}$ | rad/(s T) |
| Thomson cross section | $\sigma_{\mathrm{T}}$ | $= 0.66524616 \times 10^{-28}$ | $\mathrm{m}^2$ |
| Boltzmann constant | $k$ | $= 1.3806568 \times 10^{-23}$ | J / K |
| Stephan-Boltzmann constant | $\sigma$ | $= 5.67051 \times 10^{-8}$ | $\mathrm{W}/(\mathrm{m}^2\,\mathrm{K}^4)$ |
| Permittivity of vacuum | $\epsilon_0$ | $= 8.854187817 \times 10^{-12}$ | C/(V m) |
| Permeability of vacuum | $\mu_0$ | $= 1.2566370614 \times 10^{-6}$ | Vs/(A m) |

## C.2 Unit Conversion

### Numerical conversion factors:

**Table C.1.** Numerical conversion factors

| quantity | label | replace cgs units | by SI units |
|---|---|---|---|
| voltage | $U$ | 1 esu | 300 V |
| electric field | $E$ | 1 esu | $3\ 10^4$ V/cm |
| current | $I$ | 1 esu | $10\ c = 2.9979\ 10^9$ A |
| charge | $q$ | 1 esu | $(10c)^{-1} = 3.3356\ 10^{-10}$ C |
| resistance | $R$ | 1 s/cm | $8.9876\ 10^{11}\ \Omega$ |
| capacitance | $C$ | 1 cm | $(1/8.9876)\ 10^{-11}$ F |
| inductance | $L$ | 1 cm | $1\ 10^9$ Hy |
| magnetic induction | $B$ | 1 Gauss | $3\ 10^{-4}$ Tesla |
| magnetic field | $H$ | 1 Oersted | $1000/4\pi = 79.577$ A/m |
| force | $f$ | 1 dyn | $10^{-5}$ N |
| energy | $E$ | 1 erg | $10^{-7}$ J |

### Equation conversion factors:

**Table C.2.** Equation conversion factors

| variable | replace cgs variable | by SI variable |
|---|---|---|
| potential,voltage | $V_{\text{cgs}}$ | $\sqrt{4\pi\epsilon_0}\ V_{\text{MKS}}$ |
| electric field | $E_{\text{cgs}}$ | $\sqrt{4\pi\epsilon_0}\ E_{\text{MKS}}$ |
| current, current density | $I_{cgs}, j_{cgs}$ | $1/\sqrt{4\pi\epsilon_0}\ I_{\text{MKS}}, j_{\text{MKS}}$ |
| charge, charge density | q, $\rho$ | $1/\sqrt{4\pi\epsilon_0}\ q_{\text{MKS}}, \rho_{\text{MKS}}$ |
| resistance | $R_{\text{cgs}}$ | $\sqrt{4\pi\epsilon_0}\ R_{\text{MKS}}$ |
| capacitance | $C_{\text{cgs}}$ | $1/\sqrt{4\pi\epsilon_0}\ C_{\text{MKS}}$ |
| inductance | $L_{\text{cgs}}$ | $\sqrt{4\pi\epsilon_0}\ L_{\text{MKS}}$ |
| magnetic induction | $B_{\text{cgs}}$ | $\sqrt{4\pi/\mu_0}\ B_{\text{MKS}}$ |

Formulas are written for use of either unit. Include factors in square brackets [...] for MKS-units and omit those factors using cgs-units:

## C.3 Relations of Fundamental Parameters

fine structure constant $\quad \alpha \;=\; \dfrac{e^2}{[4\pi\epsilon_0]\,\hbar c}$

classical electron radius $\quad r_{\mathrm{c}} \;=\; \dfrac{e^2}{[4\pi\epsilon_0]\,m_{\mathrm{c}}c^2}$

electron Compton wavelength $\quad \lambda_{\mathrm{C}} \;=\; \dfrac{2\pi\hbar c}{m_{\mathrm{c}}c^2}$

## C.4 Energy Conversion

**Table C.3.** Energy conversion table

|        | calories [cal] | Joule [J] | eVolt [eV] | wavenumber [1/cm] | degKelvin [°K] |
|--------|--------------|-----------|-----------|-------------------|----------------|
| 1 cal  | 1            | 4.186     | $2.6127\ 10^{19}$ | $2.1073\ 10^{23}$ | $3.0319\ 10^{23}$ |
| 1 J    | 0.23889      | 1         | $6.2415\ 10^{18}$ | $5.0342\ 10^{22}$ | $7.2429\ 10^{22}$ |
| 1 eV   | $3.8274\ 10^{-20}$ | $1.6022\ 10^{-19}$ | 1 | 8065.8 | 11604 |
| 1/cm   | $4.7453\ 10^{-24}$ | $1.9864\ 10^{-23}$ | $1.2398\ 10^{-4}$ | 1 | 1.4387 |
| 1 °K   | $3.2984\ 10^{-24}$ | $1.3807\ 10^{-23}$ | $8.6176\ 10^{-5}$ | 0.69507 | 1 |

## C.5 Maxwell's Equations

$$\nabla E \;=\; \frac{4\pi}{[4\pi\epsilon_0]\,\epsilon_{\mathrm{r}}}\,\rho\,, \tag{C.1}$$

$$\nabla B = 0\,, \tag{C.2}$$

$$\nabla \times E \;=\; -\frac{[c]}{c}\frac{\partial B}{\partial t}\,, \tag{C.3}$$

$$\nabla \times B = \frac{4\pi}{c}\left[\frac{c}{4\pi}\right][\mu_0]\mu_{\mathrm{r}}\,\rho v + \frac{[c]}{c}[\epsilon_0\mu_0]\epsilon_{\mathrm{r}}\,\mu_{\mathrm{r}}\frac{\partial E}{\partial t}\,. \tag{C.4}$$

### C.5.1 Lorentz Force

$$F = qE + [c]\frac{q}{c}[v \times B] \tag{C.5}$$

## C.6 Wave and Field Equations

### Definition of potentials

vector potential $\boldsymbol{A}$:    $\boldsymbol{B} = \boldsymbol{\nabla} \times \boldsymbol{A}$ $\hspace{3cm}$ (C.6)

scalar potential $\varphi$ :    $\boldsymbol{E} = -\dfrac{[c]}{c}\dfrac{\partial \boldsymbol{A}}{\partial t} - \boldsymbol{\nabla}\varphi,$ $\hspace{2cm}$ (C.7)

### Wave equations in vacuum

$$\Delta \boldsymbol{A} - \frac{1}{c^2}\frac{\partial^2 \boldsymbol{A}}{\partial t^2} = \frac{4\pi}{[4\pi\epsilon_0]}\rho\boldsymbol{\beta} \tag{C.8}$$

$$\Delta\varphi - \frac{1}{c^2}\frac{\partial^2 \varphi}{\partial t^2} = -\frac{4\pi}{[4\pi\epsilon_0]}\,\rho \tag{C.9}$$

### Vector and scalar potential in vacuum

$$\boldsymbol{A}(t) = \frac{1}{[4\pi c\epsilon_0]}\frac{1}{c}\int \frac{\boldsymbol{v}\rho(x,y,z)}{R}\bigg|_{t_{\mathrm{ret}}}\,\mathrm{d}x\,\mathrm{d}y\,\mathrm{d}z \tag{C.10}$$

$$\varphi(t) = \frac{1}{[4\pi c\epsilon_0]}\frac{1}{c}\int \frac{\rho(x,y,z)}{R}\bigg|_{t_{\mathrm{ret}}}\,\mathrm{d}x\,\mathrm{d}y\,\mathrm{d}z \tag{C.11}$$

### Vector and scalar potential for a point charge $q$ in vacuum

$$\boldsymbol{A}(P,t) = \frac{1}{[4\pi c\epsilon_0]}\frac{q}{R}\frac{\boldsymbol{\beta}}{1+\boldsymbol{n\beta}}\bigg|_{t_{\mathrm{ret}}} \tag{C.12}$$

$$\varphi(P,t) = \frac{1}{[4\pi c\epsilon_0]}\frac{1}{c}\frac{q}{R}\frac{1}{1+\boldsymbol{n\beta}}\bigg|_{t_{\mathrm{ret}}} \tag{C.13}$$

### Radiation field in vacuum

$$\boldsymbol{E}(t) = \frac{1}{[4\pi\epsilon_0]}\frac{q}{cr^3}\left\{\boldsymbol{R}\times\left[(\boldsymbol{R}+\beta\boldsymbol{R})\times\dot{\boldsymbol{\beta}}\right]\right\}\bigg|_{t_{\mathrm{ret}}} \tag{C.14}$$

$$\boldsymbol{B}(t) = \frac{1}{c}\left[\boldsymbol{E}\times\boldsymbol{n}\right]_{t_{\mathrm{ret}}} \tag{C.15}$$

## C.7 Relativistic Relations

Quantities $x^*$ etc. are taken in the particle system $\mathcal{L}^*$, while quantities $x$ etc. refer to the laboratory system $\mathcal{L}$. The particle system $\mathcal{L}^*$ is assumed to move at the velocity $\beta$ along the $z$-axis with respect to the laboratory system $\mathcal{L}$.

### Lorentz transformation of coordinates

$$\begin{pmatrix} x^* \\ y^* \\ z^* \\ ct^* \end{pmatrix} = \begin{pmatrix} 1 & 0 & 0 & 0 \\ 0 & 1 & 0 & 0 \\ 0 & 0 & \gamma & -\beta\gamma \\ 0 & 0 & -\beta\gamma & \gamma \end{pmatrix}\begin{pmatrix} x \\ y \\ z \\ ct \end{pmatrix}. \tag{C.16}$$

**Lorentz transformation of frequencies (relativistic Doppler effect)**

$$\omega = \omega^* \gamma \left(1 + \beta\, n_z^*\right) \tag{C.17}$$

**Lorentz transformation of angles (collimation)**

$$\theta \approx \frac{\sin \theta^*}{\gamma(1 + \beta \cos \theta^*)}\,. \tag{C.18}$$

## C.8 Four-Vectors

Properties of 4-vectors are used in this text to transform physical phenomena from one inertial system to another.
**Space-time 4-vector**

$$\tilde{s} = (x, y, z, \mathrm{i}ct)\,, \tag{C.19}$$

**World time**

$$\tau = \sqrt{-\tilde{s}^2}. \tag{C.20}$$

*length of a 4-vector is Lorentz invariant*
*any product of two 4-vectors is Lorentz invariant*
**Lorentz transformation of time.** From (C.20)

$$c\mathrm{d}\tau = \sqrt{c^2\,(\mathrm{d}t)^2 - (\mathrm{d}x)^2 - (\mathrm{d}y)^2 - (\mathrm{d}z)^2}$$

$$= \sqrt{c^2 - \left(v_x^2 + v_y^2 + v_z^2\right)}\mathrm{d}t$$

$$= \sqrt{c^2 - v^2}\mathrm{d}t = \sqrt{1 - \beta^2}c\mathrm{d}t$$

or

$$\mathrm{d}\tau = \frac{1}{\gamma}\,\mathrm{d}t\,. \tag{C.21}$$

**Velocity 4-vector**

$$\tilde{v} = \frac{\mathrm{d}\tilde{s}}{\mathrm{d}\tau} = \gamma\frac{\mathrm{d}\tilde{s}}{\mathrm{d}t} = \gamma\left(\dot{x}, \dot{y}, \dot{z}, \mathrm{i}c\right)\,. \tag{C.22}$$

**4-acceleration**

$$\tilde{a} = \frac{\mathrm{d}\tilde{v}}{\mathrm{d}\tau} = \gamma\frac{\mathrm{d}}{\mathrm{d}t}\left(\gamma\frac{\mathrm{d}\tilde{s}}{\mathrm{d}t}\right)\,. \tag{C.23}$$

4-acceleration $\tilde{a} = (\tilde{a}_x, \tilde{a}_y, \tilde{a}_z, i\,\tilde{a}_t)$ in component form

$$\tilde{a}_x = \gamma^2 a_x + \gamma^4 \beta_x \left( \boldsymbol{\beta} \cdot \boldsymbol{a} \right) , \tag{C.24}$$

where $\boldsymbol{a}$ is the ordinary acceleration.

**Square of the 4-acceleration**

$$\tilde{a}^2 = \gamma^6 \left\{ \boldsymbol{a}^2 - [\boldsymbol{\beta} \times \boldsymbol{a}]^2 \right\} = \tilde{a}^{*2}. \tag{C.25}$$

in particle system $\beta = 0, \gamma = 1$ and therefore

$$\tilde{a}^{*2} = a^{*2}. \tag{C.26}$$

# D. Electromagnetic Radiation

**Notation** (All variables are in SI units unless otherwise noted)

| | | | |
|---|---|---|---|
| $E(GeV)$ | particle energy | $B$ | magnetic field |
| $U_0$ | energy loss/turn | $\alpha$ | fine structure constant |
| $e$ | unit of el. charge | $\rho$ | bending radius |
| $P_\gamma$ | synchrotron rad. power | $f_{rev}$ | revolution frequency |
| $I$ | beam current | $n_b$ | number of bunches |
| $N$ | number of circ. electrons | $N_b$ | electrons/bunch |
| $\dot{N}_{ph}$ | photon flux | | |
| $\omega$ | photon frequency | $\varepsilon$ | photon energy |
| $\omega_c$ | crit. photon frequency | $\varepsilon_c$ | crit.photon energy |
| $\Delta\omega/\omega$ | band width | | |

## D.1 Radiation Constants

$$C_\gamma = \frac{4\pi}{3}\frac{r_c}{(mc^2)^3} = 8.8460 \ 10^{-5} \ \frac{\text{m}}{\text{GeV}^3} , \tag{D.1}$$

$$C_P = \frac{2}{3}\frac{r_c \, c^3}{(mc^2)^3} = 379.35 \ \frac{1}{\text{s T}^2 \ \text{GeV}} , \tag{D.2}$$

$$C_\omega = \frac{2}{3}\frac{\hbar c}{(mc^2)^3} = 2.2182 \ \frac{\text{m GeV}}{\text{s}} \tag{D.3}$$

$$C_u = \frac{4\pi^2 \, r_c}{3 \, mc^2} = 7.2567 \times 10^{-20} \ \frac{\text{m}}{\text{eV}} \tag{D.4}$$

$$C_\Omega = \frac{3\alpha}{4\pi^2 e(mc^2)^2} = 1.3273 \ 10^{22} \ \frac{\text{photons}}{\text{s rad}^2 \text{GeV}^2 \text{A}} , \tag{D.5}$$

$$C_\psi = \frac{4\alpha}{9e\,mc^2} = 3.967 \ 10^{19} \ \frac{\text{photons}}{\text{s rad A GeV}} , \tag{D.6}$$

$$C_K = \frac{[c]\,e}{2\,\pi\,m\,c^2} = 0.93373 \ \frac{1}{\text{T cm}} , \tag{D.7}$$

$$C_q = \frac{55}{32\sqrt{3}}\frac{\hbar c}{mc^2} = 3.84 \ 10^{-13} \ \text{m} , \tag{D.8}$$

$$C_Q = \frac{55}{24\sqrt{3}} \frac{r_c \hbar c}{(mc^2)^6} = 2.06\,10^{-11} \frac{m^2}{GeV^5}, \tag{D.9}$$

$$C_d = \frac{c}{3} \frac{r_c}{(mc^2)^3} = 2110 \frac{m^2}{GeV^3 s}, \tag{D.10}$$

$$C_\rho = [c]\,e = 0.299792 \frac{GeV}{m\,T}. \tag{D.11}$$

## D.2 Bending Magnet Radiation

### Notation for synchrotron radiation formulas
$\theta$     angle of observation orthogonal to deflecting plane
$\psi$     angle of observation in the plane of deflection

**For isomagnetic ring:** all bending fields are equal, $\rho$ =const.
**Total radiation power**

$$P_\gamma(kW) = 14.0788\ E^4\ I \oint \frac{ds}{\rho^2} \longrightarrow \underbrace{88.460\ \frac{E^4}{\rho}\ I}_{\text{isomagnetic ring}} \tag{D.12}$$

**Energy loss per turn to synchrotron radiation**

$$U_0(keV) = 14.0788\ E^4 \oint \frac{ds}{\rho^2} \longrightarrow \underbrace{88.460 \frac{E^4}{\rho}}_{\text{isomagnetic ring}} \tag{D.13}$$

**Fundamental photon energy of synchrotron radiation**

$$\epsilon_c(keV) = 2.2181 \frac{E^3}{\rho} = 0.665\ E^2\ B \tag{D.14}$$

**Radiation power into a beam line with acceptance angle $\Delta\psi$**

$$\Delta P_\gamma(kW) = 14.079 \frac{\Delta\psi}{\rho} E^4\ I \tag{D.15}$$

### Spatial and spectral photon flux

$$\frac{d^2\dot{N}_{ph}}{d\theta d\psi}\left[\frac{photons}{s\ mrad^2}\right] = 1.3273\ 10^{16}\ E^2 I\ \frac{\Delta\omega}{\omega}\left(\frac{\omega}{\omega_c}\right)^2 K_{2/3}^2(\xi)\ F(\xi,\theta), \tag{D.16}$$

with

$$\xi = \frac{1}{2}\frac{\omega}{\omega_c}\left(1 + \gamma^2\theta^2\right)^{3/2} \tag{D.17}$$

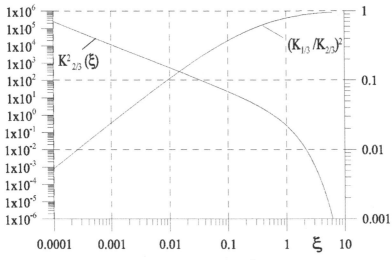

**Fig. D.1.** Functions $K^2_{2/3}(\xi)$ and $\left(K_{1/3}/K_{2/3}\right)^2$

and

$$F(\xi,\theta) = \left(1+\gamma^2\theta^2\right)^2 \left[1 + \frac{\gamma^2\theta^2}{1+\gamma^2\theta^2}\frac{K^2_{1/3}(\xi)}{K^2_{2/3}(\xi)}\right] .$$  (D.18)

**Spatial and spectral photon flux on axis ($\theta = 0$)**

$$\left.\frac{\mathrm{d}^2\dot{N}_{\mathrm{ph}}}{\mathrm{d}\theta\mathrm{d}\psi}\left[\frac{\mathrm{photons}}{\mathrm{s\ mrad}^2}\right]\right|_{\theta=0} = 1.3273\ 10^{16}\ E^2 I\ \frac{\Delta\omega}{\omega}\left(\frac{\omega}{\omega_{\mathrm{c}}}\right)^2 K^2_{2/3}\left(\frac{\omega}{2\omega_{\mathrm{c}}}\right)$$  (D.19)

**Photon flux per unit deflection angle**

$$\frac{\mathrm{d}\dot{N}_{\mathrm{ph}}}{\mathrm{d}\psi}\left[\frac{\mathrm{photons}}{\mathrm{s\ mrad}}\right] = 3.967\times10^{16}\ E\,I\ \frac{\Delta\omega}{\omega}\ S\left(\frac{\omega}{\omega_{\mathrm{c}}}\right)$$  (D.20)

**Long and short wavelength approximations are**

$$S\left(\frac{\omega}{\omega_{\mathrm{c}}}\right) = \frac{9\sqrt{3}}{8\pi}\frac{\omega}{\omega_{\mathrm{c}}}\int_{\omega/\omega_{\mathrm{c}}}^{\infty}K_{5/3}(x)\mathrm{d}x$$

$$= \begin{cases} 1.333\left(\frac{\omega}{\omega_{\mathrm{c}}}\right)^{1/3} & \text{for } \omega\ll\omega_{\mathrm{c}} \\ 0.777\sqrt{\frac{\omega}{\omega_{\mathrm{c}}}}\mathrm{e}^{-\omega/\omega_{\mathrm{c}}} & \text{for } \omega\gg\omega_{\mathrm{c}} \end{cases} .$$  (D.21)

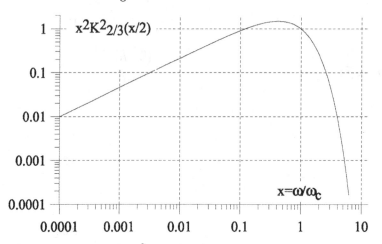

**Fig. D.2.** Function $\left(\frac{\omega}{\omega_c}\right)^2 K_{2/3}^2\left(\frac{1}{2}\frac{\omega}{\omega_c}\right)$ defining the forward photon flux

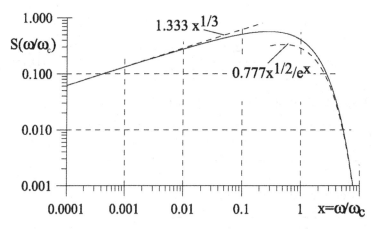

**Fig. D.3.** Universal Function $S\left(\frac{\omega}{\omega_c}\right)$

**Vertical radiation cone angle defined by** $\sqrt{2\pi}\sigma_\theta = \left(\frac{\mathrm{d}N_{\mathrm{ph}}}{\mathrm{d}\psi}\right) / \left(\frac{\mathrm{d}^2\dot{N}_{\mathrm{ph}}}{\mathrm{d}\theta\mathrm{d}\psi}\right)$

$$\sigma_\theta \ (\mathrm{mrad}) = \frac{C_\psi}{\sqrt{2\pi}C_\Omega}\frac{1}{E}\frac{S(x)}{x^2 K_{2/3}^2\left(\frac{1}{2}x\right)} = \frac{f(x)}{E(\mathrm{GeV})} \tag{D.22}$$

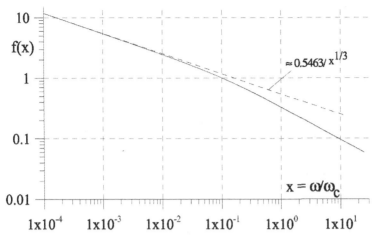

**Fig. D.4.** Scaling function $f(x) = \sigma_\theta(\text{mrad})\, E(\text{GeV})$ for the photon beam divergence

For long wavelengths $(x \ll 1)$

$$\sigma_\theta\,(\text{mrad}) \approx \frac{0.5463\,x^{1/3}}{E(\text{GeV})}. \qquad\qquad (D.23)$$

## D.3 Periodic Insertion Devices

### Notation for insertion devices

| | |
|---|---|
| $\vartheta$ | angle between observation and undulator axis |
| $\psi$ | azimuthal angle of observation about undulator axis |
| $B_0$ | maximum undulator field |
| $\theta$ | maximum deflection angle |
| $\lambda_p$ | period length (m) |
| $N_p$ | number of periods |
| $L_u$ | undulator length (m) , $L_u = N_p \lambda_p$ |

"horizontal" reflects the deflecting plane of the undulator

### D.3.1 Insertion Device Parameter

**Undulator field, on-axis**

$$B(z) = B_0 \sin \tfrac{2\pi z}{\lambda_p} \qquad\qquad (D.24)$$

**Undulator strength parameter**

$$K = 93.4\, B_0 \lambda_p \approx B_0 \lambda_p(\text{cm}) \qquad\qquad (D.25)$$

## D.3.2 Field Scaling for Hybrid Wiggler Magnets

The maximum on-axis field in a hybrid wiggler magnet depends on the gap aperture $g$ and period length $\lambda_p$ and is for $0.1 \lambda_p \lesssim g \lesssim 10 \lambda_p$ given by [34]

$$B_y(T) \approx 3.33 \exp\left[-\frac{g}{\lambda_p}\left(5.47 - 1.8\frac{g}{\lambda_p}\right)\right] \tag{D.26}$$

## D.3.3 Particle Beam Parameter

### Beam width

$$\sigma_{b,x} = \sqrt{\epsilon_x \beta_x + \left(\eta_x \frac{\sigma_E}{E}\right)^2} \tag{D.27}$$

### Horizontal beam divergence

$$\sigma_{b,x'} = \sqrt{\frac{\epsilon_x}{\beta_x} + \left(\eta'_x \frac{\sigma_E}{E}\right)^2} \tag{D.28}$$

### Beam height

$$\sigma_{b,y} = \sqrt{\epsilon_y \beta_y + \left(\eta_y \frac{\sigma_E}{E}\right)^2} \longrightarrow \underbrace{\sqrt{\epsilon_y \beta_y}}_{\text{flat ring}} \tag{D.29}$$

### Vertical beam divergence

$$\sigma_{b,y'} = \sqrt{\frac{\epsilon_y}{\beta_y} + \left(\eta'_y \frac{\sigma_E}{E}\right)^2} \longrightarrow \underbrace{\sqrt{\frac{\epsilon_y}{\beta_y}}}_{\text{flat ring}} \tag{D.30}$$

### Average drift velocity

$$\bar{\beta} = \beta \left(1 - \frac{K^2}{4\gamma^2}\right) \tag{D.31}$$

### Transverse particle coordinate

$$x(t) = \frac{K}{\gamma k_p} \cos\left(k_p \bar{\beta} ct\right) \tag{D.32}$$

### Maximum oscillation amplitude

$$a = \frac{K}{\gamma k_p} = \frac{\lambda_p K}{2\pi\gamma} \tag{D.33}$$

### Maximum deflection angle

$$\theta = \pm\frac{K}{\gamma} \tag{D.34}$$

### Longitudinal coordinate

$$z(t) = \bar{\beta} ct + \frac{K^2}{8\gamma^2 k_p} \sin^2\left(2k_p \bar{\beta} ct\right) \tag{D.35}$$

# D.4 Undulator Radiation

**Total radiation powerfrom an undulator magnet**

$$P(\text{kW}) = 0.6336\, E^2\, B_0^2\, I\, L_\text{u} \tag{D.36}$$

**Energy loss in an undulator**

$$\Delta E(\text{keV}) = 0.725\, \frac{E^2\, K^2}{\lambda_\text{p}^2(\text{cm})}\, L_\text{u}, \tag{D.37}$$

**Wavelength, frequency and photon energy for k-th harmonic**

$$\boxed{\lambda_k(\text{Å}) = 13.056\, \frac{\lambda_\text{p}(\text{cm})}{k\, E^2}\left(1 + \tfrac{1}{2}K^2 + \gamma^2\vartheta^2\right)} \tag{D.38}$$

$$\boxed{\omega_k = 1.4427 \times 10^{18}\, \frac{k\, E^2}{\lambda_\text{p}(\text{cm})\left(1 + \tfrac{1}{2}K^2 + \gamma^2\vartheta^2\right)}} \tag{D.39}$$

$$\boxed{\varepsilon_k(\text{keV}) = 0.9496\, \frac{k\, E^2}{\lambda_\text{p}(\text{cm})\left(1 + \tfrac{1}{2}K^2 + \gamma^2\vartheta^2\right)}} \tag{D.40}$$

**Undulator on-axis differential photon flux for k-th harmonic**

$$\boxed{\left.\frac{\mathrm{d}\dot{N}_\text{ph}(k\omega_k)}{\mathrm{d}\Omega}\right|_{\vartheta=0} = 1.7443\,10^{23}\, E^2 N_\text{p}^{\ 2}\, \frac{\Delta\omega}{\omega} I\, A_k(K),} \tag{D.41}$$

**Harmonic amplitude functions**

$$A_k(K) = \frac{k^2\, K^2}{\left(1 + \tfrac{1}{2}K^2\right)^2}\left[J_{\frac{1}{2}(k-1)}\left(\frac{kK^2}{4+2K^2}\right) + J_{\frac{1}{2}(k+1)}\left(\frac{kK^2}{4+2K^2}\right)\right]^2 \tag{D.42}$$

**Opening angle (diffraction limit) for on-axis photon flux**

$$\sigma_\vartheta \approx \frac{1}{\gamma}\sqrt{\frac{1 + \tfrac{1}{2}K^2}{2\, k\, N_\text{p}}} \tag{D.43}$$

**Pinhole solid angle**

$$\mathrm{d}\Omega = 2\pi\, \sigma_\vartheta^2 \tag{D.44}$$

**Pinhole photon flux**

$$\boxed{\dot{N}_\text{ph}(k\omega_1)|_{\vartheta=0} = 1.4309\,10^{17}\, N_\text{p}\, \frac{\Delta\omega}{\omega}\, G_k(K)} \tag{D.45}$$

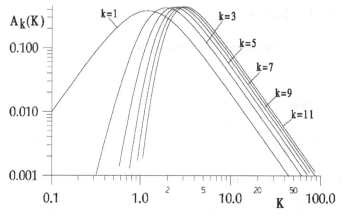

**Fig. D.5.** Functions $A_k(K)$

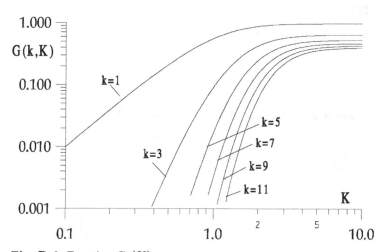

**Fig. D.6.** Function $G_k(K)$

$$G_k(K) = \frac{1 + \frac{1}{2}K^2}{k} A_k(K) \tag{D.46}$$

$$= \frac{k\,K^2}{\left(1 + \frac{1}{2}K^2\right)} \left[ J_{\frac{1}{2}(k-1)}\left(\frac{kK^2}{4 + 2K^2}\right) + J_{\frac{1}{2}(k+1)}\left(\frac{kK^2}{4 + 2K^2}\right) \right]^2$$

**Band width of radiation**

$$\frac{\Delta\omega}{\omega_k} = \frac{1}{kN_\mathrm{p}} \tag{D.47}$$

# D.5 Photon Beam Brightness

**Spectral brightness$\equiv$ 6$-$dim photon phase space density**

$$\mathcal{B}(\omega) = \frac{\dot{N}_{\mathrm{ph}}(\omega)}{4\pi^2\,\sigma_x\sigma_{x'}\sigma_y\sigma_{y'}(\mathrm{d}\omega/\omega)} \tag{D.48}$$

**Diffraction limited source size**

$$\begin{array}{ll} \text{radial source size:} & \sigma_r = \frac{1}{2\pi}\sqrt{\lambda L_{\mathrm{s}}} \\[2mm] \text{radial source divergence:} & \sigma_{r'} = \sqrt{\frac{\lambda}{L_{\mathrm{s}}}} \end{array} \tag{D.49}$$

**Source length $L_{\mathrm{s}}$**

$$L_{\mathrm{s}} = L_{\mathrm{u}} \qquad \text{for undulator,} \tag{D.50}$$

$$L_{\mathrm{s}} = 2\rho/\gamma \qquad \text{for wiggler and bending magnets} \tag{D.51}$$

**Diffraction limited brightness**

$$\mathcal{B}_{\max} = \dot{N}_{\mathrm{ph}}\frac{(4/\lambda^2)}{\mathrm{d}\omega/\omega} \tag{D.52}$$

## D.5.1 Effective Source Parameter

The beam parameters $\sigma_{\mathrm{b}0}$ are to be taken at the beginning of the source, e.g. at the entrance of the undulator and beam is assumed to be symmetric within undulator.

**Horizontal source size**

$$\sigma_{\mathrm{t},x}^2 = \tfrac{1}{2}\sigma_r^2 + \sigma_{\mathrm{b}0,x}^2 + \left(\frac{\lambda_{\mathrm{p}}K}{2\pi\gamma}\right)^2 + \tfrac{1}{12}\sigma_{\mathrm{b}0,x'}^2 L_{\mathrm{s}}^2 + \tfrac{1}{36}\theta^2 L_{\mathrm{s}}^2 \tag{D.53}$$

with increased source width due to electron path oscillations $\frac{\lambda_{\mathrm{p}}K}{2\pi\gamma}$, due to finite beam divergence $\frac{1}{12}\sigma_{\mathrm{b}0,x'}^2 L_{\mathrm{s}}^2$, and due to oblique horizontal observation angle $\frac{1}{36}\theta^2 L_{\mathrm{s}}^2$.

**Horizontal divergence**

$$\sigma_{\mathrm{t},x'}^2 = \tfrac{1}{2}\sigma_{r'}^2 + \sigma_{\mathrm{b}0,x'}^2 , \tag{D.54}$$

**Vertical source size**

$$\sigma_{\mathrm{t},y}^2 = \tfrac{1}{2}\sigma_r^2 + \sigma_{\mathrm{b}0,y}^2 + \tfrac{1}{12}\sigma_{\mathrm{b}0,y'}^2 L_{\mathrm{s}}^2 + \tfrac{1}{36}\psi^2 L_{\mathrm{s}}^2 , \tag{D.55}$$

with increased source height due to finite beam divergence $\frac{1}{12}\sigma_{\mathrm{b}0,y'}^2 L_{\mathrm{s}}^2$, and due to oblique vertical observation angle $\frac{1}{36}\psi^2 L_{\mathrm{s}}^2$.

**Vertical divergence**

$$\sigma_{t,y'}^2 = \tfrac{1}{2}\sigma_{r'}^2 + \sigma_{b0,y'}^2 \, . \tag{D.56}$$

**Effective spectral brightness**

$$\mathcal{B}(\omega) = \frac{\dot{N}_{ph}(\omega)}{4\pi^2 \, \sigma_{t,x} \, \sigma_{t,x'} \, \sigma_{t,y} \, \sigma_{t,y'}(d\omega/\omega)} \, . \tag{D.57}$$

# References

1. A. Liénard, L'Eclairage Electrique **16**, 5 (1898).
2. E. Wiechert, Archives Neerlandaises 546 (1900).
3. G. Schott, Annalen der Physik **24**, 635 (1907).
4. G. Schott, Phil. Mag.[6] **13**, 194 (1907).
5. G. Schott, *Electromagnetic Radiation* (Cambridge Univ. Press, New York, 1912).
6. A. S. N.A. Vinokurov, Preprint INP 77-59, Institute of Nuclear Physics, Novosibirsk (unpublished).
7. The author would like to thank Prof. M. Eriksson, Lund, Sweden, for introducing him to this approach into the theory of synchrotron radiation.
8. D.W. Kerst and R. Serber, Phys. Rev. **60**, 53 (1941).
9. D. Ivanenko and I.Ya. Pomeranchouk, Phys. Rev. **65**, 343 (1944).
10. J. Blewett, Phys. Rev. **69**, 87 (1946).
11. describing work of C. Sutis, Sci. News Lett. **51**, 339 (1947).
12. describing work of F. Haber, Electronics **20**, 136 (1947).
13. F. Elder, A. Gurewitsch, R. Langmuir, and H. Pollock, Phys. Rev. **71**, 829 (1947).
14. M. Sands, in *Physics with Intersecting Storage Rings*, edited by B. Touschek (Academic, New York, 1971), p. 257.
15. R. Coisson, Opt.Com. **22**, 135 (1977).
16. R. Bossart, J. Bosser, L. Burnod, R. Coisson, E. D'Amico, A. Hofmann, and J. Mann, Nucl. Instrum. Methods **164**, 275 (1979).
17. R. Bossart, J. Boser, L. Burnod, E. D'Amico, G. Ferioli, J. Mann, and F. Meot, Nucl. Instrum. Methods **184**, 349 (1981).
18. *The Large Hadron Collider in the LEP Tunnel*, edited by G. Brianti and K. Hübner (CERN, Geneva, 1985).
19. J. Jackson, *Classical Electrodynamics*, 2nd. ed. (Wiley, New York, 1975).
20. D. Ivanenko and A.A. Sokolov, DAN (USSR) **59**, 1551 (1972).
21. J. Schwinger, Phys. Rev. **75**, 1912 (1949).
22. D.H. Tomboulian and P.L. Hartman, Phys. Rev. **102**, 102 (1956).
23. G. Bathow, E. Freytag, and R. Haensel, J. Appl. Phys. **37**, 3449 (1966).
24. M. Abramowitz and I. Stegun, *Handbook of Mathematical Functions* (Dover, New York, 1972).
25. M. Born and E. Wolf, *Principles of Optics* (Pergamon, Oxford, 1975).
26. G. Airy, Trans. Cambr. Phil. Soc. **5**, 283 (1835).
27. L. Schiff, Rev. Sci. Instrum. **17**, 6 (1946).
28. T. Nakazato, M. Oyamada, N. Niimura, S. Urasawa, O. Konno, A. Kagaya, R. Kato, T. Kamiyama, Y. Torizuka, T. Nanba, Y. Kondo, Y. Shibata, K. Ishi, T. Oshaka, and M. Ikezawa, Phys. Rev. Lett. **63**, 1245 (1989).
29. E.B.Blum, U.Happek, and A.J. Sievers, Nucl. Instrum. Methods 568 (1992).
30. H. Wiedemann, P. Kung, and H.C. Lihn, Nucl. Instrum. Methods **A**, 1 (1992).

268    References

31. F. Michel, Phys. Rev. Lett. **48**, 580 (1982).
32. M. Berndt, W. Brunk, R. Cronin, D. Jensen, R. Johnson, A. King, J. Spencer, T. Taylor, and H. Winick, IEEE Trans. Nucl. Sci. 3812 (1979).
33. The author thanks T. Rabedau, SSRL, for providing this picture.
34. K. Halbach, J. Physique (1983).
35. W. Heitler, *The Quantum Theory of Radiation* (Clarendon, Oxford, 1954).
36. B. Kincaid, J. Appl. Phys. **48**, 2684 (1977).
37. W. Lavender, Ph.D. thesis, Stanford University, 1988.
38. R. Milburn, Phys. Rev. Lett. **4**, 75 (1963).
39. F.A. Arutyunian and V.A. Tumanian, Phys. Rev. Lett. **4**, 176 (1963).
40. F.A. Arutyunian, I.I. Goldman, and V.A Tumanian, ZHETF(USSR) **45**, 312 (1963).
41. I.F. Ginzburg, G.L. Kotin, V.G. Serbo, and V.I. Telnov, Preprint 81-102, Inst. of Nucl.Physics, Novosibirsk, USSR, (unpublished).
42. E.D. Cournat and H.S. Snyder, ap **3**, 1 (1958).
43. E. McMillan, Phys. Rev. **68**, 143 (1945).
44. V. Veksler, DAN(USSR) **44**, 393 (1944).
45. H. Wiedemann, *Particle Accelerator Physics I*, 2nd ed. (Springer, Heidelberg, 1999).
46. H. Wiedemann, *Particle Accelerator Physics II*, 2nd ed. (Springer, Heidelberg, 1999).
47. J.M. Paterson, J.R. Rees, and H. Wiedemann, Technical report, Stanford Linear Accelerator Center (unpublished).
48. W.K.H. Panofsky and W.A. Wenzel, Rev. Sci. Instrum. **27**, 967 (1956).
49. J. Larmor, Philos. Mag. **44**, 503 (1897).
50. J. Schwinger, Proc. Nat. Acad. of Sci. USA **40**, 132 (1954).
51. A.A. Sokolov and I.M. Ternov, *Synchrotron Radiation* (Pergamon, Oxford, 1968).
52. G. Watson, *Bessel Functions* (The Macmillan Company, New York, 1945).
53. V. Kostroun, Nucl. Instrum. Methods **172**, 371 (1980).
54. I.S. Gradshteyn and I.M. Ryzhik, *Table of Integrals, Series, and Products*, 4th ed. (Academic, New York, 1965), prepared by Yu.V. Geronimus and M.Yu. Tseytlin, translation edited by A. Jeffrey.
55. V. Baier, in *Physics with Intersecting Storage Rings*, edited by B. Touschek (Academic, New York, 1971), p. 1.
56. H. Motz, J. Appl. Phys. **22**, 527 (1951).
57. L.R. Elias, W.M. Fairbanks, J.M.J. Madey, H.A. Schwettmann, and T.J. Smith, Phys. Rev. Lett. **36**, 717 (1976).
58. W. Smythe, *Static and Dynamic Electricity* (McGraw-Hill, New York, 1950).
59. D.F. Alferov, Y.A. Bashmakov, and E.G. Bessonov, Sov. Phys.-Tech. Phys. **18**, 1336 (1974).
60. S. Krinsky, IEEE Trans. Nucl. Sci. 307 (1983).
61. A. Bienenstock, G. Brown, H. Wiedemann, and H. Winick, Rev. Sci. Instrum. **60**, 7 (1989).
62. R. Carr, NIM **306**, 391 (1991).
63. S. Sasaki, K. Kakuno, T. Takada, T. Shimada, K. Yanagida, and Y. Miyahara, Nucl. Instrum. Methods **331**, 763 (1993).
64. R. Carr and S. Lidia, in *Proc. of the SPIE* (SPIE, Bellingham, WA (USA), 1993), Vol. 2013.
65. W. Colson, Phys. Lett. **64A**, 190 (1977)

# Index

4-vector, 255
– acceleration, 32, 255
– energy-momentum, 27
– space-time, 27, 255
– velocity, 255

aberrations
– chromatic, 135
– geometric, 136
accelerating cavity, 75
acceleration
– longitudinal, 33
– transverse, 33
achromat, 131
Airy's functions, 169, 249
Ampère's law, 6
Ampère-turns, 78
approximations made
– $ct_r = \pm\rho/\gamma$, 160
– $\sin(\omega_L t_r) \approx \omega_L t_r$, 161
AS(x.y.z), see footnote, 165
asymmetric wiggler, 211

backscattered photons, 68
beam current
– circulating, 74
beam deflection, 79
beam divergence, 96
beam emittance
– equilibrium, horizontal, 109, 114
– equilibrium, vertical, 109
– minimum, 128
– scaling, 127
– vertical, 109
beam optics
– linear, 83
beam rigidity, 79
beam size, 95, 111
– height, 262
– width, 262
bend radiation
– differential photon flux, 258

– differential photon flux, on-axis, 259
– photon flux per mr, 259
– total power, 258
bending magnet, 74, 77
– radiation, 4
bending radius, 79
Bessel's functions
– modified, 160, 248
betatron
– oscillation, 83, 87, 88
– phase, 87
– tune, 87
betatron function, 51, 86, 88
– optimum in drift space, 237
– periodic, 94
– transformation, 91
Biot–Savart fields, 143
booster synchrotron, 76
brightness, 42, 207, 265
– diffraction limited, 50, 265
– effective spectral, 266
– spectral, 50, 265
bunch, 5
– length, 5
– pattern, 5
bunch length
– equilibrium, 107
bunches, 75

$C_B$, 34
$C_c$, 37
$C_d$, 114
cell, 93
$C_\gamma$, 34, 149
cgs-system, 6
Cherenkov
– angle, 19
– condition, 19
– radiation, 18, 19
chromaticity, 83, 135, 136
$C_K$, 181
coherence

– spatial, 46
– temporal, 47
coherent
– radiation power, 48
coherent radiation, 45
collimation, 3, 27
– angle, 153
collimation angle, 255
$C_\Omega$, 38, 163
Compton
– effect, 20
Compton scattering, 68
contraction
– Lorentz, 11
conversion
– energy, 253
– units, 252
coordinate system
– cartesian, 245
– cylindrical, 246
– polar, 246
Coulomb field, 142
Coulomb regime, 22, 23, 142
$C_\psi$, 39, 174
$C_Q$, 113
$C_q$, 107
$C_\rho$, 79
critical
– photon energy, 4
critical photon
– energy, 37
– frequency, 37
critical photon energy, 155, 258
$C_u$, 189

damping, 104
damping decrement, 101, 105
damping wigglers, 112, 115
dba-lattice, 131
– optimum beam emittace, 131
deflection angle, 79
diffraction, 42, 229
– Fraunhofer, 42
– intergral, Fraunhofer, 43
diffraction limit, 121
– emittance, 46
– source divergence, 47
– source size, 47
dilatation, 11
dispersion function, 92, 95
divergence
– photon beam, 41
Doppler effect, 3, 255

– relativistic, 27
dynamic aperture, 119, 132, 136

edge focusing
– wiggler magnet, 183
electromagnetic radiation, 1, 2
electron beam, 5
electron source, 75
emittance
– diffraction limited, 46, 121
energy, 13
– conservation, 17, 21
– kinetic, 9, 13
– particle, 73
– total, 13
energy conversion, 253
energy loss, 150
– per turn, 35, 149, 258
energy spread
– equilibrium, 106, 107
equation of motion, 9, 82
– analytical solution, 86
– inhomogeneous, 92
– solution, 84, 88
equilibrium
– beam emittance, horizontal, 109
– beam emittance, vertical, 109
– emittance, 108
$\eta$-function, 95

Faraday's law, 6
FEL, 5, 217
– small gain, 220
field gradient, 80
figure of eight trajectory, 183
first generation, 109
flat undulator, 65
focal length, 80
– quadrupole, 80
focal point, 80
focusing, 77
– principle of, 80
FODO cell, 86
FODO lattice, 126
FODO parameter, 94
form factor, 49
formation length, 23
forward cone, 206
forward radiation, 206
four vector
– acceleration, 255
– velocity, 255
four vectors
– space-time, 255

Fourier transform, 245
Fraunhofer
– diffraction, 42
– diffraction integral, 43
free electron laser, 5, 103, 179, 217
fringe field focusing
– wiggler magnet, 183
fundamental frequency, 178, 192
fundamental undulator radiation
– frequency, 263
– photon energy, 263
– wavelength, 263
fundamental wavelength, 62, 65, 188

gain curve, 228
Gauss's theorem, 245
Gaussian distribution, 247
GR(x.y.z), see footnote, 165

harmonic number, 5
harmonics, 63
helical undulator, 65
helicity, 210
hybrid magnet, 60
– field scaling, 262

insertion device, 55, 76, 179
integral theorems, 245
isomagnetic
– lattice, 35

$JJ$-function, 204, 223
$J_s$, 101
$J_x$, 105
$J_y$, 105

Lamor frequency, 157
Large Hadron Collider, 53
Larmor frequency, 249
lattice, 85
– cell, 93
– FODO, 85
lattice functions, 88
– optimum, 130
– periodic, 93
LHC, 35, 53
Liénard–Wiechert potentials, 140
Liénard-Wiechert potentials, 1
line spectrum, 198
– undulator, 204
linear accelerator, 2, 76
Liouville's theorem, 89, 91
LNLS, 132
Lorentz

– contraction, 3, 11
– force, 8, 77, 253
– gauge, 137
– transformation, 11, 27, 254
– – of fields, 12, 13
luminosity, 69

magnet
– dipole, 77
– excitation current, 78
matching
– photon beam, 51
matrix
– formulation, 84
– transformation, 84
Maxwell's equations, 1, 6, 137, 253
MKS-system, 6
momentum, 13
– conservation, 17
– particle, 8
momentum compaction factor, 100, 101

opening angle
– vertical, 261
optical klystron, 4
orbit, 74
– equilibrium, 83
– ideal, 83
oscillation
– phase, 100
– synchrotron, 100

Panofsky–Wenzel Theorem, 120
parallel acceleration, 148
Parseval's theorem, 145, 156, 245
particle
– energy, 73
particle beam
– emittance, 89
– envelope, 88
– focusing, 77
particle distribution
– Gaussian, 48, 89
partition number
– horizontal, 105
– synchrotron, 101
– vertical, 105
pendulum equation, 225
permanent magnet
– wiggler, 60
permeability, 7
permittivity, 7
phase

– focusing, 99
– oscillation, 100
phase ellipse, 88–90
– upright, 90
phase focusing, 100
phase space, 89
phase space ellipse, 103
phase space motion
– longitudinal, 103
photon beam
– divergence, 41
– matching, 51
– temporal structure, 55
photon beam brightness, 265
photon beam lines, 74
photon energy
– critical, 4, 37, 155
– undulator, 66
photon flux
– angular, 39, 40
– differential, 163
– per unit solid angle, 38
– spectral, 173
photon source
– parameter, 121
photon source parameters, 50
photons
– backscattered, 68
physical constants, 251
pin hole, 201
polarization, 162, 208
– elliptical, 4, 208
– $\pi$−mode, 38
– $\pi$-mode, 208
– $\sigma$−mode, 38
– $\sigma$-mode, 208
polarization states, 159
potential
– scalar, 9, 137
– vector, 9, 137
potentials
– retarded, 139, 254
Poynting vector, 17, 21–23, 143
proton
– radiation power, 35

quadrupole
– focal length, 80
– poleshape, 81
– strength, 80
quadrupole magnet, 75, 80
quantum effect, 105

radiance, 207

radiation
– bending magnet, 55, 258
– coherent, 45
– electromagnetic, 73
– forward, 206
– longitudinal acceleration, 25
– regime, 23
– shielding, 45
– spectrum, 36, 162
– spontaneous, 218
– stimulated, 217
– synchrotron, 22
– transverse acceleration, 33
radiation cone, 153
radiation constants, 257
radiation field, 141, 142, 254
– longitudinal acceleration, 25
– spectral, 161
radiation lobes, 150
radiation power, 32, 33
– instantaneous, 34, 148
– orthogonal acceleration, 148
– parallel acceleration, 148
– spatial distribution, 153, 166
– total, 35, 144
– undulator, 190
– wiggler, 60
radiation regime, 23, 142
radiation sources
– first generation, 125
– fourth generation, 125
– second generation, 125
– third generation, 125
radio antenna, 3
Rayleigh length, 229, 238
relativistic relations, 254
resonance
– integer, 88
retarded
– potentials, 1
– time, 1
retarded potentials, 139
retarded time, 138
revolution frequency, 6, 36, 74
revolution time, 74
rf-bucket, 5
rf-field, 99
rf-system, 75
Robinson criterion, 105

scalar potential, 9, 137
second generation, 109
separatrix, 103

series expansion, 243
sextupole magnet, 75
small gain FEL, 220
source length, 265
source size
– diffraction limited, 265
source sizes, 265
spatial coherence, 46
spatial distribution
– synchrotron radiation, 152
spectral brightness, 50, 207
spectral line width, 198
spectral photon flux, 173
spectral purity, 198
spectrum, 36
spontaneous radiation, 218
SPS, 35
steering magnet, 75
stimulated radiation, 217
Stokes' theorem, 245
storage ring, 5, 73, 74
– lattice, 126
strength parameter, 60
superbend, 4, 56
synchronous
– particle, 100
– phase, 99
– time, 99
synchrotron, 2
– oscillation, 99–101, 103
– radiation, 22, 179
synchrotron radiation, 25, 31, 73, 149
– angular distibution, 165
– coherent, 45
– energy loss per turn, 35, 149
– harmonics, 165
– polarization, 160
– – $\pi$–mode, 160
– – $\sigma$–mode, 160
– power per unit solid angle, 144
– spatial distribution, 152, 157, 162
– spectral distribution, 157, 162
– spectrum, 172
– total power, 144, 150
synchrotron radiation source
– first generation, 109
– second generation, 109
– third generation, 109
synshrotron radiation
– spatial distribution, 153
system of units, 6

TBA-lattice, 134

temporal coherence, 46
thin lens approximation, 85
third generation, 109
Thomson scattering, 68
– cross section, 68
time dilatation, 11, 12
transformation matrix, 84
– defocusing quadrupole, 84
– drift space, 84
– focusing quadrupole, 84
– wiggler pole, 185
transition radiation, 144
– spatial distribution, 146
– spectral distribution, 146
– total energy, 146
transverse acceleration, 148
triple bend achromat, 134
tune, 87
twin paradox, 12

undulator, 3, 177
– deflection angle, 262
– drift velocity, 262
– flat, 65
– helical, 65
– line spectrum, 3
– oscillation amplitude, 262
– period, 3
– strength parameter, 60, 261
undulator magnet, 60, 62, 76
– line spectrum, 204
undulator photon flux
– on-axis, 263
undulator radiation
– band width, 264
– energy loss, 263
– fundamental, 62, 65
– harmonic amplitudes $A_k(K)$, 263
– opening angle, 263
– period length, 3
– pin hole angle, 263
– pin hole flux, 263
– total power, 263
units, 6, 252
– conversion, 252
universal function, 39, 172, 259

vacuum system, 75
Vanadium Permendur, 60
vector
– differentials, 244
– multiple product, 244
vector potential, 9, 137

wave equations, 138, 254
wavelength
– fundamental, 62
– shifter, 4, 57, 76
– undulator, 66
wiggler magnet, 4, 58, 76, 177
– asymmetric, 211
– critical photon energy, 61
– electromagnetic, 60
– flat, 178

– fringe field focusing, 183
– hard edge, 186
– hard edge model, 186
– helical, 178
– period length, 179
– permanent magnet, 60
– strength parameter, 60, 181
wiggler pole
– transformation matrix, 185
world time, 255